SPATIAL ANALYSIS IN HEALTH GEOGRAPHY

Geographies of Health

Series Editors
Allison Williams, Associate Professor, School of Geography and
Earth Sciences, McMaster University, Canada
Susan Elliott, Professor, Department of Geography and Environmental
Management and School of Public Health and Health Systems,
University of Waterloo, Canada

There is growing interest in the geographies of health and a continued interest in what has more traditionally been labeled medical geography. The traditional focus of 'medical geography' on areas such as disease ecology, health service provision and disease mapping (all of which continue to reflect a mainly quantitative approach to inquiry) has evolved to a focus on a broader, theoretically informed epistemology of health geographies in an expanded international reach. As a result, we now find this subdiscipline characterized by a strongly theoretically-informed research agenda, embracing a range of methods (quantitative; qualitative and the integration of the two) of inquiry concerned with questions of: risk; representation and meaning; inequality and power; culture and difference, among others. Health mapping and modeling has simultaneously been strengthened by the technical advances made in multilevel modeling, advanced spatial analytic methods and GIS, while further engaging in questions related to health inequalities, population health and environmental degradation.

This series publishes superior quality research monographs and edited collections representing contemporary applications in the field; this encompasses original research as well as advances in methods, techniques and theories. The *Geographies of Health* series will capture the interest of a broad body of scholars, within the social sciences, the health sciences and beyond.

Also in the series

The Afterlives of the Psychiatric Asylum
Recycling Concepts, Sites and Memories
Edited by Graham Moon, Robin Kearns and Alun Joseph

Geographies of Health and Development
Edited by Isaac Luginaah and Rachel Bezner Kerr

Soundscapes of Wellbeing in Popular Music
Gavin J. Andrews, Paul Kingsbury and Robin Kearns

Mobilities and Health
Anthony C. Gatrell

Spatial Analysis in Health Geography

Edited by

PAVLOS KANAROGLOU
McMaster University, Canada

ERIC DELMELLE
University of North Carolina at Charlotte, USA

ANTONIO PÁEZ
McMaster University, Canada

ASHGATE

Published by
Ashgate Publishing Limited
Wey Court East
Union Road
Farnham
Surrey, GU9 7PT
England

Ashgate Publishing Company
110 Cherry Street
Suite 3A
Burlington, VT 05401-3818
USA

www.ashgate.com

British Library Cataloguing in Publication Data
A catalogue record for this book is available from the British Library.

The Library of Congress has cataloged the printed edition as follows:
Spatial analysis in health geography / [edited] by Pavlos Kanaroglou, Eric Delmelle and Antonio Páez.
    pages cm. -- (Ashgate's geographies of health series)
  Includes bibliographical references and index.
  ISBN 978-1-4724-1619-3 (hardback) -- ISBN 978-1-4724-1620-9 (ebook) -- ISBN 978-1-4724-1621-6 (epub) 1. Medical geography--Statistical methods. I. Kanaroglou, Pavlos. II. Delmelle, Eric. III. Páez, Antonio.
  RA791.S64 2015
  614.4'2--dc23

                                                                2014036510

ISBN 9781472416193 (hbk)
ISBN 9781472416209 (ebk – PDF)
ISBN 9781472416216 (ebk – ePUB)

MIX
Paper from
responsible sources
FSC
www.fsc.org   FSC® C013985

Printed in the United Kingdom by Henry Ling Limited,
at the Dorset Press, Dorchester, DT1 1HD

Contents

List of Figures

List of Tables

Notes on Contributors

Matthew D. Adams is a PhD student in the School of Geography and Earth Sciences at McMaster University, Canada. His research interests include the application of spatial analysis to the assessment of air pollution. He focuses on the monitoring and assessment of both environmental and human health impacts, including the modeling of air pollution, the design of pollution monitoring networks, and the use of biomonitors for assessing air pollution.

Linda Beale is a spatial analytics researcher at Esri specializing in developing applied approaches to spatial analysis through real examples and applications. Previously she was at Imperial College London with responsibility for the development of methodologies for spatial analysis, modeling and programming Geographical Information Systems (GIS) applications to explore geographical variation in chronic diseases and environmental risk for epidemiological study. Linda has published numerous papers, book chapters, presented at national and international conferences and been invited to deliver workshops on spatial analysis and environmental epidemiology. She remains an Honorary Research Fellow at Imperial College London.

Brian H. Bossak is an Associate Professor and Department Chair of Health Sciences at Florida Gulf Coast University. He is a medical geographer and spatial epidemiologist whose research interests include climate change and health, modeling coastal hazards utilizing GIS and remote sensing, the spatial epidemiology of vector-borne diseases, and water quality investigations. He holds BS and MA degrees in Geography from the University of Georgia, a PhD in Geography from Florida State University, and an MPH in Environmental and Occupational Health from Emory University.

Margaret Carrel is an Assistant Professor in the Department of Geographical & Sustainability Sciences and the Department of Epidemiology at the University of Iowa. As a medical geographer, her research centers on exploring geographic patterns of health and disease, focusing in particular on how complex interactions between people and environments result both in disease outcomes and the progressive evolution of human pathogens. She has studied H5N1 influenza in Vietnam, H1N1 in China, malaria in the Democratic Republic of Congo, MRSA in Iowa and HIV drug resistance in North Carolina.

Irene Casas is an Associate Professor in the Department of Social Sciences at Louisiana Tech University and has conducted research in Cali, Colombia with spatial epidemiologists to better understand health care provision and access.

Christopher J. Coutts is an Associate Professor of Urban and Regional Planning and a Research Associate in the Center for Demography and Population Health at Florida State University. He has published numerous articles on the environmental determinants of disease in urban planning, public health, and geography outlets. He holds a Bachelor of Community Health and Masters of Public Health, both from New Mexico State University, and a Ph.D. in Urban, Technological, and Environmental Planning from the University of Michigan. His research focuses on the influence of the built environment and ecologically-sensitive land use practices on community health and health behavior.

Eric Delmelle is an Associate Professor in the Geography and Earth Sciences Department at the University of North Carolina at Charlotte. His interests are in the development of spatial modeling techniques and visualization techniques to the field of spatial epidemiology and infectious diseases in particular.

Patrick DeLuca is a researcher and lecturer at McMaster University. He conducts a wide range of research that relies heavily on GIS and spatial analysis, with interests in quantitative methods, air quality monitoring, modeling and mapping, regional variations in the quality of health and health care and environment and health. He has extensive experience with geocoding in many settings and has been invited to public health conferences to discuss geocoding and its limitations.

Coline Dony is a graduate student in Geography at the University of North Carolina at Charlotte. Her primary research interests are into health accessibility and inequity.

Cong Fu is a PhD student in the Department of Geography and Anthropology, Louisiana State University. His research interests include the applications of GIS and spatial analysis.

Sandro Galea is a physician and an epidemiologist. He is Dean and Professor at the Boston University School of Public Health. Prior to his appointment at Boston University, Dr Galea served as the Anna Cheskis Gelman and Murray Charles Gelman Professor and Chair of the Department of Epidemiology at the Columbia University Mailman School of Public Health where he launched several new educational initiatives and substantially increased its focus on six core areas: chronic, infectious, injury, lifecourse, psychiatric/neurological, and social epidemiology. He previously held academic and leadership positions at the University of Michigan and at the New York Academy of Medicine. In his own scholarship, Dr Galea is centrally interested in the social production of health of urban populations, with a focus on the causes of brain disorders, particularly common mood-anxiety disorders and substance abuse. He has long had a particular interest in the consequences of mass trauma and conflict worldwide, including as a result of the September 11 attacks, Hurricane Katrina, conflicts in sub-Saharan Africa, and the American wars in Iraq and Afghanistan. This work has been principally funded by the National Institutes of Health, Centers for Disease Control and Prevention, and several foundations. He has published over 500 scientific journal articles, 50 chapters and commentaries, and nine books and his research has been featured extensively in current periodicals and newspapers. His latest book, co-authored with Dr Katherine Keyes, is an epidemiology textbook, *Epidemiology Matters: A New Introduction to Methodological Foundations*. Dr Galea has a medical degree from the University of Toronto, and graduate degrees from Harvard University and Columbia University. He was named one of TIME magazine's epidemiology innovators in 2006. He is past-president of the Society for Epidemiologic Research and an elected member of the American Epidemiological Society and of the Institute of Medicine of the National Academies of Science. Dr Galea serves frequently on advisory groups to national and international organizations. He has formerly served as chair of the New York City Department of Health and Mental Hygiene's Community Services Board and as member of its Health Board.

Maxime Goovaerts is a senior student in the Computer Science and Engineering program at Michigan State University. He is interested in the numerical analysis of geospatial data and conducted some of the analysis described in that Chapter 10 during a summer internship with BioMedware, Inc.

Pierre Goovaerts is a Chief Scientist for the R&D Company, Biomedware, Inc, where he conducts NIH funded research on the development of geostatistical methodology for the analysis of health and environmental data. He has been developing and implementing in the software SpaceStat new techniques for mapping, cluster and boundary analysis of health outcomes, with a particular focus on cancer. He has authored more than 130 refereed papers in the field of theoretical and applied geostatistics, and he is a reviewer for 50 international journals. He created in 2001 his own consulting company, PGeostat, LLC. He now acts as a consultant for the Environmental Protection Agency, the Nuclear Regulatory Commission, and he is bringing his expertise to numerous projects in U.S. and Europe dealing with the characterization of air, soil, and water pollution and its impact on human health.

Daniel A. Griffith is Ashbel Smith Professor of Geospatial Information Sciences, University of Texas/Dallas. He earned degrees in mathematics, statistics and geography. He holds awards from the Fulbright and John Simon Guggenheim Memorial Foundations, Association of American Geographers, American Statistical Association, Leverhulme Trust, and Pennsylvania Geographical Society, is an elected Fellow of the American Association for the Advancement of Science, the Regional Science Association International, the New York Academy of Sciences, the Spatial Econometrics Association (a founding fellow), and Fitzwilliam College (University of Cambridge), a past president of the North American Regional Science Council and the Syracuse Chapter of Sigma Xi, and received a Doctor of Science, honoris causa, degree from Indiana University of Pennsylvania. He has published 17 books/monographs and more than 200 papers, given more than 200 invited talks, and a past editor of *Geographical Analysis*. His recent spatial statistical research focuses on urban public health.

Tony H. Grubesic is the Director of the Center for Spatial Reasoning and Policy Analytics and a Professor of Policy Analytics within the College of Public Service and Community Solutions at Arizona State University. His research and teaching interests are in geographic information science, regional development, spatial epidemiology and public policy evaluation. Author of over 100 research publications, his recent work focuses on urban health disparities, critical infrastructure systems, broadband deployment and air transportation systems. Grubesic obtained a B.A. in Political Science from Willamette University, a B.S. in Geography from the University of Wisconsin-Whitewater, a M.A. in Geography and Planning from the University of Akron, and a Ph.D. in Geographic Information Science from the Ohio State University.

Oliver Gruebner is a health geographer and postdoctoral research fellow at Harvard Medical School, Department of Population Medicine, Boston, MA. Working on a spatial epidemiological approach on well-being in urban slums, he received his PhD in 2011 from Humboldt University in Berlin, Germany. He is co-directing the International Summer School "Spatial Epidemiology, Climate and Health: Concepts and Modeling" in Bielefeld, Germany since 2009. His research focuses on applying spatial epidemiological approaches to the complex relationships between the socio-ecological environment and urban population health.

Mark W. Horner is a Professor in the Department of Geography, Florida State University. He currently serves as a U.S. editor for *Transportation* (Springer) and on the editorial boards of *Computers, Environment, and Urban Systems* (Elsevier), *Travel Behavior and Society* (Elsevier), and the *Journal of Transportation and Land Use*. Mark also chairs Transportation Research Board Standing Committee ADD20, Social and Economic Factors of Transportation. He holds a B.S. in geography and regional planning (Salisbury University [MD]), an M.A. In geography (University of North Carolina at Charlotte) and a Ph.D. in geography (Ohio State University). His research and teaching interests are in urban and regional analysis, geographic information science, transportation, and sustainability.

Patrick Hostert is a geographer with a PhD in Remote Sensing. He is head of the Geomatics Lab at the Geography Department at Humboldt University of Berlin, Germany. His major research topics focus around Land System Science, that is, on monitoring rates and spatial patterns at which land use change is occurring. His particular interest lies in areas where little is known about recent changes. Assessing the consequences of land change for ecosystem services and human well-being are his prime research interests. Methodologically, Hostert's focus is on spatially-explicit data analyses, which support a better understanding of rapidly changing systems from local to regional scales and with a strong background in remote sensing.

Michael Jerrett is an internationally recognized expert in Geographic Information Science for Exposure Assessment and Spatial Epidemiology. He is a professor in and the chair of the Department of Environmental Health Science, Fielding School of Public Health, University of California, Los Angeles, and a professor-in-residence, School of Public Health, University of California, Berkeley. Dr. Jerrett earned his PhD in Geography from the University of Toronto (Canada). For the past 16 years, Dr. Jerrett has researched how to characterize population exposures to air pollution and built environmental variables, how to understand the social distribution of these exposures among different groups (e.g., poor vs. wealthy), and how to assess the health effects from environmental exposures. Over the decade, Dr. Jerrett has also studied the contribution of the built and natural environment to physical activity behavior and obesity. In 2009, the United States National Academy of Science appointed Dr. Jerrett to the Committee on "Future of Human and Environmental Exposure Science in the 21st Century." The Committee recently concluded its task with the publication of a report entitled *Exposure Science in the 21st Century: A Vision and a Strategy*. In 2013, the U.S. Environmental Protection Agency appointed Dr. Jerrett to the Clean Air Scientific Advisory Sub-Committee for Nitrogen Oxides. In 2014, Dr. Jerrett was named to the Thomson Reuters List of Highly-Cited Researchers, indicating he is in the top 1% of all authors in the fields of Environment/Ecology in terms of citation by other researchers.

Meijuan Jia is a graduate student in Geography at the University of North Carolina at Charlotte. Her interests are in computational modeling (for example, agent-based model) to infectious diseases.

Pavlos S. Kanaroglou holds a Tier 1 Canada Research Chair in Spatial Analysis and is Professor in the School of Geography and Earth Sciences at McMaster University. His research interests include the development of methods in spatial analysis and the application of such methods to urban transportation and to understanding the relationship between environmental pollution and health.

Mobarak Hossain Khan is an Assistant Professor at the Department of Public Health Medicine at Bielefeld University since early 2007. He obtained his B.Sc. and M.Sc. in Statistics in 1989 and 1990 from Jahangirnagar University, Bangladesh with first class first position in both examinations. He also obtained M.Sc. in Community Health and Health Management in Developing Countries from Medical School of Heidelberg University in 2000. In 2007 he received his Ph.D. in Public Health from Sapporo Medical University School of Medicine, Japan. Currently he is finalizing his habilitation on the tentative topic entitled Urban health and healthcare utilization in a rapidly urbanizing Bangladesh. His teaching and research interest mainly includes public health and health care services in megacities and urban areas, health inequity, climate stressors and migration and health, environmental health including water contamination and arsenicosis problem in Bangladesh, applied statistical modeling of health and demographic outcomes and interdisciplinary health research. He attended many national and international conferences and already published about 90 research articles in peer-reviewed journals and books.

Petros Koutrakis has conducted a number of comprehensive air pollution studies in the United States, Canada, Spain, Chile, Kuwait, Cyprus, and Greece that investigate the extent of human exposures to gaseous and particulate air pollutants. Other research interests include the assessment of particulate matter exposures and their effects on the cardiac and pulmonary health. He is the Director of the EPA/Harvard University Clean Air Research Center and is a Professor in Harvard University's School of Public Health.

Alexander Krämer, MD, PhD had cofounded the first School of Public Health in the German-speaking region of Europe at the University of Bielefeld, Germany, in the year 1994, and since then held the chair of its Department of Public Health Medicine. Krämer's research interests comprise the fields of infectious disease epidemiology, migrant health, global and national burden of disease studies, setting projects like "Health in Megacities" and "Health at Institutions of Higher Education" and the topics "violence and health" and the Impact of Climate Change on Health. He is author of over 300 articles published in peer-reviewed international journals and editor of several books. Krämer has experience in the management and successful scientific conduct of multinational studies within Europe and further internationally, particularly in South Asia and China.

Tobia Lakes is an Assistant Professor for Geoinformatics at the Humboldt University since 2007. She has studied Geography in Duisburg, Belfast and Bonn and has received her PhD on Geoinformation science in urban ecological planning at the Technical University Berlin. Her major research interests are in the fields of developing and applying spatiotemporal analysis and modeling techniques to study human–environment interactions. In current projects she and her team work on land use change and climate change impacts on ecosystem services and particularly on human livelihoods in Berlin, the former Soviet Union, Brazil, Bangladesh, Indonesia, and Malaysia.

Sven Lautenbach is a Geographer and System Scientist by training and holds a position as an Assistant Professor for land use modeling and ecosystem services at the Department of Urban Planning & Real Estate Management, Institute of Geodesy & Geoinformatics, Bonn University, Germany. He obtained his PhD at the mathematics/computer science department of the University of Osnabrück, Germany. From 2005 until 2012 he worked as a researcher at the Department for Computational Landscape Ecology, Helmholtz Centrum for Environmental Research—UFZ, Leipzig Germany. In 2009 he acted as a substitute for the Assistant Professorship for Geoinformatics at the Humboldt University. His research focuses on effects of land use on ecosystem services and human well-being, statistical data analysis of human–environment interactions and decision support systems.

Daniel J Lewis is a researcher in the field of Health and Medical Geography based at London School of Hygiene and Tropical Medicine, UK. His current work applies spatial analysis to understanding how the London 2012 Olympic Games impacted the health and well-being of people in East London over time. Daniel completed his PhD in 2012 at University College London, UK, specializing in spatial measures of access to health care; his PhD, supervised by Prof. Paul Longley, explored the provision of local services in the UK National Health Service (NHS). Daniel also holds a Master's degree in Geographic Information Science, again from UCL; and a Bachelor's degree in Geography from the London School of Economics and Political Science (LSE), UK.

Fernando A. López-Hernández is an Associate Professor in the Department of Quantitative Methods and Computing in the Technical University of Cartagena, Spain. His research focuses on spatial econometrics and economic geography.

Thomas E. McKone, is a Senior Staff Scientist at the Lawrence Berkeley National Laboratory (LBNL) and an Adjunct Professor at the University of California, Berkeley School of Public Health. His research focuses on the development, use, and evaluation of models and data for human health and ecological risk assessments and the health and environmental impacts of energy, industrial, and agricultural systems. McKone has served on the US EPA Science Advisory Board, been a member numerous National Academy of Sciences committees, and worked with the World Health Organization. He is a fellow of the Society for Risk Analysis and a former president of the International Society of Exposure Science (ISES). He earned his PhD in engineering from the University of California, Los Angeles.

Daniel Müller is an agricultural economist and serves as Deputy Head of Department and Senior Research Associate at Leibniz Institute of Agricultural Development in Transition Economies (IAMO), Halle (Saale), Germany. His research focuses on understanding landsystem change, including its multiple repercussions on human welfare, food production, carbon balance and biodiversity. Müller approaches this by applying a range of approaches including econometrics, spatial analysis and artificial intelligence. The current geographic focus of his work is in Central and Eastern Europe, the former Soviet Union, Southeast Asia, and Western China.

Alan T. Murray is a Professor in the College of Computing and Informatics, Professor in the School of Public Health, and Director of the Center for Spatial Analytics and Geocomputation, all at Drexel University. He is currently an editor of *International Regional Science Review*, and former editor of *Geographical Analysis*. His formal training includes a B.S. in mathematical sciences (emphasis in operations research), an M.A. in statistics and applied probability (emphasis in operations research) and a Ph.D. in geography, all from the University of California at Santa Barbara. His research and teaching interests are in: optimization; geographic information science; urban/regional planning and development; and, transportation.

Juan A. Ortega-García coordinates the Pediatric Environmental Health Speciality Unit in the Childrens Hospital University Virgen de la Arrixaca of Murcia (Spain). He is a pediatrician, with expertise in environmental health and public health, and with speciality boards in Prevention of Pediatric Cancer and Environmental Health.

Antonio Páez is Professor of Geography in the School of Geography and Earth Sciences at McMaster University. His fields of expertise include spatial analysis and urban processes. He serves as Editor-in-Chief of the *Journal of Geographical Systems*.

Colleen E. Reid is a Ph.D. candidate in Environmental Health Sciences at the University of California, Berkeley's School of Public Health. Colleen's research interests focus on the health impacts of climate change, specifically using novel spatial exposure assessment methods and causal inference epidemiology to estimate the health impacts of natural air pollutants and extreme heat. Colleen also piloted the development of the heat vulnerability index, a way to map populations vulnerable to extreme heat events across the US. Prior to her doctoral studies, Colleen was an Environmental Health Fellow at the Global Change Research Program within the U.S. EPA.

Manuel Ruiz is an Associate Professor in the Department of Quantitative Methods and Computing at the Technical University of Cartagena, Spain. His fields of expertise include spatial econometrics, nonparametric statistic and symbolic analysis.

Xun Shi is an Associate Professor in the Department of Geography, Dartmouth College. His research interests include the applications of GIS and spatial analysis in public health, epidemiology and socioecnomic systems.

Umaporn Siangphoe is a PhD student in Biostatistics at the Virginia Commonwealth University, where she earned a master's degree in Biostatistics. Her research interests include statistical computing, statistical methods for high-throughput genomic data, and spatial epidemiology.

Wenwu Tang is an Assistant Professor in the Geography and Earth Sciences Department at the University of North Carolina at Charlotte. His research interests are in the domain of Agent-Based Models and Spatio-Temporal Simulation, high-performance and parallel computing as well as GIS & Spatial Analysis and Modeling.

Fahui Wang is a James J. Parson Professor and department chair of the Department of Geography and Anthropology, Louisiana State University. His research interests include the applications of GIS and spatial analysis in public health, crime, and urban and regional development.

Mary Ward is a Senior Investigator in the Occupational and Environmental Epidemiology Branch at the National Cancer Institute. She received an M.S. in ecology from the University of Tennessee and a Ph.D. in epidemiology from The John Hopkins School of Hygiene and Public Health. Her research focuses on environmental and occupational causes of cancer, with special emphasis on pesticides and nitrates in relation to the etiology of childhood leukemia, non-Hodgkin's lymphoma, gastrointestinal cancers, and thyroid cancer. She has expertise in environmental epidemiology and using geographic information systems (GIS) for visualization and analysis of environmental exposures. She participates in interdisciplinary collaborations to develop new methods of exposure assessment for epidemiologic studies of cancer risk in relation to drinking water contaminants and agricultural pesticides.

Mark R. Welford is a physical geographer and an Associate Professor in the Department of Geology & Geography at Georgia Southern University. His research interests include the geodynamics, biogeography and conservation of tropical montane environments and the spatial dynamics of modern and historical pandemics. He holds a BSc from Coventry Polytechnic, an MS from the University of Idaho, and a PhD from the University of Illinois, Urbana-Champaign.

David C. Wheeler is Assistant Professor in the Department of Biostatistics in the School of Medicine at Virginia Commonwealth University. He was previously a Cancer Prevention Fellow at the National Cancer Institute. He has graduate degrees in statistics and geography from The Ohio State University and a graduate degree in public health from Harvard University. His primary research activities are in the areas of spatial epidemiology and cancer control and prevention. His current research interests include modeling spatialtemporal variability in cancer risk and applying statistical methods to identify and predict risk factors for cancer. His recent research projects include examining the association between ultraviolet exposure and risk of cancer in the AARP Diet and Health Study and developing a data mining approach to predict occupational exposure assessments for diesel exhaust in the New England Bladder Cancer Study.

Acknowledgments

Chapter 3 (DeLuca and Kanaroglou)

The authors would like to thank Ervin Ruci from Geocoder.ca who provided one of the geocoded datasets used in this analysis free of charge. The authors also would like to acknowledge the very helpful comments provided by the reviewers.

Chapter 4 (Páez, López-Hernández, Ortega-García, and Ruiz)

Manuel Ruiz was partially supported by MINECO (Ministerio de Economíay Competitividad) and FEDER (Fondo Europeo de Desarrollo Regional) projects ECO2012-36032-C03-03 and MTM2012-35240; and COST Action IS1104 "The EU in the new economic complex geography: models, tools and policy evaluation".

Chapter 5 (Bossak and Welford)

The authors would like to thank Kathryne Henderson, Justin Marsh, and Adam Middleton for their assistance with portions of this research project.

Chapter 6 (Delmelle, Jia, Dony, Casas, and Tang)

The authors thank Dr. Alejandro Varela and Dr. Diego Calero, former Ministers of Health for the municipality of Cali for providing access to the data. We also thank Dr. Jorge Rojas, epidemiologist for the Public Health Municipality of City of Cali for his valuable feedback throughout the project.

Chapter 8 (Wheeler and Siangphoe)

We would like to thank the following collaborators for providing consultation and access to the data from the National Cancer Institute-Surveillance, Epidemiology, and End Results Interdisciplinary Case-Control Study of NHL: Mary Ward, Lindsay Morton, Patricia Hartge, Anneclaire De Roos, James Cerhan, Wendy Cozen, and Richard Severson. We also acknowledge Robin Bliss for sharing computer code to estimate generalized additive models.

Chapter 9 (Gruebner, Khan, Lautenbach, Müller, Krämer, Lakes, Hostert, and Galea)

This study was funded by the German Research Foundation (DFG, HO 2568/5-3 and GR 4302/1-1).

Chapter 10 (Goovaerts and Goovaerts)

This research was funded by grant 1R21 ES021570–01A1 from the National Cancer Institute. The views stated in this publication are those of the author and do not necessarily represent the official views of the NCI.

Chapter 12 (Griffith)

This work was supported by a grant from the National Science Foundation: award BCS-0552588. Dust and soil assays were completed by Drs. A. Hunt and D. Johnson.

Chapter 1
Introduction: Spatial Analysis and Health

Eric Delmelle and Pavlos S. Kanaroglou

Medical Geography or Spatial Epidemiology is concerned with two fundamental questions: (1) where and when do diseases tend to occur? and (2) why do such patterns exist? The field has experienced substantial growth over the last decade with the widespread recognition that the concept of "place" plays a significant role in our understanding of individual health (Kwan, 2012) while advances in geographical modeling techniques have made it easier to conduct spatial analysis at different granularities, both spatially and temporally (Cromley and McLafferty, 2011). Several journals (for example, *Health and Place*, *Spatial and Spatio-Temporal Epidemiology*, *International Journal of Health Geographics*, *Geospatial Health* and *Environmental Health*) have a long tradition to publishing research on topics in Spatial Epidemiology.

This introductory chapter reviews some contemporary themes and techniques in medical geography. Specifically, we discuss the nature of epidemiological data and review the best approaches to geocode and map information while maintaining a certain level of privacy. Analytical and visualization methods can inform public health decision makers of the reoccurrence of a disease at a certain place and time. Clustering techniques, for instance, can inform on whether diseases tend to concentrate around specific locations. We examine the role of the environment in explaining spatial variations of disease rates. Next, we address the importance of accessibility models, travel estimation and the optimal location of health centers to reduce spatial inequalities when accessing health services. We also review the increasing contributions of volunteered geographic information and social networks, helping to raise public awareness of the risk posed by certain diseases, especially vector-borne diseases following a disaster. The concepts of scale and uncertainty are discussed throughout as they are known to affect the suitability of certain methods and consequently impact the stability of the results. Some of the concepts set forth are illustrated with a data set of a 2010 dengue fever outbreak in Cali, Columbia. We conclude this chapter by discussing the layout and contributions of this volume to Spatial Epidemiology.

Mapping Epidemiological Data

Medical geography studies the relationship between place and health; specifically it evaluates how the physical and social environments shape the health and well-being of different individuals (Cromley and McLafferty, 2011). Geographical Information Systems (GIS) and spatial analysis provide unique tools to determine where and when a particular disease has occurred and could resurface in the future. Accurate spatial (and temporal) data is thus critical to identifying such patterns.

Epidemiological data comes at different scales (disaggregated or aggregated data) and different levels of accuracy. Addresses can be transformed into geographic coordinates by means of geocoding (Goldberg, Wilson, and Knoblock, 2007), but the process may be sensitive to the completeness of the addresses and the quality of the underlying network (Zandbergen, 2009; Jacquez, 2012). Scatter maps are used to display geocoded, disaggregated data; for example, in Figure 1.1(a), each dot is an occurrence of a reported dengue fever[1] case in Cali, an urban area of Colombia, during an outbreak in 2010 (Delmelle, and Casas et al. 2013). Besides cartographic outputs, GIS can link spatially-explicit data to environmental and census data using one of the available spatial join algorithms. This approach facilitates our understanding on the role that the physical and social environment may play on health and well-being.

Due to *privacy* concerns, epidemiological data may be geomasked,[2] or be *aggregated* at a certain level of census geography, for instance at the county or postal and zip code level. Figure 1.1(c) uses a proportional symbology to map the variation of dengue cases per neighborhood, suggesting an uneven pattern. Other techniques, such as choropleth mapping, are widely used to display disease rates across an area. Figure 1.1(e) suggests that dengue fever rates are not randomly distributed, possibly owing to population density, shown in Figure 1.1(d).

A concept that has received significant attention in medical geography is the level of *spatial scale* at which the analysis is conducted (Diez Roux, 2001). As pointed out by Root (2012), "the impact of neighborhoods on health is uniquely geographic." Spatially aggregating data, however, give rise to the modifiable areal unit problem (MAUP). This is because the basic assumption of any aggregation scheme is that there is uniformity within but sharp contrast among the defined geostatistical areas (Cromley and McLafferty, 2011). Using different boundaries an analysis may lead to significantly different results. It has thus become clear that it is increasingly important to conduct analysis at several granularities of scale.

Visualizing Disease Patterns and Clustering Techniques

Clustering techniques help identify whether disease events are randomly distributed and if not, where clusters may be located. Delimiting the extent of those clusters is important for the determination of areas potentially at risk. In this context, the contributions of exploratory spatial data analysis (ESDA), including kernel density estimation (KDE), are well documented in the literature (Delmelle et al., 2011; Delmelle, 2009; Cromley and McLafferty, 2011; Kulldorff, 1997). Eisen and Eisen (2011) and Vazquez-Prokopec et al. (2009) underline the importance of GIS and ESDA to monitor vector-borne diseases, where prompt space-time monitoring techniques are critical for timely detection and mitigation purposes. Spatial analytical methods can generate disease distribution maps revealing significant information in terms of direction, intensity of a disease, as well as its likelihood to spread to inhospitable areas.

---

1   Dengue fever is a vector-borne disease transmitted from one individual to another by the the the *Aedes Aegypti* mosquito (Gubler, 1998).

2   Geomasking is a process which explicitly introduces a small perturbation in the spatial coordinates of the events when those are presented in the form of a map (Kwan, Casas, and Schmitz, 2004).

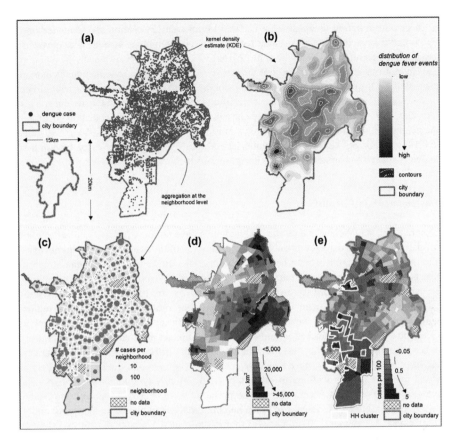

Figure 1.1   Dengue fever cases for the city of Colombia, 2010 (geocoded at the street intersection level), in (a). Kernel density estimation in (b), aggregated dengue cases per neighborhood in (c), population density in (d) and dengue fever rates in (e)

The ESDA techniques are used traditionally to identify spatial and more recently spatio-temporal patterns. The statistical significance of identified clusters is tested by Monte Carlo simulations. Kulldorff et al. (1998) and Levine (2006) have developed spatial analytical tools (SatScan and CrimeStat, respectively) to detect clusters of point events and then to conduct simulations for the evaluation of the statistical significance of those clusters. Such tools are now incorporated into commercial GIS packages and are available to the common GIS user (Fischer and Getis, 2009). An example of an ESDA technique is the kernel density estimation, illustrated in Figure 1.1(b) for monitoring hot spots of dengue fever. In essence, the map shows areas with greater expectation of dengue fever occurrences. Contours reinforce the extent of such hot spots.

Space-time clustering techniques are still in their development phase, partly due to their computational challenges (Jacquez, Greiling, and Kaufmann, 2005; Robertson et al., 2010). Research on space-time clustering tests has focused mainly on uncertainty, which is introduced through biased or incomplete data, perhaps because of incorrect addresses or inaccurate reported diagnosis (Lam, 2012; Malizia, 2012). Within the limits imposed by computational requirements, much recent research attempts to remedy weaknesses in visualization techniques (Delmelle et al., 2014a).

Nearby observations may exhibit similarity (Tobler, 1970). *Spatial autocorrelation*, estimated by a global Moran's I statistic (Moran, 1950), measures whether nearby data (generally aggregated) are dependent on one another, while its local statistic counterpart (Anselin, 1995) informs on where those clusters of similar observations tend to occur. For the neighborhood data, for example, shown in Figure 1.1(e), the estimated Moran's I value is 0.14, indicating a weak autocorrelation. The Moran's I statistic can be extended in time to detect space-time autocorrelation (Goovaerts and Jacquez, 2005).

Environment and Health

Geographers, statisticians and public health experts have not only focused on the detection of spatial clusters of diseases, but also on the evaluation of the association of natural factors and the built environment with health and individual well-being. The hypothesis here is that geographic behaviors and outcomes of health (that is, *health disparities*) cannot only be explained by *individual* factors; *neighborhood factors* are likely to play a contributing role (Diez Roux, 2001; Krieger et al., 2003). For instance, individuals living in rural regions will experience geographic barriers in traveling to health services, given that the numbers of facilities that can be reached within a certain time budget is much smaller than in urban areas. Women living in poor areas may find it particularly difficult to access mammography facilities when they do not have a vehicle and must rely on public transportation (Peipins et al., 2011). Children walking to schools or living in an environment where parks and playgrounds are readily accessible may be prone to be more active than others (Cooper et al., 2010). Clusters of violence in urban neighborhoods may be related to alcohol outlets (Grubesic and Pridemore, 2011). These examples illustrate the breadth of pathways through which environmental factors give rise to *health disparities* over space.

Other non-neighborhood factors may play an important role in shaping our understanding of the potential for outbreaks of certain diseases. As suggested by Comrie (2007), climatic variation and weather-related factors is likely to create particularly suitable conditions for certain vectors to thrive and potentially increase the geographical extent of vector-borne diseases. Spatial regression and multilevel modeling are examples of some of the key methods that were developed for the evaluation of the impact of neighborhoods on health (Cromley and McLafferty, 2011). Variation in the dependent variable (disease rate, accessibility) can be explained by a set of individual characteristics (age, gender, income and education for instance), environmental factors (neighborhood characteristics) and spatial terms accounting for the presence of spatial autocorrelation. Geographically Weighted Regression quantifies the spatial importance of each explanatory variable on the dependent variable (Fotheringham, Brunsdon, and Charlton, 2003).

What defines a neighborhood and the concept of scale will affect which methodology is used and ultimately the results. Krieger et al. (2003) underline that the geographic scale of secondary data, such as socio-economic characteristics, may determine the level of aggregation at which a study is conducted. Evaluating the effect of different artificial boundaries is thus necessary by repeating those analyses at different scales. Using only the local scale of a neighborhood may not account for the entire activity space of an individual (Cummins, 2007). GPS and GIS technologies appear to be particularly useful in mapping the daily activity of individuals and determining the extent of an individual's neighborhood (Kwan, 2004). Also, in studies of exposure analysis it is important to take account of the residential history of subjects under study, although relevant data are not always available (Root, 2012).

Health Care Provisions and Accessibility

Accessibility is a critical element of any health care system. In an ideal system, every member of a community should have similar access to health care professionals; however, a perfect match between supply and demand is not possible, leading to spatial inequalities (Cromley and McLafferty, 2011; Parker and Campbell, 1998). In rural areas, for instance, access to care is constrained due to longer travel distances and scarcity of providers.

A critical objective of a health care system is to guarantee a minimum level of geographic access to primary care services. Accessibility below that level can make the difference between life and death or between a controlled outbreak and an epidemic (Higgs, 2004). *Travel impedance* is thus a contributing factor in the utilization of health care services (Lovett et al., 2002; Delamater et al., 2012). Impedance can be evaluated with different metrics such as travel distance (Euclidean or network), or travel time. The latter may be a more precise measure since it accounts for en-route conditions (Cromley and McLafferty, 2011). Delmelle and Cassell et al. (2013) propose a GIS-based methodology to estimate travel impedance for children with birth defects, suggesting that children living in urban areas have a much lower travel burden than children in rural areas. Having to rely on public transportation, urban residents of low-income areas may be at a disadvantage. Several internet-based providers (Open Street Map, GoogleMaps) can estimate travel impedance; however, when using those providers, careful attention must be paid to confidentiality issues, the accuracy of the travel estimates themselves and the restriction in the number of queries that can be submitted to those providers.

One way of visualizing health care accessibility is by means of KDE, as discussed in previous sections of this introduction. In this case, one can estimate the density of service providers over space, revealing differences in access (Lewis and Longley, 2012; Casas, Delmelle, and Varela, 2010). Another popular approach is the *two-step floating catchment area* (Luo and Wang, 2003) which evaluates the availability of health services in regards to population need. Methods based on gravity models can capture the interaction of an individual with a health facility, using several of its characteristics, including size and quality of service. Nevertheless, these aforementioned approaches remain theoretical in nature. Although more difficult to obtain due to confidentiality concerns, *revealed accessibility* provides actual information on the utilization of health services, allowing the identification of facilities that are underutilized or overutilized while delimiting the catchment area of any facility. It is therefore desirable for researchers to obtain information on the utilization of health services at a disaggregated level.

Disparities in geographical accessibility can be reduced by selecting the *optimal location and capacity* for new health centers or when existing facilities are to be upgraded or their size calibrated (Wang, 2012). Operations Research and Location-Allocation Modeling are proven techniques that effectively answer questions as to where new facilities should be opened and of what capacity in order to maximize coverage and minimize travel. More behavioral research is needed coupled with simulation modeling regarding the utilization of health care services following a change in the structure of a network of facilities.

Volunteered Geographic Information

Boulos et al. (2011) discuss the increasing interest among health researchers to disseminate analytical functionality over the internet, partly due to the massive epidemiological datasets that are becoming available through social networks, such as twitter (Freifeld et al. (2010); Chunara, Andrews, and Brownstein (2012)). However, the development of analytical methods over the web are computationally challenging; for instance commonly used spatial analytical functions, such as the KDE, are time consuming as web-based GIS services (Dominkovic et al., 2012; Delmelle et al., 2014b). Adding the dimension of time, for the development for example of space-time clustering techniques, presents serious research challenges.

Participation of *volunteers* in mapping health information has the inherent potential to foster community involvement, ultimately improving the well-being of individuals (Skinner and Power, 2011). This is critical especially for developing countries where there are limited financial resources and GIS expertise (Fisher and Myers, 2011; Kienberger et al., 2013). Following the Haiti earthquake (Zook et al. 2010), for example, volunteers over the internet helped to create a geospatial database that proved to be very useful for the allocation of resources to places of higher need.

Structure of the Book

Previous sections in this introduction highlight that there is an established tradition of the application of existing spatial analytic methods and techniques to public health. The reverse, however, is also true. Public health issues pose new challenges for spatial analysts forcing them to innovate and develop new methodologies, thus enhancing the field of spatial analysis. The symbiotic relationship between spatial analysis and health is the subject matter of this book. The 16 chapters included in the volume discuss methods and techniques that are applied to substantive issues of health. In dividing the material into sections we had two choices. The first was to group the chapters by type of methodology and the second by the substantive area where the methods are applied. Although both ways have their drawbacks and advantages, we selected the latter method since many researchers in the field are interested in specific areas of health and this way the book will be of better service to the research community.

We divided the material into five sections. Section 1 covers purely methodological issues in spatial analysis that have wide applicability in a variety of public health issues. The four sections that follow focus on methods as they are applied to: infectious disease (Section 2), chronic disease (Section 3), exposure (Section 4), and accessibility (Section 5). A more detailed description of the contents in the five sections follows.

Section 1: Methods

This section consists of three chapters. In Chapter 2 by Linda Beale sets the stage for subsequent chapters by discussing the benefits from using GIS in Spatial Epidemiology. She covers substantial ground by describing the uses of GIS in visualizing, exploring and modeling methods that have been developed specifically for the use and exploitation of the spatial properties of epidemiological data. One issue that Beale brings to the forefront is that for all these methods to yield fruits it is important for the data to be properly and correctly geo-referenced.

The theme of Chapter 2 ties in with Chapter 3 by DeLuca and Kanaroglou which evaluates three popular and commercially available methods of automatic geocoding. These are (1) ESRI's Online geocoding available through ArcGIS.com, (2) an online geocoding service provider, GeoCoder.ca and (3) the freely available Yahoo geocoding API. The data set used for the evaluation is residential addresses in the city of Hamilton, Canada, using the parcel fabric of the city as a baseline. Results indicate that a disturbing proportion of the geocodes can be off significantly and in some cases by as much as six kilometers. Errors introduces through geocoding may have severe implication for studies in health geography. These include exposure misclassification or erroneous assessment of accessibility.

Chapter 4 by Páez, López-Hernández, Ortega-García and Ruiz compares two seemingly overlapping techniques that are used to detect the concentration of events over space. These are the techniques of clustering and co-occurrence. The chapter uses simulations as well as data on events of pediatric cancer from Murcia, Spain, to demonstrate the concepts. The results indicate that the two are not competing techniques but can provide complementary results for the better understanding of the process that generates the events.

Section 2: Infectious Disease

Chapter 5 by Bossak and Welford opens the theme on infectious disease by examining one of the most lethal epidemics ever that killed 30–50% of the population in certain parts of Europe within four years. This was the mid-14th century Medieval Black Death. The authors examine the spatial and temporal aspects of the disease bringing historical but also environmental and socio-economic data within a modern spatial analytical framework offering a fresh look at the epidemic.

Chapter 6, by Delmelle, Jia, Dony, Casas and Tang, discusses space-time visualization methods to examine and detect infectious disease outbreaks. The method proposed in this chapter is the well-known kernel density estimation, a spatial technique, extended to include the temporal dimension. The proposed method is applied to dengue fever data from Cali, Columbia, for the year 2010.

In Chapter 7, Carrel discusses methods for the exploration and identification of spatial patterns in the changing genetic character of infectious disease pathogens, such as viruses, bacteria and protozoa. She claims that understanding where and when the pathogen genetic changes are taking place is crucial in preventing or containing infectious disease outbreaks. Several exploratory methods are discussed in this context, including interpolation and clustering techniques, as well as modeling such as geographically weighted regression.

Section 3: Chronic Disease

This section consists of three chapters. Chapter 8 by Wheeler and Siangphoe discusses a family of models that derive from the generalized additive model, as they relate to the analysis of the spatial variation of disease risk. Several modeling approaches are compared within a simulation framework. The methods are applied to data from Los Angeles County for the investigation of the spatial variation of risks for non-Hodgkin lymphoma.

Chapter 9 by Gruebner, Khan, Lautenbach, Müller and Krämer examines the mental well-being in urban areas. More specifically, the focus is in urban slums. Using generalized linear regression models and spatial autocorrelation the authors analyze a cross section of survey data collected in the slums of Dhaka. The WHO-5 Well-being Index was used to assess mental well-being. The authors test the hypothesis that this metric is related to the socio-ecological environment in the slums.

In Chapter 10 Goovaerts and Goovaerts introduce a variety of methods for the visualization and exploration through spatial analysis of a time series of health data. The techniques include 3D displays, binomial kriging, joinpoint regression and cluster analysis. The methods are applied to incidences of late-stage breast cancer diagnosis for counties in Michigan Lower Peninsula over the period 1985–2007.

Section 4: Exposure

Exposure analysis is the subject of three chapters in Section 4 of the book. Chapter 11 by Adams and Kanaroglou explores a recurring theme in air quality exposure that relates to assigning outdoor exposure estimates to subjects that are calculated as long-term mean concentrations of ambient air quality from incomplete time series data sets. A method is proposed that appears to produce better estimates of long-term mean concentrations. The method is evaluated through simulations and data from Paris, France.

Chapter 12 by Griffith, examines the correlation between metal concentrations found in yard soil and in dust from inside residences to blood lead levels of children less than six years of age living in those residences. Taking into account socio-demographic characteristics, the relationship is examined at various geographic scales using canonical correlation analysis. The primary data used were collected from Syracuse, New York, in the time period 1992–96.

Chapter 13 by Jerrett, Reid, McKone, and Koutrakis is adapted from a lengthy 2012 National Academy of Science report on exposure science. The chapter provides an overview of the state of the art in exposure science and highlights potential future directions, especially with the emergence of new technologies for the collection of more accurate exposure data. New concepts, such as "ubiquitous," "embedded" and "participatory" sensing, are discussed that are to have substantial relevance for exposure science in the 21st century.

Section 5: Accessibility and Health

The last section of the book consists of four chapters. Chapter 14 by Murray and Grubesic provides an overview of optimization models developed in location analysis to support strategic decisions for the siting of hospitals, clinics and health care facilities. The objective of such models is to ensure that given a spatial distribution of the population the number, size, and location of facilities are sufficient to guarantee adequate accessibility to health care. The chapter serves as an introduction to the other chapters that follow in this section.

Chapter 15 by Wang, Fu and Shi focuses on the accessibility of cancer centers in the United States, as proposed by the National Cancer Institute. The authors use spatial optimization methods, such as integer and quadratic programming to evaluate two scenarios of improving population accessibility to the centers. The first scenario is the allocation of additional resources to existing centers and the second is the establishment of additional centers.

Lewis in Chapter 16 digs deeper into the concept of accessibility to healthcare services by examining how it is conceptualized, qualified, quantified and modeled. He focuses on the spatial dimensions after first describing a holistic view of conceptualizing access.

Important is the distinction he draws between an epidemiological and a spatial framework within which access is conceptualized and analyzed.

The last chapter in this section by Coutts and Horner examines the relationship between the accessibility of people to green space and premature mortality. The specific study employs regression analysis using death certificates for the state of Florida in the time period 2000–2012. Proximity to green space was estimated with the help of GIS. The developed model controls for social and demographic characteristics of subjects. The results, although exploratory, indicate that distance of place or residence from green space increases the likelihood of premature death.

Concluding Remarks

In this chapter we have discussed current and emerging research themes in spatial analysis as they apply to issues of public health. We then provided an overview of the contributions to those issues through the volume in hand. Although the majority of the chapters in the volume are heavily methodological in nature, two are focusing on conceptual contributions. Chapter 7 by Carrel proposes that the timing and location of genetic mutations of pathogens are crucial in understanding the spread of infectious disease. To test this proposition she recommends the use of well-known methods in spatial analysis, including interpolation, clustering and regression. Also, Lewis in Chapter 16 discusses the conceptualization of accessibility and he proposes several measures for it derived from spatial analysis methods.

For the rest of the chapters, while all dwell on methods, one can classify them into three types. The first group of chapters deals with spatial data and proposes ways to enhance the accuracy of georeferencing or to combine different databases into a single spatial framework that allows a richer analysis of public health phenomena (DeLuca and Kanaroglou, Bossak and Welford). The second group reviews methods that are suitable for specific problems of public health and examine new and innovative technologies that are expected to play a significant role in Spatial Epidemiology for the years to come (Jerrett, Reid, McKone and Koutrakis; Murray and Grubesic). All the rest of the chapters in the volume can be considered to form a third group that proposes the use of a combination of known or new spatial analysis methods to study various types of problems in public health. It is interesting that in some of the chapters the application of a combination of well-established spatial analytic methods can provide insights to phenomena that are not clear with the use of a single method.

The substantive issues discussed in the volume go beyond to identifying relationships between disease and socio-demographic factors and into conceptual or institutional issues. In some instances old problems, such as the mid-14th century Medieval Black Death in Europe, are analyzed within a GIS using modern spatial analytic modeling methods (Bossak and Welford). We believe that academics as well as practitioners in the field will find the material interesting and informative and will make use of the methodologies discussed in this book in their own research.

References

Anselin, L. 1995. Local indicators of spatial association—LISA. *Geographical Analysis*, 27(2): 93–115.

Boulos, K.M., B. Resch, D. Crowley, J. Breslin, G. Sohn, R. Burtner, W. Pike, E. Jezierski, and K-Y. Chuang. 2011. Crowdsourcing, citizen sensing and sensor web technologies for public and environmental health surveillance and crisis management: Trends, OGC standards and application examples. *International Journal of Health Geographics*, 10(1): 67.

Boulos, K.M. and S. Wheeler. 2007. The emerging Web 2.0 social software: An enabling suite of sociable technologies in health and health care education. *Health Information & Libraries Journal*, 24(1): 2–23.

Casas, I., E. Delmelle, and A. Varela. 2010. A space-time approach to diffusion of health service provision information. *International Regional Science Review*, 33(2): 134–56.

Chunara, R., J.R. Andrews, and J.S. Brownstein. 2012. Social and news media enable estimation of epidemiological patterns early in the 2010 Haitian cholera outbreak. *The American Journal of Tropical Medicine and Hygiene*, 86(1): 39–45.

Comrie, A. 2007. Climate change and human health. *Geography Compass*, 1(3): 325–39.

Cooper, A.R., A.S. Page, B.W. Wheeler, P. Griew, L. Davis, M. Hillsdon, and R. Jago. 2010. Mapping the walk to school using accelerometry combined with a global positioning system. *American Journal of Preventive Medicine*, 38(2): 178–83.

Cromley, E. and S. McLafferty. 2011. *GIS and Public Health*. New York: Guilford Press.

Cummins, S. 2007. Commentary: Investigating neighbourhood effects on health—avoiding the 'local trap.' *International Journal of Epidemiology*, 36(2): 355–7.

Delamater, P.L., J.P. Messina, A.M. Shortridge, and S.C. Grady. 2012. Measuring geographic access to health care: Raster and network-based methods. *International Journal of Health Geographics*, 11(1): 15.

Delmelle, E. 2009. Point pattern analysis. *International Encyclopedia of Human Geography*, 8: 204–11.

Delmelle, E., I. Casas, J. Rojas, and A. Varela. 2013. Modeling spatio-temporal patterns of dengue fever in Cali, Colombia. *International Journal of Applied Geospatial Research*, 4(4): 58–75.

Delmelle, E., C. Cassell, C. Dony, and E.T. Radcliff, J-P., Siffel, C. Kirby, R.S. 2013. Modeling travel impedance to medical care for children with birth defects using geographic information systems. *Birth Defects Research Part A: Clinical and Molecular Teratology*.

Delmelle, E., E. Delmelle, I. Casas, and T. Barto. 2011. H.E.L.P: A GIS-based health exploratory analysis tool for practitioners. *Applied Spatial Analysis and Policy*, 4(2): 113–37.

Delmelle, E., C. Dony, I. Casas, M. Jia, and W. Tang. 2014a. Visualizing the impact of space-time uncertainties on dengue fever patterns. *International Journal of Geographical Information Science* (ahead-of-print): 1–21.

Delmelle, E.M., H. Zhu, I. Casas and W. Tang 2014b. A web-based geospatial toolkit for the monitoring of Dengue Fever. *Applied Geography*, 52: 144–52.

Diez Roux, A.V. 2001. Investigating neighborhood and area effects on health. *American Journal of Public Health*, 91(11): 1783–89.

Dominkovics, Pau et al. 2011. Development of spatial density maps based on geoprocessing web services: Application to tuberculosis incidence in Barcelona, Spain. *International Journal of Health Geographics*, 10(1): 62.

Eisen, L. and R. Eisen. 2011. Using geographic information systems and decision support systems for the prediction, prevention, and control of vector-borne diseases. *Annual Review of Entomology*, 56(1): 41–61.

Fischer, M.M. and A. Getis. 2009. *Handbook of Applied Spatial Analysis: Software Tools, Methods and Applications*. Berlin, Heidelberg and New York: Springer.

Fisher, R.P. and B.A. Myers. 2011. Free and simple GIS as appropriate for health mapping in a low resource setting: A case study in eastern Indonesia. *International Journal of Health Geographics*, 10: 15.

Fotheringham, A.S., C. Brunsdon, and M. Charlton. 2003. *Geographically Weighted Regression: The Analysis of Spatially Varying Relationships*. New York: John Wiley & Sons.

Freifeld, C.C., R. Chunara, S.R. Mekaru, E.H. Chan, T. Kass-Hout, A.A. Iacucci, and J.S. Brownstein. 2010. Participatory epidemiology: Use of mobile phones for community-based health reporting. *PLoS Medicine*, 7(12): e1000376.

Goldberg, D.W., J.P. Wilson, and C.A. Knoblock. 2007. From text to geographic coordinates: The current state of geocoding. *Journal of the Urban and Regional Information Systems Association*, 19(1): 33–46.

Goovaerts, P. and J. Jacquez. 2005. Detection of temporal changes in the spatial distribution of cancer rates using local moran's I and geostatistically simulated spatial neutral models. *Journal of Geographical Systems*, 7: 137–59.

Grubesic, T.H. and W.A. Pridemore. 2011. Alcohol outlets and clusters of violence. *International Journal of Health Geographics*, 10(1): 30.

Gubler, D.J. 1998. Dengue and dengue hemorrhagic fever. *Clinical Microbiology Reviews*, 11(3): 480–96.

Jacquez, G. 2012. A research agenda: Does geocoding positional error matter in health GIS studies? *Spatial and Spatio-temporal Epidemiology*, 3(1): 7–16.

Jacquez, G., D. Greiling, and A. Kaufmann. 2005. Design and implementation of a space-time intelligence system for disease surveillance. *Journal of Geographical Systems*, 7(1): 7–23.

Kienberger, S., M. Hagenlocher, E. Delmelle, and I. Casas. 2013. A WebGIS tool for visualizing and exploring socioeconomic vulnerability to dengue fever in Cali, Colombia. *Geospatial Health*, 8(1): 313–16.

Krieger, N., J.T. Chen, P.D. Waterman, D.H. Rehkopf, and S. Subramanian. 2003. Race/ethnicity, gender, and monitoring socioeconomic gradients in health: A comparison of area-based socioeconomic measures-the public health disparities geocoding project. *American Journal of Public Health*, 93(10): 1655–71.

Krieger, N., P. Waterman, K. Lemieux, S. Zierler, and J.W. Hogan. 2001. On the wrong side of the tracts? Evaluating the accuracy of geocoding in public health research. *American Journal of Public Health*, 91(7): 1114.

Kulldorff, M. 1997. A spatial scan statistic. *Communications in Statistics—Theory and Methods*, 26(6): 1481–96.

Kulldorff, M., K. Rand, G. Gherman, G. Williams, and D. DeFrancesco. 1998. SaTScan v 2.1: Software for the spatial and space-time scan statistics. Bethesda, MD: National Cancer Institute.

Kwan, M-P. 2004. GIS methods in time-geographic research: Geocomputation and geovisualization of human activity patterns. *Geografiska Annaler B*, 86: 205–18.

Kwan, M-P. 2012. Geographies of health. *Annals of the Association of American Geographers*, 102(5): 891–2.

Kwan, M-P., I. Casas, and B. Schmitz. 2004. Protection of geoprivacy and accuracy of spatial information: How effective are geographical masks? *Cartographica: The International Journal for Geographic Information and Geovisualization*, 39(2): 15–28.

Lam, N.S-N. 2012. Geospatial methods for reducing uncertainties in environmental health risk assessment: Challenges and opportunities. *Annals of the Association of American Geographers*, 102(5): 942–50.

Levine, N. 2006. Crime mapping and the crimestat program. *Geographical Analysis*, 38(1): 41–56.

Lewis, D.J. and P.A. Longley. 2012. Patterns of patient registration with primary health care in the UK national health service. *Annals of the Association of American Geographers*, 102(5): 1135–45.

Lovett, A., R. Haynes, G. Sünnenberg, and S. Gale. 2002. Car travel time and accessibility by bus to general practitioner services: A study using patient registers and GIS. *Social Science & Medicine*, 55(1): 97–111.

Luo, W., and F. Wang. 2003. Measures of spatial accessibility to health care in a GIS environment: Synthesis and a case study in the Chicago region. *Environment and Planning B: Planning and Design*, 30(6): 865–84.

Malizia, N. 2013. The effect of data inaccuracy on tests of space-time interaction. *Transactions in GIS*, 17(3): 426–51.

Moran, P.A. 1950. Notes on continuous stochastic phenomena. *Biometrika*, 37(1–2): 17–23.

Parker, E.B. and J.L. Campbell. 1998. Measuring access to primary medical care: Some examples of the use of geographical information systems. *Health & Place*, 4(2): 183–93.

Peipins, L.A., S. Graham, R. Young, B. Lewis, S. Foster, B. Flanagan, and A. Dent. 2011. Time and distance barriers to mammography facilities in the Atlanta metropolitan area. *Journal of Community Health*, 36(4): 675–83.

Robertson, C., T.A. Nelson, Y.C. MacNab, and A.B. Lawson. 2010. Review of methods for space–time disease surveillance. *Spatial and Spatio-temporal Epidemiology*, 1(2): 105–16.

Root, E.D. 2012. Moving neighborhoods and health research forward: Using geographic methods to examine the role of spatial scale in neighborhood effects on health. *Annals of the Association of American Geographers*, 102(5): 986–95.

Skinner, M.W. and A. Power. 2011. Voluntarism, health and place: Bringing an emerging field into focus. *Health & Place*, 17(1): 1–6.

Tobler, W.R. 1970. A computer movie simulating urban growth in the Detroit region. *Economic Geography*, 46: 234–40.

Vazquez-Prokopec, G.M., S.T. Stoddard, V. Paz-Soldan, A.C. Morrison, J.P. Elder, T.J. Kochel, T.W. Scott, and U. Kitron. 2009. Usefulness of commercially available GPS data-loggers for tracking human movement and exposure to dengue virus. *International Journal of Health Geographics*, 8(1): 68.

Wang, F. 2012. Measurement, optimization, and impact of health care accessibility: A methodological review. *Annals of the Association of American Geographers*, 102(5): 1104–12.

Zandbergen, P. 2009. Geocoding quality and implications for spatial analysis. *Geography Compass*, 3(2): 647–80.

Zook, M., M. Graham, T. Shelton, and S. Gorman. 2010. Volunteered Geographic Information and Crowdsourcing Disaster Relief: A Case Study of the Haitian Earthquake. *World Medical & Health Policy*, 2(2): 7–33.

SECTION 1
Methods

Chapter 2
Effective Use of GIS for Spatial Epidemiology

Linda Beale

This chapter describes ways in which geographical information systems (GIS) can be used for epidemiological analysis.

Geography clearly has a key role to play in health analysis with spatial variations in environmental hazards, population distribution, population characteristics (including susceptibility) and health outcomes. Furthermore, management and policy can benefit when geographically targeted. Effectively defining the sources of hazards can lead to prevention, mitigation and amelioration by defining sources of hazards, hotspots of exposure and disease clusters and excesses. These goals form the basis of spatial epidemiology.

There is an increasing need to address public concerns about perceived environmental health risks. Although routinely collected geocoded data on health, population distribution and environmental pollutants are becoming more available, these datasets are rarely collected specifically for spatial epidemiological analysis. Despite improved statistical and geographical information science functionality in commonly used software; effective spatial analysis has to be underpinned by sound domain knowledge and an understanding of analytical techniques. Using GIS, disparate data can be integrated into a common geographic form where it can be validated. Furthermore unmeasured characteristics can be modeled, other models can be linked or integrated and of course maps of disease, environmental hazards and socio-economic factors can be generated.

A number of limitations impact epidemiological analysis, such as inconsistent geography, data scale/resolution, health and population data issues, exposure misclassification, selection and ecological bias, however, there is enormous potential to taking a geographical approach to epidemiology and health research. Using effective approaches, GIS offers valuable functions and tools for health analysis, for example to define exposed populations and effectively communicate results. Crucial to effective communication of spatial information is the use of suitable mapping techniques that convey the results effectively.

Geographical Information Science

Geographical Information Science (GI Science) or Geoinformatics combine the disciplines of the geosciences and technology for the study of spatial data and spatial issues. A Geographical Information System is the software that can be used to capture, store, process, analyze and visualize spatial information.

Data in a GIS must be geo-referenced (that is, linked to some systematic form of spatial referencing) and uses location and topology for the basis for inference. Any data that occurs at a known location can be captured and stored in a GIS and, increasingly with mobile and web technologies, real-time data capture is possible in the field. Data is mapped in a GIS and maps can form the output, however, they are not the only form of output, nor in all cases, the best means of representation.

The spatial properties of data, such as topology, can be used in geographical analysis. Access to healthcare, for example, can be found using travel distance or travel time of a population to the nearest health service (Delamater et al., 2012) or these spatial properties combined can be incorporated with other characteristics, as gravity models, that define distance-dependent interactions between health services and population demands (Wand and Roisman, 2011).

A GIS provides the ability to integrate disparate data from many sources and display and analyze the information together to generate meaningful results. Unmeasured characteristics can be modeled either directly in a GIS or through linkage or integration with other models for example for exposure assessment. Using a GIS spatial analysis can be automated and so calculations can be made across large areas, over time, involving large populations. The advantages of extending analysis through time and space give increased statistical confidence, with larger numbers, together with the ability to evaluate patterns in a broader context.

Using a GIS, analysis includes a visual component. Visualizing data is a very powerful component that can be used in a number of ways. Viewing data, together with its geographical component, often reveals new information such as where data is unexpectedly missing and, therefore, highlighting possible errors in data collection. The ability to visualize events in situ, particularly together with other geographical data, can help inform the analyst. Furthermore, the ability to map the results provides a powerful and effective form of reporting. Conversely, the advantages that visualizing results can bring means the analyst must be wary not be led by the data display (as opposed to a designed map) and clear study design be followed. Care should be taken throughout the analysis process and mapped results should be carefully designed to ensure it conveys the intended message. Any map can mislead a reader so map design should form an important component of any study.

Spatial Epidemiology

Spatial epidemiology encompasses a number of key areas including disease mapping, spatial exposure assessment, disease risk assessment and cluster detection.

Disease mapping

Disease mapping involves both the analysis and display of mortality or morbidity rates that show spatial patterns of health outcomes. The spatial component of health data can play a crucial part in helping to explain variability in risk, for example in relation to environmental risk factors. The suitability of any analytical approach is defined in part by the study design, for example whether it is a case, case-control or cohort study and also, perhaps inappropriately, by data availability. The conventional approach to disease mapping is to use either directly or indirectly standardized rates (Elliott et al., 2000).

Standardization (or adjustment) is a crucial component in spatial epidemiology to ensure the valid comparison of groups (for example, age-adjustment). In many cases variables included in the analysis are correlated. For example, when analyzing the impact of environmental exposure, individuals with low income or of low socio-economic status are likely to be correlated with both the exposure and the disease outcome.

Directly standardized rates (DSR) are calculated as the ratio between observed and expected cases for a specified health outcome. The age-specific mortality rates from the study area's population are used to derive expected deaths in the standard or reference population.

$$DSR_i = \frac{\sum_j N^*_j r_{ij}}{\sum_j N^*_j} * 100,000$$

where:

DSR_i is the directly standardized rate in the study area i
N_j^* is the population-years at risk in age strata j in the reference population
r_{ij} is the rate of disease D in age strata j in the study area i
(result is multiplied by 100,000 to obtain rate per 100,000 person years)

Standardized mortality/morbidity rate (SMR) are calculated in a similar way except, the age-specific rates of a standard population are applied to the age distribution of the study population.

$$SMR_i = \frac{O_i}{E_i}$$

where:

SMR are standardized mortaility/morbidity/incidence ratios (using indirect standardization)
O_i the number of observed cases in area i
E_i the number of expected cases in area i

and:

$$E_i = \sum_j N_{ij} r^*_{ij}$$

where:
N_{ij} is the population-years at risk in age/sex/covariate strata j in area i
r^*_j is the rate of disease in strata j in the reference region

Directly standardized rates can be directly compared between different exposure groups, however, they can be statistically unstable in areas of small populations and/or when studying rare diseases. Indirectly standardized risks or standardized morbidity/mortality ratios (SMRs) offer a relatively stable approach as the rates are calculated based on the larger comparison population rates. The results, however, are not directly comparable between different exposure groups, particularly in regions where population structures significantly differ.

Disease mapping provide a means of visualizing spatial patterns of disease, using rates and risks. Furthermore, it provides a basis for spatial modeling and risk assessment (for example, through exposure assessment) and allows hypotheses to be developed about potential causal factors and associations.

Exposure Assessment

The evaluation of factors from the natural and built environment that affect the incidence, prevalence, and geographic pattern of health conditions are key components in spatial epidemiology. Exposure assessment comprises a number of different approaches including both direct and indirect methods. Direct methods such as biomonitoring, personal exposure measures and self-reporting provide individual measures and can be very accurate; however, data collection is both costly and difficult. Indirect methods offer a different approach using proximity, environmental monitoring or modeling and provide broader ranging estimates (both temporally and spatially), however accuracy can be a concern. Invariably, any broader scale analysis will involve some form of modeling or interpolation to obtain estimates across a continuous surface (for example, soil, water or air pollution).

Co-location, spatial overlap, contiguity, and proximity are often used as proxies for exposure (Briggs, 2005). In the simplest case distance from a potential hazardous source can be used to define potential exposed and un-exposed populations (Beale et al., 2010; Elliott et al., 2001) with standard functions (such as buffer and overlay) being available in GIS. These types of analyses are relatively simple, fast, easy to understand and have very few data requirements making them valuable in areas where data collection is limited. In many cases, however, the oversimplifications assumed in these approaches will lead to unreliable estimates of who is exposed or exposure. For example, in the case of air pollution, dispersion is complex and the influence by factors in the natural and built environment (for example, wind, buildings, etc.) is crucial (Kimborough et al., 2013).

Spatial interpolation can be used to estimate a value of a variable at an un-sampled location from measurements made at other sites and is based on the notion that points close together in space tend to have similar attributes (Tobler, 1979). A number of different approaches are available in GIS including deterministic methods that directly use the input sample points and stochastic or geostatistical methods that use the sample points together with their statistical properties. Geostatistical modeling arose from the idea that spatial variation of a continuous attribute is often too irregular to be modeled by a simple, smooth mathematical function. Kriging offers a powerful way to model surfaces breaking it down into three steps: 1) modeling the general trend of the data (the drift) that is, deterministic variation, 2) the smaller local variations (that are spatially auto-correlated but physically difficult to explain) and 3) the uncorrelated noise.

Cluster Detection

The area of cluster detection often forms an important part of surveillance and is concerned with looking for spatial patterns of disease that vary from the expected (Knox, 1989). A number of approaches are used to define clusters in data, from traditional methods that can be used to test hypothesis (for example, chi-squared statistic) to those that use proximity such as autocorrelation statistics and K-functions, kernel functions to calculate event intensity or density and finally, moving window approaches (for example, scan statistics). To avoid issues inherent with assuming a distribution, the significance of clusters can be found using a permutation approach, such as Monte Carlo simulation (Rushton and Lolonis, 1996) but beyond simply identifying clusters in data, effective cluster detection for spatial epidemiology relies on an understanding of the expected risks (for example, risk surface). Moving window methods superimpose a number of circular regions onto the study area in which the significance of the number of cases are estimated. The circular areas are defined in terms of distance (Openshaw, 1987), the number of cases (Besag and Newell, 1991) and, population size (Turnbull et al., 1990; Kulldorff and Nagarwalla, 1995). Automating analysis in a GIS means that more complex methods, such as kernel methods and geostatistical approaches can now be extended to large databases. This has huge advantages for public health response. Cluster detection often arises in response to public fears and, as such, is at risk of focusing on a pre-defined, often small area. This approach to pre-defining a cluster often leads to boundary shrinkage, also known as the 'texas-sharpshooter' effect, as he fires his bullet and then draws the target around it. With increased computing and software capabilities it is now possible to extend the analysis spatially and temporally to analyze whether the initial area of concern is indeed a potential cluster or area that differs from others, given a reasonable comparison that represents the bigger picture.

Scale of Analysis

Since geographic analysis is never independent of scale or location, this poses significant challenges for epidemiological study. Ideally, individual level health data would be used for such analysis but this is often not available and more so, descriptive data that can be used as confounders within analysis, are rarely available at individual level. As the number of biobanks increases over time, where individual biological samples are archived for use in research, the opportunity for more and larger sized individual level analysis will increase in the future.

Ecological Analysis

Using aggregated data, or ecological analysis as it is termed within epidemiology (Robinson, 1950), brings with it a number of issues that impact upon the results and, therefore, validity of a study. Spatial units are derived based on number of factors such as administrative boundaries, historical events, and ease of data collection. Spatial boundaries are rarely developed with research in mind, moreover, area boundaries often change over time, as populations increase or decline census areas are subdivided and aggregated, political boundaries can change with governments and specific studies will usually differ by design for each unique project.

Censuses, commonly undertaken every ten years and in some countries five years, are used to collect population and other statistics essential to those who have to plan and allocate resources. Census geography should provide an appropriate framework for the publication of small area census statistics which meet users' needs (Martin, 2002). A number of factors are usually taken into consideration for census boundaries such as confidentiality thresholds for publication, internal social homogeneity and the regularity/conformity with other boundaries and that each level of boundary should nest in the next hierarchy. These are ideals and population densities can dramatically change with the construction of new estates, new physical boundaries are built and the social mix of an area can alter. Census boundaries can change over time in an attempt to meet these ideals, making it difficult for studies that span multiple censuses as many health studies do.

The modifiable areal unit problem (Openshaw, 1984) means that analytical results, using the same data for the same study area, can vary enormously when aggregated in different ways (Figure 2.1). Much of the data used in spatial epidemiology with rely on data collected for other purposes and requires collating from various sources, cleaning and, in some cases, transforming to common boundaries. A number of different methods exist that can be used to transform and aggregate data but effects due to aggregation (or zonation) and scale will be inherent.

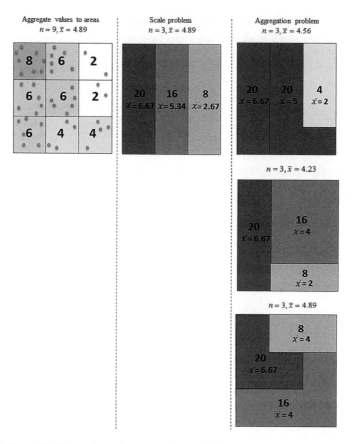

Figure 2.1   MAUP: scale and aggregation problem

Changes that occur as digital data is aggregated, disaggregated or even just transformed to a different data type will have an effect upon the results. Data is often from a number of different sources and is brought into a common framework for analysis. Even at the simplest level, the patterns we see differ when aggregated in different ways. As visualization of data becomes a part of analysis, with increased used of GIS, this should perhaps be included as the third problem of the modifiable areal unit problem, together with scale and aggregation.

Data Errors

Spatial representation is usually the decision of the analyst, not an inherent characteristic of the feature. The locations of sites of interest, for example industrial sites, can for some analysis, be equally valid as points (for example, representing the center of the site) or areas (using the site boundary). Indeed the data formats may already have been altered since initial data collection. Since the spatial units and data structure affects analysis, metadata is a crucial component in data manipulation and storage.

Data issues are an important factor within spatial analysis. All the advantages that the automation of analysis brings (such as large temporal and spatial analysis with big population numbers) also bring associated issues. As data is aggregated over time and space and databases grow, incompleteness or inaccuracies in the data become lost. There may be local variations in ascertainment of health data either due to specialist centers or from errors in data collection. Exploring the data visually in a GIS can help to highlight some data errors (for example, missing data or extreme outliers from miscoding); however local knowledge or additional data should be used at the initial exploratory phase to evaluate data before analysis. Equally, changes in health event recording over time (for example, adoption of a new ICD revision) should not be overlooked in studies that span a number of years. Errors in population data, possibly from migration, or incomplete/inaccurate geocoding of either health or population data can vary spatially, particularly due to geocoding errors (for example, there tend to be greater positional errors for rural than for urban addresses) (Ward et al., 2005; Krieger et al., 2002). Population data is often based on census data, so accuracy will decline as you move away from the census year, furthermore, where census boundaries have changed over time long-term ecological analysis becomes more complex as error will likely be introduced by an increased need to model data. Moving area data between different boundaries requires some form of areal interpolation. Simple area-weighting is a volume preserving method and uses the percentage of overlap between zones to apportion the population while assuming that data is evenly distributed across space. Mask area-weighting (also called binary mapping) builds on the volume preserving approach but attempts to improve on the inherent simplification by initially masking out areas of zero population and distributing the population across the remainder of the area. More advanced methods, based on dasymetric modeling, employ ancillary data to more intelligently apportion data according to where the population actually exist in space (Eicher and Brewer, 2001; Mennis and Hultgren, 2006). All areal interpolation methods should follow the pycnophylactic principle (Tobler, 1979); that is, the derived output data should still sum to the total in the input data to ensure that information is not being lost. Although error may be introduced through such modeling, it is imperative that the same population groups are used throughout a study.

Mapping Health Data

The ability to map and visualize spatial data helps reveal and understand spatial distributions and relations (Kraak, 2005). With the ever increasing use of spatial data and geographical information systems, data visualization is not only easily altered but is also used at many stages during spatial analysis. Indeed, geovisualization has now become integral to the process of handling spatial data (Kraak, 2005). Spatial data analysis offers enormous potential for spatial epidemiology, however, interpretation must be informed and should not be led by the geovisualization method. The ability of non-cartographers to map data, which has been primarily driven by the increased use of GIS, has greatly assisted in understanding spatial patterns and process. It has also led to development of alternative methods of display for example animated maps, three-dimensional diagrams and web maps (Hardy and Field, 2012). Worryingly, though as the use of GIS has increased, so the number of people with an understanding of cartographic theory and design has become proportionally smaller. Maps are never neutral or passive and care must always be taken when designing and interpreting them (Briggs and Field, 2000; Monmonier, 1996). Statistical maps produced within a GIS are characteristically thematic (Dent, 2008), so that they show specific information connected with its relevant geographic area. Thematic maps use base data as a geographic point of reference for the phenomenon being mapped. In showing information by location, they show spatial patterns and comparisons can be drawn between patterns in different maps. There are a number of different methods for thematic mapping, with the most common including: choropleth, proportional symbol, isopleths and dot density.

Choropleth Mapping

The most commonly used form in spatial epidemiology is the choropleth map, indeed they appear to be the 'default' map type. Disease maps show rates and risks for each area of analysis, usually a census region, with different colors or hues representing different value ranges. Choropleth maps are most effective for discrete data that is specific to a known area. Spatial epidemiology is most often concerned with ecological data for census regions, due to data availability (Briggs and Field, 2000). Data classification is a crucial component of choropleth mapping allowing raw data to be transformed into a number of classes, each with a unique single symbol on a map. There are many different methods for classifying data. None are universally perfect and the chosen scheme should fit the intended map use and the data values, with breaks between categories allowing identification of areas above or below a meaningful value, such as a national, provincial or state average rate (Dent, 2008). Classifications of map data in spatial epidemiology have tended to use quantiles, typically quintiles, or equal breaks. A quantile classification scheme is an ordinal ranking of the data values, dividing the distribution into intervals that have an equal number of data values. Quantile classification ensures maps are easily comparable and can be 'easy to read' (Brewer and Pickle, 2002), however, in using the ranked distribution of the data and, not the numerical distribution, important spatial patterns may be hidden if the values are too close. One of the most obvious problems with quantile classification occurs where data are skewed such that, for instance, four of five classes have narrow class intervals at one end of the data distribution and the fifth includes values across a much wider class interval. If the purpose of a choropleth map is to represent areas with similar characteristics in the same symbol (to aid visual comparison) then such a classification scheme can be inherently misleading. Some data values would be much closer to the lower end of the data distribution yet be classified in the upper quantile. Nevertheless, past research has shown the accuracy (in terms of a wide range of epidemiological map reading tasks) showed quantile classifications to be reasonably interpretable (Brewer and Pickle, 2002). Any classification scheme will substantially alter the subsequent view of the data when mapped and it is therefore a vital component in effective data visualization for spatial epidemiology in the same way as it is for any thematic map data presented using a choropleth (Figure 2.2).

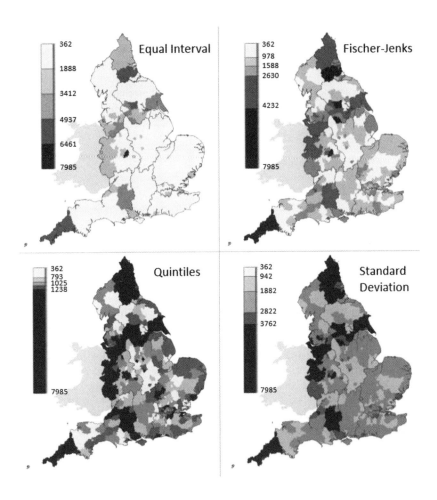

Figure 2.2   The effect of classification: Directly standardized mortality rates for 2011, London, UK

Alternative Thematic Mapping Techniques

Although rarely used in spatial epidemiology proportional symbols are an effective method for displaying values for areas, where the geographical area does not have a visually dominating effect. The lack of continuity in these maps seem to make them more challenging to any form of pattern recognition, however, in some cases the lack of neighborhood influence may be a better representation of reality that is, where physical or political boundaries exist. Dot density maps, where a dot may be used to locate each occurrence of a disease case, are not practical for ecological studies such as those carried out for disease mapping. A dot may also be used to indicate any number of entities, for example, 100 disease cases, however, the use of a single point location to represent a population group must be done with caution. For disease maps that are to be used for environmental epidemiological interpretation, patterns will be an outcome of representative location, which has the visual appearance of spatial accuracy. Isopleth maps use isolines, use connected lines between locations of equal value. They are effective for continuous and smooth phenomena, such as elevation values (contours) and precipitation. The isolines can be 'filled' color lines, or simple line features. Such maps could usefully be used to display general spatial trends in health outcomes across space where the visual association with an enumeration area is unnecessary.

Discussion

Ecological Issues

Ecological analysis investigates the rate or risk of disease by area or in relation to exposure in groups of people within the group, rather than the individual, being the unit of observation and analysis. In many cases, data defining exposure and health outcomes are more readily available for areas (for example, census areas) and, therefore, ecological studies can be carried out relatively quickly and cheaply. As such, ecological studies are very valuable for hypothesis generation. In many cases, covariate data (such as an index of deprivation) is only available at ecological level. Additionally, ecological studies can be used to evaluate the effectiveness health policies that have been implemented across defined regions. It is important to note, however, that results from ecological analysis may only hold true at the adopted scale of analysis (for example, the defined group of people). Results of analysis at the population or group level cannot be extrapolated to individuals. Any attempts to infer associations at the individual level from those observed at the group level are subject to ecological bias. Small-area studies are less susceptible to ecological bias created by within-area heterogeneity; however, unless the analysis is carried out at the individual level it is impossible to rule out this bias entirely. Consequently, ecological analysis cannot be used to infer a causal relationship between an environmental factor and a disease.

Bayesian modeling

Although small area studies are less prone to issues of within area heterogeity, issues with small numbers can make such analysis problematic. In areas with low numbers of observed cases and/or small populations both rates and SMRs can become numerically unstable (rates even more than SMRs). Extreme risks, together with wide confidence intervals are associated with areas of low population. In some cases disease maps will be very uneven with adjacent areas showing completely opposite risks. Using a Bayesian approach allows the area risk estimates to be statistically stabilized by smoothing the area values to a global or local mean. Empirical Bayes estimation is a non-spatial method of smoothing or Poisson-Gamma model by Clayton and Kaldor (1987). It offers a hierarchical approach that provides more precise estimates of relative risk and more accurate assessments of significant changes than standard methods, by accounting for differential variability in the data. Specifically, if λ_i represents the relative risk in area i, the Poisson-Gamma model assumes:

$$O_i \text{ Poisson}(\lambda_i E_i)$$

$$\lambda_i \text{ Gamma}(\alpha, \beta)$$

A full Bayesian model can take a number of forms, specifically the conditional autoregressive (CAR), Heterogeneity (HET) and Besag, York and Mollie (BYM) models. These methods smooth the areas to a local mean and therefore, require estimation of the mean from all neighboring areas. Areas of highest uncertainty of SMR will show the highest shrinkage.

In order to effectively deal with some of inherent problems with spatial epidemiological analysis a number of advanced statistical approaches have been developed. Rates and risks are usually calculated together with confidence intervals, rarely, however, is any attempt made to include these values that, significantly add to the interpretation, on any maps that are produced. Using empirical Bayes and full Bayesian models measures of uncertainty are produced, in the form of credible levels with empirical Bayes and posterior probabilities with the full Bayes approach. Posterior probabilities were proposed by Richardson et al. (2004) as an effective method to identify areas with an excess risk. In areas where relative risk is greater than 1, a decision rule based on a cutoff probability of 0.8 was found to achieve a good balance between specificity and sensitivity. Sensitivity measures the proportion of actual positives which are correctly identified that is, areas that show 'true' high risk whereas specificity measures the proportion of negatives which are correctly identified that is, areas of 'true' low risk. Using cartographic methods, such as transparency and hatching, in combination with statistical methods can ensure that areas of numerical instability are subdued do not dominate a map.

Exposure Misclassification

Exposure assessment can easily be done using proxy measures (for example, proximity based measures) in a GIS, however, these simple measures, as opposed to more sophisticated modeling techniques can lead to incorrectly assigning individuals or populations to a exposure group (for example, non-exposed or exposed). Such exposure misclassification can also occur when environmental levels do not accurately represent actual exposure. Latency periods between exposure and disease onset and population migration in and out of the area under study can also cause exposure misclassification.

Data Confidentiality

Issues of confidentiality have long been a concern for those working in epidemiology and public health, however, it has taken on greater significance as publicly available spatial data and spatial analysis becomes more commonplace. Futhermore, data collection methods are improving and the quality and or scale of available datasets is increasing. Methods of preserving data confidentially by perturbing the data have been proposed (Armstrong, 1999; Kwan, 2004). These approaches alter the geographic coordinates of individual level data, thereby adding stochastic or deterministic noise, however, in the context of spatial analysis, this can have serious implications. Spatial analysis uses location as a key component of the analysis and so any changes to location will impact results. Within the confines of a specific analysis the impact of any spatial changes may be statistically acceptable, however, in the context of GIS analysis the results being displayed will be erroneous. Secondly, the perturbed spatial dataset will contain unknown spatial errors that would simply propagate and become magnified in subsequent analysis by the unwitting or careless. Clearly, high quality metadata and warnings to guide interpretation can help ameliorate the way such results and maps might be interpreted but ultimately the output would remain spatially inaccurate.

Analysis of health data in a GIS places an increased emphasis on effectively displaying results without disclosing any individual data or breaching confidentiality laws. This is an area where the tools in GIS and cartographic design can be combined to ensure that effective analysis and presentation of results can be achieved. It is important that data is presented in a way that does not allow the reader to be able to determine or infer original data values. The problem here is that spatial data by its nature reveals information about 'where' and in some cases this allows 'who' to be determined. This is particularly acute when dealing with point locations, where the location is precise, but the issue is not resolved by using aggregated data is all situations. In cases of rare diseases and/or areas of low population, it may be possible to determine individuals when results are mapped at small area level. Ecological analysis is subject to a number of issues that can introduce bias and lead to inaccurate results, therefore, increased accuracy of data for analysis is usually advantageous. The role of cartography in effectively displaying the results, without data disclosure, becomes paramount in such cases, to ensure that effective analysis will be possible in the future. A number of interpolation techniques can be used to display the data which allow the confidentiality of the data to be protected. Areas where the numbers are sufficiently low to risk disclosure should be visually suppressed in any map, just as the data would be in tabular form. Such areas should be displayed in a way that differentiates them from other areas where results can be shown, such as a different color (for example, using transparency) or cross-hatched.

Visualization

The statistical methods used for disease mapping take into account population density, however, spatial patterns can be difficult to interpret given area extent of many census regions that are commonly used for disease mapping. Census regions, in attempting to apportion the areas into regions of similar population density, are by their nature significantly larger in area within rural areas. Furthermore, this apportionment is carried out within tolerances of minimum and maximum population totals, to ensure a balance between other factors (for example, socially homogenous, physical boundaries). This means that in many cases the rural areas contain slightly smaller populations and therefore lower and statistically less stable disease rates. The urban areas with higher rates of disease and often more reliable estimates are found in the smallest and least visible areas on the map. In areas where spatial units differ greatly in size, for example where both urban and rural areas are found, it may be difficult to interpret spatial pattern as larger areas will tend to dominate the map and the smallest areas may be indistinguishable. Using inset maps allows readers to see the spatial patterns in the smallest areas, which in many cases, form the most important areas in terms of patterns of disease. Alternatively, multi-scale web maps allow readers to see a broad overview at small scales and zooming in to see detail in such areas at large scales. In many cases these high-density urban areas are those with a high population turn over and high deprivation where higher rates of poor health may be present (Meijer et al., 2012), as wealth tends to bring outward migration to areas of lower density and overall lower disease rates.

One method proposed to overcome the small area problem is the use of cartograms (Dorling, 1996; Selvin et al., 1988). Cartograms, or density-equalizing maps, re-size each area according to the variable that is being displayed. Until recently the spatial distortion and computational demands have meant that they have not been widely used but, improved methods for cartogram construction address these earlier problems of complexity and distortion by retaining shape as far as possible (Gastner and Newman, 2004). Their approach provides an alternative to the original choropleth maps where spatial units differed significantly in size but, there still seems to be some resistance in commonly adopting this approach. This may be in part because this mapping method is not available directly in most GIS packages, despite the relative efficiency of the algorithm.

Conclusions

Increasingly, data collection and analysis contains both a spatial and temporal element. Spatio-temporal analysis has almost been inherent in spatial epidemiology but increased data availability and technology means that analysis and display of results is becoming ever more sophisticated. The ability to show and interpret changing spatio-temporal patterns of health offers great potential for the understanding of health patterns. Time is an important factor in analyzing disease outbreaks. Variations over time in disease frequency can provide valuable information about the determinants of that disease in a given population. Infectious disease surveillance can usefully analyze the spatial entities and relationships using common GIS functions such as proximity analysis and statistical methods to observe when and where a disease appears, however, improved spatio-temporal analysis would permit more effective analysis of spread of disease. GI science, commercial GI systems, statistics and cartography must soon meet the challenges that dealing with such data will bring, as it will no doubt also bring many opportunities for understanding.

The use of GIS as part of epidemiological analysis means that spatial display of data is becoming a part of analysis, as we unconsciously start to interpret the spatial patterns that we see. Mapping uncertainty, such as the posterior probabilities, together with relative risks will allow readers to interpret the results more accurately, in the same way as it is usual to include confidence intervals when reporting the results in tables.

As web technologies become more pervasive and GIS analysis moves into the cloud domain, this not only opens up access to many new opportunities, for example extending analysis that have previously been the domain of the expert to a new audience and, at the same time providing new opportunities to map and share analysis. The ability to map data at multiple scales, when results are scale dependant increases analysis requirements and brings challenges to cartographic representation. Interactive multi-scale maps bring a host of new opportunities to disseminate information but research is needed to ensure this is done effectively. This will require a consideration of the role of design in cartographic information products to meet the needs of cloud based publishing (Field and Demaj, 2012)

As individual level data becomes available, different approaches and tools for analyzing and mapping health data will be required. These will also place more demands on effective visualization as confidentiality of results will be paramount. Estimating relative risks using density estimation methods where kernels are placed at the centers of areas and weighted by their population size have been showed to be effective for individual case and control data

(Bithell, 1990; Kelsall and Diggle, 1998) while more recent developments in geostatistical methods now allow point or area data to be more effectively handled. Although relatively unused to date, both these methods offer the ability to effectively display individual, potentially disclosure data, as a surface that preserves anonymity.

Recent advances in technology and GIS offer many advantages for spatial epidemiology. Spatial epidemiological analysis can be effective as a first-cut surveillance tool. GIS offers a number of valuable functions and tools both for spatial analysis and for effective communicate of results and results from spatial analysis have the potential to be used by public health agencies to implement prevention and control measures to reduce disease burdens. There are however a number of inherent issues with spatial epidemiology that technology can offer some solutions, however, only in combination with knowledge of epidemiological principles, statistics, spatial analysis and cartography. The power of effective spatial epidemiological analysis with GIS lies in a multidisciplinary approach. Digital data will increasingly be collected and data quality is likely to improve over time but the validity of using a particular dataset or method for now, remains something that should be analyzed on a case-by-case basis.

References

Armstrong, M.P., Rushton, G. and D.L. Zimmerman. 1999. Geographically masking health data to preserve confidentiality. *Statistics in Medicine*, 18(5): 497–525.

Beale, L., Hodgson, S., Abellan, J.J., LeFevre, S., and L. Jarup. 2010. Evaluation of spatial relationships between health and the environment: The rapid inquiry facility. *Environmental Health Perspectives*, 118(9), September: 1306–12.

Beale, L., Abellan, J., Hodgson, S., and L. Jarup. 2008. Risk assessment using spatial epidemiological methods. *Environmental Health Perspectives*, 116(8): 1105–10.

Besag, J. and J. Newell. 1991. The detection of clusters in rare diseases. *J R Stat Soc Ser A.*, 154: 143–55.

Bithell, J.F. 1990. An application of density estimation to geographical epidemiology. *Stat Med*, 9: 691–701.

Briggs, D. 2005. The role of GIS: Coping with space (and time) in air pollution exposure assessment. *Journal of Toxicology and Environmental Health* Part A 68: 1243–61.

Briggs, D.J. and K.S. Field. 2000. Using GIS to link environment and health data, in *Decision-Making in Environmental Health: From Evidence to Action*, edited by C. Corvalan, D. Briggs, and G. Zielhuis. London: E&FN Spon.

Breslow, N.E. and N.E. Day. 1987. *Statistical Methods in Medical Research, Vol. II.* International Agency for Research on Cancer. Lyon.

Breslow, N.E. and N.E. Day. 1988. *Statistical Methods in Cancer Research, Volume II- The Design and Analysis of Cohort Studies.* International Agency for Research on Cancer. New York: Oxford Press.

Brewer, C. and L.W. Pickle. 2002. Evaluation of methods for classifying epidemiological data on choropleth maps in series. *Annals of the Association of American Geographers,* 92: 662–81.

Clayton, D. and J. Kaldor 1987. Empirical Bayes estimates of age-standardised relative risks for use in disease-mapping *Biometrics*, 43: 671–81.

Delamater, P.L., Messina, J.P., Shortridge, A.M., and S.C. Grady. 2012. Measuring geographic access to health care: Raster and network-based methods. *International Journal of Health Geographics*, 11: 15.

Dent, B.D., Torquson, J. and T. Hodler. 2008. *Cartography: Thematic Map Design*, 6th Edition, McGraw-Hill.

Dorling, D. 1996. Area cartograms: Their use and creation. *Concepts and Techniques in Modern Geography (CATMOG)* 59. Geo Books: Norwich.

Elliott, P., Wakefield, J.C., Best, N.G., and D.J. Briggs. 2000. *Spatial Epidemiology: Methods and Applications*. Oxford: Oxford University Press.

Elliott, P., Briggs, D., Morris, S., De Hoogh, K., Hurt, C., Kold, J.T., Maitland, I., Richardson, S., Wakefield, J., and L. Jarup. 2001. Risk of adverse birth outcomes in populations living near landfill sites. *BMJ* August 18; 323(7309): 363–8.

Field, K.S. and Demaj, D. 2012. Reasserting design relevance in cartography: Some concepts. *The Cartographic Journal*, 49(1): 70–76.

Gastner, M.T. and M.E.J. Newman. 2004. Diffusion based method for producing density-equalizing maps. *Proceedings of the National Academy of Sciences*, 101: 7499–504.

Hardy, P. and K.S. Field. 2012. Portrayal and Cartography, in *Handbook of Geographic Information*, edited by Kresse W and Danko, DM eds. Dordrecht: Springer Verlag.

Kelsall, J.E. and P.J. Diggle. 1998. Spatial variation in risk: A nonparametric binary regression approach. *Applied Statistics*, 47: 559–73.

Kimbrough, S., Baldauf, R.W., Hagler, G.S.W., Shores, R.C., Mitchell, W., Whitaker, D.A., Croghan, C.W., and D.A.Vallero. 2013. Long-term continuous measurement of near-road air pollution in Las Vegas: Seasonal variability in traffic emissions impact on local air quality. *Air Quality, Atmosphere & Health*, 6(1): 295–305.

Kraak, M-J. 2005. Visualising spatial patterns, in *Geographical Information Systems* (Abridged): *Principles, Techniques, Management and Applications*, Second Edition, edited by P. Longley, M.F. Goodchild, D.J. Maguire, and D.W. Rhind. Chichester: John Wiley & Sons, .

Kulldorff, M. and N. Nagarwalla. 1995. Spatial disease clusters: Detection and inference. *Stat Med.*, 14: 799–810.

Krieger, N., Waterman, P., Chen, J.T., Soobader, M., Subramanian, S.V., and R. Carson. 2002. Zip Code Caveat: Bias Due to Spatiotemporal Mismatches Between Zip Codes and US Census–Defined Geographic Areas—The Public Health Disparities Geocoding Project. *American Journal of Public Health*, 92(7): 1100–102.

Kwan, M., Casas, I. and B.C. Schmitz. 2004. Protection of geoprivacy and accuracy of spatial information: How effective are geographical masks? *Cartographica*, 39(2): 15–28.

Martin, D. 2002. Geography for the 2001 Census in England and Wales. *Population Trends*, 108: 7-15.

Meijer, M., Kejs, A.M., Stock, C., Bloomfield, K., Ejstrudd, B., and P. Schlattmann. 2012. Population density, socioeconomic environment and all-cause mortality: A multilevel survival analysis of 2.7 million individuals in Denmark. *Health and Place*, 18: 391–9.

Monmonier, M. 1993. *Mapping It Out*. Chicago, IL: University of Chicago Press.

Monmonier, M. 1996. *How To Lie With Maps*. Chicago, IL: University of Chicago Press.

Openshaw, S. 1984. Ecological fallacies and the analysis of areal census data. *Environ Plan A*, 16: 17–31.

Openshaw, S., Charlton, M.E., Wymer, C., and Craft, A. 1987. A Mark 1 Geographical Analysis Machine for the automated analysis of point data sets. *International Journal of Geographical Information Science*, 1: 335–58.

Pang, A. 2001. Visualizing Uncertainty in Geo-spatial Data. Workshop on the Intersections between Geospatial Information and Information Technology, prepared for the National Academies Committee of the Computer Science and Telecommunications Board 1–14.

Robinson, W.S. 1950. Ecological correlations and the behavior of individuals. *American Sociological Review*, 15(3): 351–7.

Rushton, G. and P. Lolonis. 1996. Exploratory spatial analysis of birth defect rates in an urban population. *Statistics in Medicine*, 15: 717–26.

Selvin, S., Merrill, D., Schulman, J., Sacks, S., Bedell, L., and L. Wong. 1988. Transformations of maps to investigate clusters of disease. *Social Sciences and Medicine*, 26(2): 215–21.

Wang, L. and D. Roisman. 2011. Modeling spatial accessibility of immigrants to culturally diverse family physicians. *Professional Geographer*, 63: 73–91.

Ward, M.H., Nuckols, R., Giglierano, J., Bonner, M.R., Wolter, C., Airola M, Mix W, Colt JS, Hartge. 2005. Positional accuracy of two methods of geocoding. *Epidemiology*, 16: 542–7.

Chapter 3

An Assessment of Online Geocoding Services for Health Research in a Mid-Sized Canadian City

Patrick DeLuca and Pavlos S. Kanaroglou

Historically, geographers have been able to demonstrate the importance of location in health research. Spatial models are employed in diverse applications such as cluster analysis (DeLuca and Kanaroglou, 2008; Zimmerman and Lin, 2010), disease surveillance (Gumpertz et al., 2006; Waller and Gotway, 2004), exposure assignment (Dolk et al., 2000; Finkelstein et al., 2003), accessibility to health care (Passalent et al., 2013), allocation of subsidies to underserviced areas (Sorensen et al., 2000) and studies of obesity and the built environment (Merchant et al., 2011). These applications are often made possible through some form of automated geocoding of addresses. Geocoding is the process whereby one matches a text-based address to positional information (for example, latitude and longitude, or some other form of coordinates). This process can be carried out in a variety of ways including the use of postal code conversion files (PCCFs), address-based matching found in GIS packages, value added utilities, like DMTI's GeoPinpoint Suite (DMTI Spatial, 2007) and internet-based approaches, such as ArcGIS Online geocoding service (ESRI, 2012) or Google or Yahoo's geocoding API (application programming interface) either directly, or through sites such as www.batchgeo.com or www.spatialepidemiology.net which are built on Google. All approaches involve the input of a standardized address or postal code (source data), and a reference data set (typically a street network file) for which an iterative comparison of the address to the reference data can take place to calculate geographic coordinates. The calculation generally is based on interpolation along a street segment for which the geographic coordinates of the beginning and end points are known, and/or areal interpolation within a parcel, ZIP code, or city polygon (Jacquez, 2012). The accuracy of the geocode then is directly related to the quality of both the source data supplied and the quality of the reference data utilized. Quite often, full address information is incorrect, missing, or suppressed due to confidentiality. In situations like these, postal codes are often used either as the sole source of address information or as additional information in a multi-staged geocoding approach using postal codes to increase the number of matches (McElroy et al., 2003).

Generally, quality is assessed by examining the match rates, scores, and types as well as the spatial accuracy of the resulting geocode. The match (or hit) rate is the percentage of records capable of being geocoded, and depending on the geocoding approach, may be reported explicitly as part of the output, otherwise it is easily computed. The match score is a number that represents the degree of confidence associated with the output geocode. For example, if the user supplies the street number, street address, city, state or province, and zip or postal code a perfect score of 100 would indicate that a complete match was made to the reference data set. Numbers less than 100 would indicate that one or more of the user supplied address components could not be matched exactly. The greater the number of components that cannot be matched, the lower the score. The match type is the level of geography that the address is matched to. This may vary from one application to the next, and might include descriptions such as rooftop, street address, street name, ZIP or Postal Code level, City, and Province match. Spatial accuracy is a measure of how close the computed geocode is to the actual position on the surface of the earth (Goldberg and Jacquez, 2012).

Errors related to geocoding, have the potential to bias results, regardless of the application for which they are intended. This means that the ability to identify any effect may be limited. On the other hand, it is also possible for associations to be identified which are purely artifacts of systematic geocoding errors (DeLuca and Kanaroglou, 2008). In response to this, attempts to quantify errors from various geocoding processes have been prevalent in the literature. For example, some studies have assessed accuracy by examining differences in the Euclidean distance between point locations ascertained by automated geocoding, and corresponding locations obtained through the use of more intensive and accurate techniques, such as GPS (Bonner et al,. 2003) or aerial photography (Cayo and Talbot, 2003).

Aside from Burra et al. (2002), Bow et al. (2004) and DeLuca and Kanaroglou (2008), little has been done to assess the accuracy of geocoding products readily available to public health practitioners in Canada. Although there are several American examples, gaps still exist in the literature, particularly pertaining to the quality of the newer online geocoding services which are growing in popularity. The primary aim of this research is to evaluate three current online geocoding products in terms of positional accuracy in both urban and rural settings. We assess this in two ways. First, we aggregate the geocoded address points to their respective census tract boundaries in Hamilton, Ontario, Canada and compare them to the parcel fabric centroids aggregated in the same manner. This is accomplished via an observed (geocoded addresses) to expected (parcel fabric centroids) ratio for each census tract. Second, the differences in Euclidean distance between the geocoded points and corresponding parcel fabric centroids will be computed for both rural and urban parts of the city. These are also aggregated and mapped in order to explore the spatial variation of these differences.

Study Area

The City of Hamilton, Ontario (Figure 3.1) is a mid-sized industrial Canadian city located at the western end of Lake Ontario (43.3°N, 79.9°W). It is an amalgamation of five suburban communities (Ancaster, Dundas, Flamborough, Glanbrook, and Stoney Creek) and a central urban core (the old City of Hamilton). In addition, there are numerous smaller communities scattered throughout the city boundary. For example, Flamborough is comprised of 20 former townships and communities that all maintain a very strong identity, along with their community name as part of their mailing address. Examples of these communities include Waterdown, Carlisle, Rockton, Sheffield and Freelton. The total population of the city is 519, 949 (Statistics Canada, 2012), the majority of which reside in the old city (Table 3.1). A large part of the surrounding area is rural (Figure 3.1). This urban/rural dichotomy makes Hamilton a useful example since it has been well documented that the levels of positional accuracy of geocoded events differ significantly between these two types of areas (Vine et al. (1997), Bonner et al. (2003), Cayo and Talbot (2003)).

Table 3.1   Population and dwelling counts for the different communities of Hamilton

Community Name	Population	Number of Dwellings
Ancaster	36,911	12,513
Dundas	24,907	10,191
Flamborough	40,092	14,192
Glanbrook	22,438	8,379
Hamilton	330,481	145,463
Stoney Creek	65,120	23,804
Total	519,949	214,542

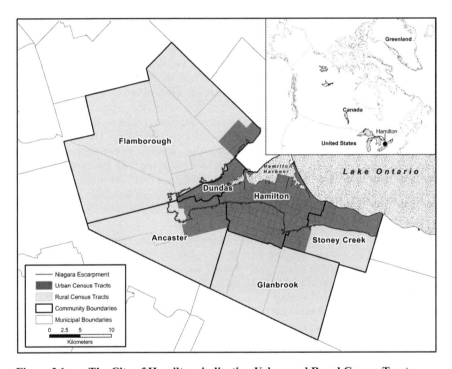

Figure 3.1   The City of Hamilton, indicating Urban and Rural Census Tracts

Data and Methods

The full set of residential addresses for the city of Hamilton (N = 139,792) was extracted from the city's 2011 parcel fabric dataset. This data set consists of individual property boundaries and attributes pertaining to each parcel such as tax roll number, street address, land use type and four fields indicating dwelling type: single family, semi-detached, townhouse and apartment. Parcel centroids were derived from this dataset and act as the base case for comparing the various geocoding approaches described in the following section. Dwelling centroids would be preferable, but those were not available for the entire city. In a study examining accessibility to public resources, Healy and Gilliland (2012) illustrated that next to dwelling centroids, parcel centroids are the second most accurate address proxy regardless of whether the area was urban or rural. While this dataset is not an explicit health dataset, it is the universe of residential records from which any health data would be sampled from for the city of Hamilton. The advantage of using a dataset such as this for demonstration is that it is free from biases that may arise from selection of observations and from oversampling or under-sampling from particular parts of a city. This is especially important since there are known issues with geocoding accuracy in rural places.

Online Geocoding Services

Three different geocoding services were utilized in this study: GeoCoder.ca, Yahoo's geocoding API and ESRI's online geocoding available through ArcGIS.com. For each service, batch geocoding was employed due to the size of the database. In batch geocoding, several observations are geocoded at the same time, in this case, based on the full street address with the city name set to Hamilton, the Province name set to Ontario and the country name set to Canada. The remainder of this subsection briefly describes each online service and specific steps employed to complete the geocoding process.

Geocoder.ca (Geolytica, 2013) provides many different geocoding services including real-time geocoding, batch geocoding, and postal code geocoding. The reference data used in geocoding is provided by various agencies which gather GIS data including GeoBase (GeoBase, 2013) for the National Road Network files for Canada and the U.S. Census Bureau for TIGER files (US Census Bureau, 2013). The data they collect is filtered for inaccuracies and corrected using a proprietary algorithm. The geocoding is done by building a line or curve through the data points representing a street then interpolating based on the street number of the location or finding the intersection of two streets. With respect to accuracy, they claim that average error is 9 meters in urban areas, with rural areas being described as having more "rough approximations." The addresses were sent to the operator of Geocoder.ca as a text file with the results of the batch geocoding operation returned in a single text file with the latitude and longitude appended to the addresses originally sent.

The second online service used was Yahoo's Geocoding API (known as Yahoo! BOSS Geo Services), which was accessed through www.gpsvisualizer.com (Schneider, 2012).[1] This website has an easy to use front-end for the API so that no programming is involved for the user. It accepts many address formats and has the limitation that only 1000 addresses can be geocoded at a time and only 50,000 per day per IP address. In this example, 140 separate queries were executed to geocode all 139,792 addresses in the data set. This process took several days to complete and upon completion the results were collated into one file for analysis. The reference data for geocoding is based on the map data that is available through Yahoo! Maps, whose providers are NAVTEQ, TeleAtlas, iCubed and also other unspecified data that is in the public domain (Yahoo! Maps, 2013). The website also offers the Google

---

1 Initial geocoding took place on June 14, 2013. During the review process, gpsvisualizer.com switched from using the Yahoo API to Bing Maps for geocoding. As of April 24, 2014, the Yahoo geocoding API can still be accessed at developer.yahoo.com/boss/geo.

geocoding API, however, its use is limited in that any geocodes produced with the Google API must be displayed using a Google product like GoogleEarth or GoogleMaps and that the X,Y coordinate information cannot be extracted and analyzed in any other software package (Google, 2013).

The third approach utilized ESRI's online geocoding service, found embedded in ArcGIS 10.0 and only available online from version 10.1 onwards. Specifically, the North American Address locater was employed with reference data from NAVTEQ based on their 3rd quarter, 2011 update. This address locator is a composite geocoder supporting multiple levels of geocoding. For Canada, the following levels of geocoding are supported: address point, street address, street name, postal code and city/province. The geocoding service requires a subscription to ArcGIS Online and requires credits that can be purchased to utilize the geocoding service. The online documentation suggests that 25,000 records at a time can be batch geocoded, however, in practice after approximately 12,000 records the service would halt. As a result, the addresses were subdivided based on the six communities that exist in Hamilton (listed in Table 3.1). Depending on the number of records in each community it was further subdivided to have no more than 12,000 records in each batch. The dataset took several hours to complete the geocoding process.

Analytical Methods

In order to assess the accuracy of each of the online services, they were compared to the parcel fabric in two ways. First each approach's resultant geocodes were aggregated to the census tract level of geography along with the parcel fabric centroids. Then a ratio was computed with the number of geocodes in the numerator and the number of parcels in the denominator to determine how many of the geocodes were correctly coded to the appropriate census tract. Secondly, the Euclidean distance between each address and its corresponding geocode from each of the automated approaches was computed using ESRI's ArcGIS 10.1. Basic summary statistics, such as the mean distance, minimum and maximum distances and the standard deviation were reported. Similar to Cayo and Talbot (2003) all computations were completed for urban and rural census tracts separately. These were then mapped by census tract to show how the errors varied across the city.

Results

The quality of a geocode is assessed by examining the match rates, scores, type of geocode as well as the spatial accuracy (Jacquez, 2012). With respect to the match rate, almost all records geocoded for each approach (Table 3.2), however some of them were coded to locations outside of the Hamilton border, and as such are not considered successful.

Only the results from the geocoding in ArcGIS included match scores and type of geocode to help assess the overall quality, the other two methods only included the latitude and longitude in the result. For the ArcGIS Online result, 42,327 records could not be matched to the street address, but only to the city name, resulting in the observations being located at the centroid of the city. A further 5,261 could be matched to the street, but not the street number, resulting in the geocoded address being placed at one end of a link with the particular street name. The rest of the observations were matched to the full street address.

Table 3.2   Match rates of each of the geocoding approaches

Type of Geocode	Number of Unmatched Addresses	Number Coded Outside of Hamilton	Number Coded Within Hamilton	Success Rate (%)
Yahoo API	5	6	139,781	99.99
Geocoder.ca	4	164	139,624	99.88
ArcGIS Online	0	1,442	138,350	98.96

Of those, 6,356 were not perfect matches as evidenced through the match score, which was as low as 80.05%.

To assess the spatial accuracy, the geocodes were first aggregated to the 2011 census tract boundaries and then compared to the parcel fabric centroids, which were aggregated in the same fashion. Figure 3.2a is the result for Geocoder.ca, which shows that for the outlying areas of Hamilton, with two exceptions, there were less addresses geocoded than what was expected. In the worst census tract, there was a difference of 1,030 geocodes less than what was expected. At the other end of the spectrum, there is one census tract with almost five times more geocodes than expected (a difference of 6,045 addresses), while two census tracts had the same number of geocodes as expected. The average rate was 1.01 with a standard deviation of 0.37. Figure 3.2b presents the result for the Yahoo API. The pattern here is similar to Geocoder.ca, except that parts of Dundas and Stoney Creek have more geocodes than expected, and there are now six census tracts that match what is expected. The worst census tract has 7.5 times as many geocodes compared to the expected case. The average rate is 1.04 with a standard deviation of 0.6. Figure 3.2c presents the result for ArcGIS Online. The pattern here is considerably different from the other two in that the census tracts in the areas outside of the old city of Hamilton all have less than half of the geocodes that are expected in the census tract. The worst census tract has 72 times as many geocodes as expected. The average rate was 0.06 with a standard deviation of 0.179.

Table 3.3 shows the average distance between the parcel fabric centroid and its corresponding geocode from an online geocoding service. Yahoo's API has the lowest average distance separating the centroid from the resultant geocode of 835 m. ArcGIS had the highest average distance at approximately 6 km. With respect to the maximum distance between a parcel centroid and its respective geocode, Geocoder had the highest value at 221 km, while Yahoo had the lowest maximum distance at 51 km. The higher standard deviations for Geocoder and ArcGIS are attributed to the events that have been geocoded outside of Hamilton (64 and 1,442 respectively).

Figure 3.2 **Observed geocodes versus what is expected based on the parcel fabric centroids for (A) Geocoder.ca, (B) Yahoo API and (C) ArcGIS Online**

(a)

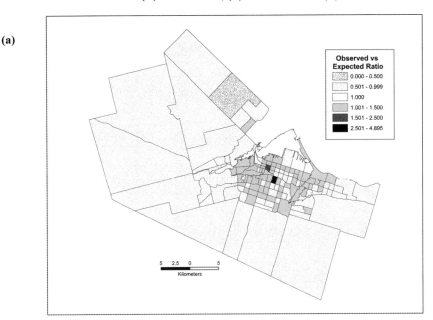

(b)

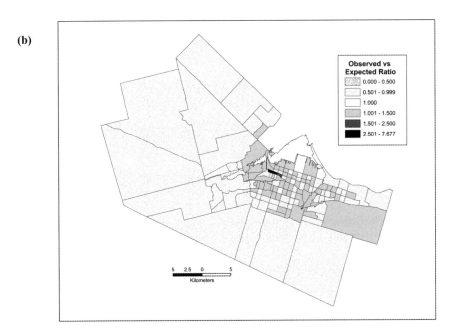

(c)

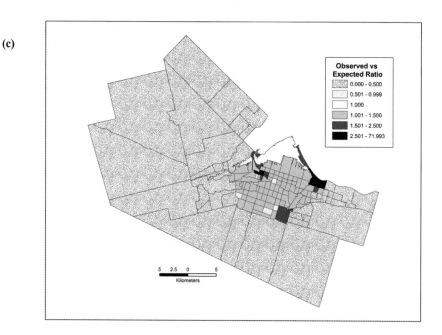

Table 3.3   Distance between parcel centroids and each online geocoding approach

Distance (m)	Geocoder	Yahoo	ArcGIS
Mean	1.132	835	6,096
Std. Dev	5,873	3,872	16,733
Minimum	0.11	5	1
Maximum	221,011	51,320	181,232

Table 3.4 provides summary statistics for the urban and rural differences for the different types of geocodes. In general, Yahoo had a lower average distance separating the parcel centroid and its respective geocode than both Geocoder and ArcGIS in both urban and rural settings. The average difference in each case was less for urban areas than for rural, which is consistent with the literature (for example, Cayo and Talbot, 2003; Bonner et al., 2003).

Table 3.4   Urban and rural differences (in meters) between parcel centroids and resulting geocodes from each online geocoding approach

Distance (m)	Geocoder Urban	Geocoder Rural	Yahoo Urban	Yahoo Rural	ArcGIS Urban	ArcGIS Rural
Mean	578	5,408	430	3,996	4,798	16,087
Std. Dev	4,895	9,746	2,618	8,166	16,203	17,362
Minimum	0.11	1	5	13	1	11
Maximum	221,011	169,802	51,320	41,070	178,457	181,232

Figure 3.3 shows the variation in the average distance separating parcel centroids and the respective geocodes by census tract. In general, all three approaches produce reasonable results for most census tracts in the old city of Hamilton, with 100m or less separating the centroid and the respective geocode on average. In suburban areas, Yahoo's API produced the best results, with only three census tracts having an average separation distance of 6000m or greater. Geocoder was slightly worse with five census tracts in the same category. Both Yahoo and Geocoder produced higher average separation distances in the rural parts of Flamborough and Ancaster. These distances decreased towards the eastern part of the city in Glanbrook and Stoney Creek. ArcGIS was the poorest with all tracts outside of the old city of Hamilton having average separation distances greater than 6000m.

Figure 3.3   Average distance separating geocode and parcel fabric centroid by census tract for (A) Geocoder.ca, (B) Yahoo API and (C) ArcGIS Online

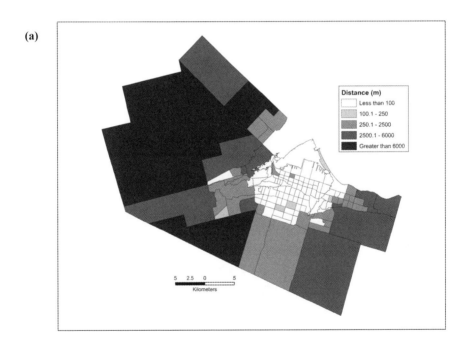

(a)

(b)

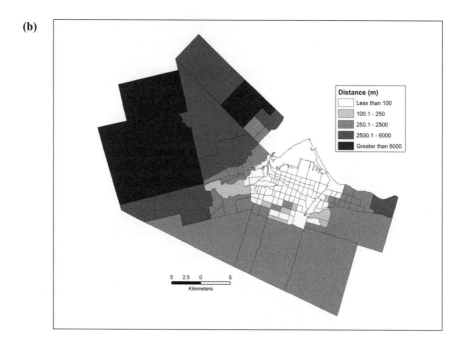

(c)

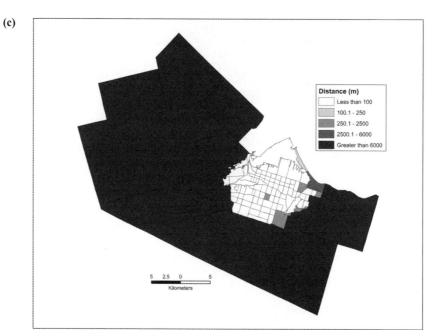

Discussion

The results indicate that on a local scale, the positional accuracy of the online geocoding tools are not very good, with the average ranging from 835m to over 6 km away from its respective parcel-based centroid. As expected, when examining the positional accuracy of rural cases alone, the distance separating the online geocode and its corresponding parcel based geocode is much greater. This poor accuracy is well documented in the literature (see Zimmerman et al. 2007; Bonner et al. 2003; Cayo and Talbot 2003 for example).

Zandbergen (2009) identified four main types of errors that arise from address matching. These include:

1. Error arising from geocoding an incorrect address. This generally arises from typographic errors or an error in street type (street, road, avenue, etc.), street direction (N, S, E, W) or street prefix (these can be directions like E 17th Street or road types like Highway 8 or County Rd).
2. Errors related to inaccurate interpolation along a street segment. These errors are less important and tend to be small in urban areas where street segments are shorter (Cayo and Talbot, 2003).
3. Error resulting from incorrect offsets from the road segment. These errors tend to be very small.
4. Error in the reference data street segments within the street network file. These errors relate to missing streets, incorrect placement of streets, incorrect street names or types, incorrect street directions, incorrect city name. Additionally, the reference data may be missing alias names for streets or cities.

All four of these issues may be involved with online geocoding services, with the only one that can be controlled by the user to some degree being errors arising from geocoding of an incorrect address. However, even if the user has the correct address, free of typographic errors, the reference data may be incorrect or coded differently producing either an incorrect location or failing to produce a match altogether. Geocoder.ca and the Yahoo API, like most online geocoding systems in use today, are considered black boxes, where the details describing the algorithms and/or the reference data used are not provided in any detail, other than perhaps listing the source of the reference data. Many also do not provide any way of checking the overall accuracy. Of the approaches examined here, only ArcGIS affords the user some diagnostics to assess the resulting geocodes through the interactive rematch that can be used to examine any geocode that did not return a perfect score.

Through this rematch process the user can compare their full address information to different examples available in the reference data, so issues like incorrect street types, directions or incorrect place names in the reference data can be matched to yield the proper location. For example, in this dataset, upon completion of the geocoding with ArcGIS online, it was observed that none of the observations in the rural parts of Hamilton were geocoded correctly so those were rematched, this time changing the specific community name used for the city field as opposed to using Hamilton. This switch significantly improved the geocoding, although there were still gaps in the areas where the smaller communities within Flamborough, Ancaster and Glanbrook are located. These data were rematched a third time using the smaller community name as the city which successfully filled in the gaps from the second geocoding attempt. This rematch process yielded positive improvements over the initial matching (which yielded an average of 0.06 with a standard deviation of 0.179). The rate of observed to expected changed considerably with 39 census tracts having the same number of observed as expected addresses with an average of 0.99 (standard deviation of 0.095). The maximum rate was 1.315 which is a positive improvement when compared to the original geocoding result. The average distance separating the parcel centroid and the geocode changed from 6,096m to 170m and spatially, the pattern presented in Figure 3.3C did not exist with many census tracts in the suburban communities having less than 100m separating the geocode from the parcel fabric on average (Figure 3.4). These improvements suggest that it is absolutely critical to have detailed knowledge of the geography of the area for which geocodes are computed.

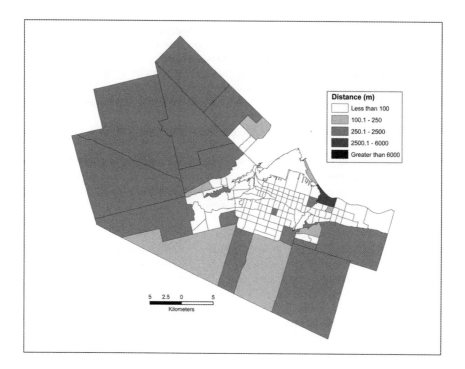

Figure 3.4   Average distances based on the interactive rematch in ArcGIS

The implications of errors in geocoding range from minor to severe, depending on the application. If the objective is to produce maps of disease rates at an aggregate level of geography, such as census tracts, then the effects can be significant, with some tracts over-reporting disease incidence with others under-reporting. The results for Hamilton in Figure 3.2 illustrate systematic under-reporting in suburban and/or rural parts of the city, and over-reporting in the central core. If the objective is exposure assessment based on location, as in the case of Bonner et al. (2003), and Gilboa et al. (2006), then having observations coded near a point source of pollution when the correct location is several kilometers away (and vice versa), is a major problem potentially leading to incorrect exposure assignment.

Cluster studies are another research area where geocoding errors may have a significant impact on the results. For example, DeLuca and Kanaroglou (2008) found that Kulldorf's

Scan Statistic (1997) is very sensitive to errors in geocoding, yielding opposite results in the same geographic area for reproductive health outcomes geocoded using two different products. Using a variant of Ripley's K function, however, they found that the type of geocoding made little difference in the overall conclusions. Although that research was based on postal code geocoding, in theory something similar could happen using an address-based approach. Malizia (2013) extended this to show how inaccuracies in geocoding also play a role in tests of spatial-temporal clustering, while Delmelle et al. (2014) illustrated how these errors can be mapped in space and time using a space-time kernel density estimate.

There are also issues with accessibility to health care practitioners (McLafferty et al., 2012; Delmelle et al., 2013) and allocation of resources. For example, in Ontario, Canada, which has publicly funded health care, there are a number of financial incentives aimed at increasing medical services in underserviced areas. In some cases there are payment adjustments, if the practice is located in an underserved census subdivision (CSD), in other cases there is direct compensation for education costs (up to $100,000) providing the graduating physician will practice in an underserviced area. These areas are in part defined by a physician-to-population ratio within a CSD or city boundary. If physicians are geocoded to an incorrect CSD, the resulting ratio would be incorrect and could lead to either over or under compensation. In this case, the latitude and longitude of the practice is very important since many physicians, particularly in the sparsely populated north may report the closest town as the location of practice even though their practice may be in a smaller neighboring community (OPHRDC, 2011).

Conclusion and Recommendations

For any of the above examples, the incorrect locations could lead to erroneous associations in any spatial analysis exercise. Therefore, it is important to pay careful attention when using online geocoding tools or providers. Based on the example presented here, there are several recommendations that any user should keep in mind. First, ensure that your address information is free of typographic errors to the extent possible. Second, it is important to have as complete an address as possible, including street addresses, directions, types, city names and, if the reference data includes it, the postal or zip code. Also, it is critical to have a detailed knowledge of the geography of the area which you are trying to geocode. As was demonstrated by the example here, the geocoding with ArcGIS Online improved dramatically when the city name was changed to match the smaller communities within the city of Hamilton. Lastly, if possible, verify the geocoded locations within a GIS. This could be done by applying a spatial join of the geocode locations with a complete street network file. Using the spatial join functionality, the closest street name and other information contained in the street network file will be joined to the geocoded points. Once this is complete, either manually or through a script, the address used to geocode can be compared to the address in the street network file and the location can be manually changed as required.

The fact that geocoding using the Yahoo API performed better than the initial geocoding with ArcGIS Online and Geocoder.ca in this exercise is an important finding given that higher trust is often placed with geocoding providers. The main advantage that online geocoders such as Yahoo, Google and OpenStreetMap provide is the simplicity of approach, particularly when it acts as a backend for an interface such as batchgeo.com or gpsvisualizer.com. An additional advantage is that they are free, unlike geocoding providers, but they are often capped with regards to how many geocodes can be executed daily from a specific IP address, as is the case with Yahoo. However, these caps can be circumvented with a commercial account. For example, Yahoo! Boss Geo offers tiered pricing for different amounts of queries per day including unlimited geocoding at $3.00 per 1,000 queries per day provided that the number of observations to be geocoded is 35,000 or more (Yahoo!, 2013).

The main disadvantages of online geocoders include the precision of result and issues around confidentiality. The reality is that there does not exist a unified standard for geocoding precision, meaning that what is produced by Google, Bing, and Yahoo will all be slightly different depending on the source data and the types of interpolation involved. A useful website to compare source data that can be used for online geocoders is Map Compare (Geofabrik GmbH, 2014). This webpage is split into four panes where the user can compare OpenStreetMap to Bing, and Google at the same time (unfortunately, Yahoo Maps is not currently part of this system). Perhaps a more important issue is confidentiality. Health records such as a database of cancer or AIDS patients are considered confidential and submitting them to a site such as Google or Yahoo would violate any confidentiality agreement. Google explicitly states as part of their terms of service that the user gives them "a perpetual, irrevocable, worldwide, royalty-free, and non-exclusive license to reproduce, adapt, modify, translate, publicly perform, publicly display and distribute Your Content" (Google, 2013). This violation of confidentiality would potentially raise the benefits of using a commercial geocoder, like Geocoder.ca, depending on agreements signed between the provider and the customer. Alternatively, a database of health observations can be mixed in with a much larger set of random addresses making it very difficult to identify any of the health observations contained within the combined data set.

The ability to display and analyze health data, while appealing, must be balanced with a profound understanding of the limitations embodied in the geocoding method applied. In addition, it is necessary to understand how these limitations may potentially impact the results of any potential spatial analysis. As a result, a researcher should not be seduced by the simplicity of geocoding online and carefully examine the results of any geocoding exercise, not taking for granted that the online geocoding services produce perfect geocoded locations.

References

Bonner, M.R., Han, D., Nie, J., Rogerson, P., Vena, J.E., and J.L. Freudenheim. 2003. Positional accuracy of geocoded addresses in epidemiologic research. *Epidemiology*, 14: 408–12.

Burra, T., Jerrett, M., Burnett, R.T., and M. Anderson. 2002. Conceptual and practical issues in the detection of local disease clusters: A study of mortality in Hamilton, Ontario. *Canadian Geographer*, 46: 160–71.

Bow, C.J.D., Waters, N.M., Faris, P.D, Seidel, J.E., Galbraith, P.D., Knudtson, M.L., Ghali, W.A., and the APPROACH Investigators. 2004. Accuracy of city postal code coordinates as a proxy for location of residence. *International Journal of Health Geographics*, 3(5).

Cayo, M.R. and T.O. Talbot. 2003. Positional error in automated geocoding of residential addresses. *International Journal of Health Geographics*, 2(10).

Delmelle, E.M., Cassell, C.H., Dony, C., Radcliff, E., Tanner, J.P., Siffel, C., and R.S. Kirby. 2013. Modeling travel impedance to medical care for children with birth defects using Geographic Information Systems. *Birth Defects Research Part A: Clinical and Molecular Teratology*, 97(10): 673–84.

Delmelle, E.M., Dony, C., Casas, I., Jia, M., and W. Tang. 2014. Visualizing the impact of space-time uncertainties on dengue fever patterns. *International Journal of Geographic Information Science*, 28(5): 1107–27.

DeLuca, P.F. and P.S. Kanaroglou. 2008. Effects of alternative point pattern geocoding procedures on first and second order statistical measures. *Journal of Spatial Science*, 53(1): 131–41.

DMTI Spatial. 2007. GeoPinpoint Suite Windows, Toronto, Ontario, Canada.

Dolk, H., Pattenden, S., Vrijheid, M., Thakrar, B., and B. Armstrong. 2000. Perinatal and infant mortality and low birth weight among residents near cokeworks in Great Britain. *Archives of Environmental Health*, 55: 26–30. ESRI. 2012. North American Address Locator [online]. Available at: http://www.arcgis.com/home/item.html?id=8b980709e0534bb39784dc42f550d554 [Accessed February 21, 2013].

Finkelstein, M.M., Jerrett, M.L., DeLuca, P.F., Finkelstein, N., Verma, D., Chapman, K., and M. Sears. 2003. Environmental justice: A cohort study of income, air pollution and mortality. *Canadian Medical Association Journal*, 169(5): 397–402.

Geobase. 2013. *Geobase.* Available at: http://www.geobase.ca/geobase/en/index.html [Accessed June 14, 2013].

Geofabrik GmbH. 2014. Map Compare. Available at: http://tools.geofabrik.de/mc [Accessed, April 25, 2014].

Geolytica. 2013. *Services—Products—Solutions.* Available at: http://geocoder.ca/?services=1 [Accessed June 14, 2013].

Gilboa, S.M., Mendola, P., Olshan, A.F., Harness, C., Loomis, D., Langlois, P., Savitz, D., and A.H. Herring. 2006. Comparison of residential geocoding methods in population-based study of air quality and birth defects. *Environmental Research*, 101: 256–62.

Goldberg, D.W. and M.G. Cockburn. 2012. The effect of administrative boundaries and geocoding error on cancer rates in California. *Spatial and Spatio-temporal Epidemiology*, 3(1): 39–54.

Goldberg, D.W. and G.M. Jacquez. 2012. Advances in geocoding for the health sciences. *Spatial and Spatio-temporal Epidemiology*, 3(1): 1–5.

Google. 2013. *GeoGuidelines.* Available at: http://www.google.com/permissions/geoguidelines.html [Accessed June 14, 2013].

Gumpertz, M.L., Pickle, L.W., Miller, B.A., and B.S. Bell. 2006. Geographic patterns of advanced breast cancer in Los Angeles: Associations with biological and sociodemographic factors (United States). *Cancer Causes Control*, 17(3): 325–39.

Healy, M.A. and J.A. Gilliland. 2012. Quantifying the magnitude of environmental exposure misclassification when using imprecise address proxies in public health research. *Spatial and Spatio-temporal Epidemiology*, 3(1): 55–67.

Jacquez, G.M. 2012. A research agenda: Does geocoding positional error matter in health GIS studies? *Spatial and Spatio-temporal Epidemiology*, 3(1): 7–16.

Kulldorff, M. 1997. A spatial scan statistic. *Communications in Statistics: Theory and Methods*, 26: 481–1496.

Malizia, N. 2013. The effect of data inaccuracy on tests of space-time interaction. *Transactions in GIS*, 17(3): 426–451.

McElroy, J.A., Remington, P.L., Trentham-Dietz, A., Robert, S.A., and P.A. Newcomb. 2003. Geocoding addresses from a large population-based study: Lessons learned. *Epidemiology*, 14: 399–407.

Merchant, A.T., DeLuca, P.F., Shubair, M., Emili, J., and P. Kanaroglou. 2011. Local data for obesity prevention: Using national datasets. *Professional Geographer*, 63(4): 550–59.

OPHRDC. 2012. *2011 Annual Report—Physicians in Ontario*. Available at: https://www.ophrdc.org/Public/Report.aspx [Accessed June 18, 2013].

Passalent, L., Borsy, E., Landry, M.D., and C. Cott. 2013. Geographic information systems (GIS): An emerging method to assess demand and provision for rehabilitation services. *Disability and Rehabilitation*, DOI: 10.3109/09638288.2012.750690.

Schnieder, A. 2013. *GPS Visualizer*. Available at: http://www.gpsvisualizer.com/geocoder/ [Accessed June 14, 2013].

Sorensen, R.J. and J. Grytten. 2000. Contract design for primary care physicians: Physician location and practice behavior in small communities. *Health Care Management Science*, 3: 151–7.

Statistics Canada. 2012. Hamilton, Ontario (Code 3525005) and Hamilton, Ontario (Code 3525) (table). Census Profile. 2011 Census. Statistics Canada Catalogue no. 98 316-XWE. Ottawa. Released October 24, 2012. http://www12.statcan.gc.ca/censusrecensement/2011/dp-pd/prof/index.cfm?Lang=E [accessed: March 8, 2013].

US Census Bureau. 2013. *Tiger Products*. Available at: http://www.census.gov/geo/maps-data/data/tiger.html [accessed: June 14, 2013].

Vine, M.F., Degnan, D., and C. Hanchette. 1997. Geographic information systems: Their use in environmental epidemiologic research. *Environmental Health Perspectives*, 105: 598–605.

Waller, L.A. and C.A. Gotway. 2004. *Applied Spatial Statistics for Public Health Data*. New Jersey: Wiley Inter-Science.

Yahoo! 2013. *Yahoo! BOSS Geo Services*. Available at: http://developer.yahoo.com/boss/geo/ [Accessed June 14, 2013].

Zandbergen, P.A. 2009. Geocoding quality and implications for spatial analysis. *Geography Compass*, 3(2): 647–80.

Zimmerman, D.L., Fang, X., Mazumdar, S., and G. Rushton. 2007. Modeling the probability distribution of positional errors incurred by residential address geocoding. *International Journal of Health Geographics*, 6(1).

Zimmerman, D. and J. Li. 2010. Spatial autocorrelation among automated geocoding errors and its effects on testing for disease clustering. *Stat Med*, 29: 1025–36.

Chapter 4

Clustering and Co-occurrence of Cancer Types: A Comparison of Techniques with an Application to Pediatric Cancer in Murcia, Spain

Antonio Páez, Fernando A. López-Hernández,
Juan A. Ortega-García, and Manuel Ruiz

Spatial epidemiology, the geographical analysis of health-related information, is a field of research that harkens back to the simple mapping of disease outcomes (Snow, 1855), and that has become increasingly sophisticated and useful to understand the etiology of disease (Elliott and Wartenberg, 2004). Mapping, use of statistical methods, and other forms of visualization for spatial health data are valuable tools to identify similar or overlapping sets of unobservable risk factors (RF) that influence the occurrence of diseases (Jin et al., 2005) or locations where common genetic traits may be present (Abbaszadeh et al., 2010).

Exploratory spatial data analysis in particular expands the domain of classical exploratory data analysis to situations where geographical information is appended to each datum. Its main objective is to detect the spatial properties of data sets, rather than trying to explain the processes that led to such properties (Haining et al., 1998; Bailey and Gatrell, 1995). As such, exploratory spatial data analysis provides a valuable set of tool to describe the data (an important component of fact-finding, as for instance in the discovery of high geographical concentrations of cancer), and also to generate hypotheses (ideas about the underlying factors) to be tested in further research. Furthermore, these tools serve to target interventions (Pickle et al., 2006) and to effectively communicate with policy makers and the public (Bell et al., 2006).

In recent years, the spatial epidemiology of cancer has been a particularly active area of research (see inter alia Yin et al., 2011; Tian et al., 2010; Kloog et al., 2009; Shi, 2009; Cassetti et al., 2008; Pollack et al., 2006; Mather et al., 2006; Kulldorff et al. 1997; Rigby and Gatrell, 2000; Timander and McLafferty, 1998; Gatrell et al., 1996). A recent review of methods used in the spatial analysis of cancer is due to Goovaerts (2010). A vast majority of research that studies the geography of cancer has concentrated on the incidence in adults, but increasingly as well attention is paid to pediatric cases (for example, Amin et al., 2010; McNally et al., 2009; Rainey et al., 2007; Wheeler, 2007; Knox and Gilman, 1996; Gilman and Knox, 1998). Pediatric cancer (PC) is a rare disease with a low incidence rate of 10–15 cases per hundred thousand people less than 15 years old in various geographical settings (Ferris-Tortajada et al., 2004; Grupp et al., 2011; Bao et al., 2010). Like many multi-factorial diseases, pediatric cancers are likely the result of complex interactions between constitutional, genetic, and environmental risk factors. Vulnerability to carcinogenic factors, it is known, tends to be higher during the early years of life, with the risk for infants being 10 times higher compared to adults during the first two years of life, and three times as high compared to ages 3–15 (Preston, 2004). A better understanding of the geographical incidence of cases can provide valuable insights into the etiology of the disease in children.

In terms of research approaches, existing work in exploratory spatial epidemiology has focused mainly on the analysis of a single disease (Cassetti et al., 2008). A reason for this is that, as noted by Jung et al. (2010), there are few methods useful for exploring the spatial pattern of categorical variables with more than two outcomes (p. 1910). The relevance of multinomial variables in spatial epidemiology is that their analysis can help to generate and/ or validate hypotheses concerning the possible relationships of various diseases with respect to risk factors, as shown by recent studies of diabetes and leukemia in the UK and Denmark (Manda et al., 2009; Schmiedel et al., 2011), and other similar investigations of multiple types of disease (for example, Alfo et al., 2009; Knorr-Held and Best, 2001; Held et al., 2005; Downing et al., 2008; Dreassi, 2007; Gelfand and Vounatsou, 2003; Song et al., 2006).

The methods available for joint spatial analysis of diseases include the shared component model whereby the underlying risk surface is decomposed into portions comprised of shared and disease-specific variation (for example, Dreassi, 2007; Knorr-Held and Best, 2001; Held et al., 2005). Multivariate random fields are used to study overall specific risks and the correlation of prevalence between diseases while controlling for covariates (for example, Gelfand and Vounatsou, 2003; Song et al., 2006). More recently other techniques such as a combination of Bayesian estimation and Poisson kriging (Cassetti et al., 2008), and finite mixture models (Alfo et al., 2009) have also been proposed for analysis of multiple disease types. A common thread of these approaches is that they are used for applications with aggregated data where the exact location of individual cases is not known (Held et al., 2005). Analysis of individual case data, however, has been identified as a priority to advance cancer control efforts (Pickle et al., 2006). A limiting factor, besides data privacy and confidentiality, is that techniques useful to analyze individual cases, such as kernel estimation, the K function, and nearest neighbor techniques (Goovaerts, 2010; Wheeler, 2007), are less well-suited for the analysis of multiple disease types.

Our objective in this chapter is to compare two recently developed techniques for the exploratory spatial analysis of individual level multinomial variables. One of the most widely used tools in spatial epidemiology is the spatial scan statistic (Kulldorff, 1997). A multinomial version of the spatial scan statistic was recently proposed by Jung et al. (2010) with the explicit objective of detecting geographical clusters when there are multiple disease types. The $Q(m)$ statistic of Ruiz et al. (2010) was also recently developed in the spatial analysis literature for the exploration of multinomial variables, but with a focus on the co-occurrence of different types of events. The two methods deal with separate aspects of spatial pattern: clustering is the spatial concentration of events in specific regions, whereas co-occurrence is the location pattern of various types of events with respect to each other. The techniques are compared using simple simulated datasets and then in an application to PC types in the region of Murcia in Spain.

Definitions and Methods

Spatial analysis of disease cases can be conducted from different perspectives. A common task is the study of the geographical incidence of a disease, either by itself or in reference to a population at risk (Gatrell et al., 1996). The specific approach adopted depends on the nature of data and the purpose of the analysis. For instance, analysis can be based on regional count data or case-control point data. In the case of regional count data, space is discretized using a system of zones, so that each case is assigned to a pre-defined zone (for example, a municipality or postal code) for which the population at risk is known (for example, Amin et al., 2010; Schmiedel et al., 2010). Frequently, the objective of this type of analysis is to identify adjacent zones that display a level of incidence that is higher than the expectation under some explicitly stated null hypothesis, for instance, that the disease is equally likely to be observed in any member of the population. Analysis of disease data at the zonal level is often an imposition of the availability of sources of information, and there are reasons why preserving the original continuous setting of the data is desirable (Gatrell et al., 1996, pp. 257–8). When geographical information of cases is recorded in continuous space so that the exact location (as a point) of the disease is known, the analysis of point patterns is an attractive alternative (Gatrell et al., 1996). In addition, when a set of controls (non-cases) is also available, analysis of cases-controls is possible, which allows the formulation of a null hypothesis of the form: cases and controls are independently and randomly sampled from the same population. There are numerous examples of case-control analysis in the literature (for example, Schmiedel et al., 2011; Wheeler, 2007). From the perspective of the analysis of point data in continuous space, the use of controls is not without issues. Controls are by design sampled, and thus do not constitute a complete enumeration of events (a mapped pattern; see Bailey and Gatrell, 1995, p. 76), which is problematic as the results are conditional on the location of the sampled controls.

For the purpose of this chapter, we concentrate on the spatial pattern of disease cases only. This is partly due to the nature of the data available for our empirical example, but also because such are the terms of the techniques selected for comparison. In general terms, our focus is on the identification of anomalous geographical patterns of diagnosed cases. The two techniques that form the basis for our comparison, the multinomial scan statistic and $Q(m)$, are exploratory spatial data analysis methods.

The two techniques deal with separate aspects of spatial pattern. A commonly investigated property of spatially distributed events is their *clustering*, defined as a high intensity of events (number of events per unit area) with respect to the intensity of a pre-defined spatially random process. Clustering is of interest because it may be indicative of attraction between events or of common environmental factors that propitiate their emergence. The alternative to clustering is *regularity* (or hyperdispersion; see Schmiedel et al., 2011), of interest because it may indicate repulsion or regular features of the environment. In the case of the multinomial spatial scan statistic of Jung et al. (2010), the objective is to determine the existence spatial clusters of several types of events. This is accomplished by means of a likelihood ratio that maximizes the probability of event type j being inside a spatial window, for all types of events simultaneously. The null hypothesis is that for at least one type of event, the probability of being inside the window is different from that of being outside.

A different property of spatial pattern that we are interested in is *co-occurrence*, which we define as the pattern of co-location of events of different types with respect to each other, irrespective of the intensity of the process. This property has seldom before been identified as a separate property of spatial pattern, although the term has occasionally been used as a synonymous of clustering (for example, Cassetti et al., 2008). There are few examples of analysis of this property. One relatively isolated example is the work of Knox and Gilman (1996) where pairs and triads of childhood leukemia and cancers in Britain are investigated. The null hypothesis in the analysis of co-occurrence when using the $Q(m)$ statistic of Ruiz et al. (2010) is that no combination of events within a pre-defined neighborhood is observed more or less frequently than expected by chance.

Three simulated data sets help to exemplify the differences between clustering and co-occurrence (see Figure 4.1). The examples consist of a discrete and spatially distributed variable with three possible outcomes—the actual techniques, to be described next, generalize to an arbitrary number of outcomes. In Example 1, there is no obvious pattern of clustering, or a particularly high intensity of events in any segments of the region. In contrast, it can be seen that some types of events tend to co-occur: triangles appear to be co-located, even if they do not form part of a cluster, and this could be the case as well for circles and squares. In Example 2, the pattern gives a stronger impression of clustering, perhaps accompanied of co-occurrence of triangles and squares. In Example 3 there is some appearance of clustering near the central part of the region, with perhaps co-occurrence of triangles and squares in combinations. The statistics are described next and illustrated in reference to these simulated examples.

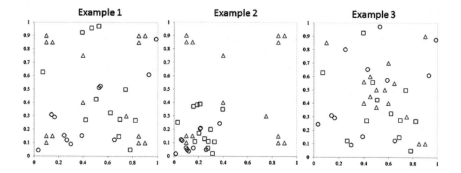

Figure 4.1   **Three examples of spatial patterns using simulated data. The x- and y-axis represent geographical coordinates (easting and northing). Triangles, squares, and circles represent different types of diseases. The scale in the maps has been normalized to create a unit square**

The multinomial spatial scan statistic (Jung et al., 2010)[1]

The spatial scan statistic (Kulldorff, 1997) is currently one of the most widely used tools to investigate clustering of spatial events. The statistic is developed based on the simple yet powerful idea of using moving windows of varying size to "scan" a study area—an idea that can be traced back to the Geographical Analysis Machine of Openshaw et al. (1987). As the area is swept, the number of events within the window is counted. Initially the windows were circular, but further extensions now allow the use of other shapes (Tango and Takahashi, 2005; Yiannakoulias et al., 2007). The likelihood of the observed spatial pattern (the number of observed events within the window) is compared to the likelihood of the pattern under the null hypothesis (the expected number of events). Maximization of the ratio of these two likelihoods as a function of window size allows an analyst to identify the most likely spatial cluster, the second most likely cluster, and so on. Furthermore, it is possible to attach a probability value to each cluster in order to support inferential statements.

---

1   First published online on May 7, 2010.

In its original formulation for spatial data, the spatial scan statistic was developed based on the Poisson distribution (Kulldorff and Nagarwalla, 1995; Kulldorff, 1997), but it has since been extended to a number of other distributions, including most recently to the multinomial distribution by Jung et al. (2010).

The multinomial version of the spatial scan statistic is useful to investigate clustering when a discrete spatial variable can take one and only one of k possible outcomes that lack intrinsic order information. If the region defined by the moving window is denoted by Z, the null hypothesis for the statistic can be stated as follows:

$$H_0 : p_1 = q_1; p_2 = q_2; \cdots; p_j = q_j; \cdots; p_k = q_k \qquad (1)$$

where p_j is the probability of being of event type j inside the window Z, and q_j is the probability of being of event type j outside the window. In other words, under the null hypothesis of no clustering, for any given window the probability of an event being inside or outside of the window is the same. The alternative hypothesis is that for at least one type event the probability of being of that type is different inside and outside of the window.

The statistic is built as a likelihood ratio, and takes the following form after transformation using the natural logarithm:

$$\Lambda = \max_Z \left\{ \sum_j \left[S_j^Z \log\left(\frac{S_j^Z}{S^Z}\right) + (S_j - S_j^Z)\log\left(\frac{S_j - S_j^Z}{S - S^Z}\right) \right] - \sum_j S_j \log\left(\frac{S_j}{S}\right) \right\} \qquad (2)$$

In equation (2), S is the total number of events in the study area and S_j is the total number of events of type j. The superscript Z denotes the same but for the sub-region defined by the moving window. The theoretical distribution of the statistic under the null hypothesis is not known, and therefore significance is evaluated numerically by simulating neutral landscapes (obtained using a random spatial process) and contrasting the empirically calculated statistic against the frequency of values obtained from the neutral landscapes. The results of the likelihood ratio serve to identify the most likely cluster, which is followed by secondary clusters by the expedient of sorting them according to the magnitude of the test. As usual, significance is assigned by the analyst, and the cutoff value for significance reflects the confidence of the analyst, or tolerance for error.

When implementing the statistic, the analyst must decide the shape of the window and the maximum number of cases that any given window can cover. Currently, analysis can be done using circular or elliptical windows (see www.satscan.com). Elliptical windows are more time consuming to evaluate but provide greater flexibility to contrast the distribution of events inside and outside the window, and are our selected shape in the analyses to follow. Furthermore, it is recommended that the maximum number of cases entering any given window does not exceed 50% of all available cases. We follow this recommendation.

The Q(m) statistic (Ruiz et al., 2010)[2]

The $Q(m)$ statistic was introduced in the spatial analysis literature by Ruiz et al. (2010) as a tool to explore geographical co-occurrence of qualitative data. The statistic is developed based on the principle of symbolic entropy, and can be applied to a variable X that is the outcome of a qualitative and discrete spatial process indexed at s (the coordinates of the event), and as such can take one of $j=1, \ldots, k$ different values.

We can denote the possible outcomes of X as $a_1, a_2, \ldots, a_k$. These could be, for instance, different types of cancers. In order to assess the spatial property of co-occurrence, the observed value of the variable at a specific location s, call it s_0, is spatially embedded—in other words, the spatial relationships of X at s_0 can be defined. This is done (see Ruiz et al., 2010) by selecting the $m-1$ nearest neighbors of s_0 to give a neighborhood, called an m-surrounding, of size m. Nearest neighbors are selected based on a distance criterion, and in the case of ties, recourse can be made to secondary criteria, such as direction. In the case of irregularly arranged data, secondary criteria are seldom invoked. Other possible distance criteria can be used as appropriate, but are not implemented in this study, including network distance or the presence of natural barriers.

In general, the m-surrounding of spatial index $s0$ can be represented in the following terms:

$$X_m\left(s_0\right) = \left(X_{s_0}, X_{s_1}, \cdots, X_{s_{m-1}}\right)$$

The m-surrounding is a string that collects the values of the variable in the neighborhood of s_0, and therefore gives the local spatial configuration of the variable, with each observation receiving identical weight. For instance, consider the event in Example 1 located at coordinate $s_0=[0.10,0.90]$ (a triangle). The two nearest neighbors are the events at coordinates [0.10,0.85] and [0.15,0.85], both triangles. The spatial embedding of size $m=3$ (the m-surrounding) for s_0 in this case is a string containing three triangles, and so the triangle at s_0 is said to co-occur with two triangles. Consider now in the same example the event located at coordinate [0.75,0.30]. If we set this as s_0, the two nearest neighbors are two squares, and therefore the spatial embedding of size $m=3$ for the variable at this site is a string that contains one triangle and two squares. The event then is said to co-occur with two squares.

Since in general there are k outcomes, there are exactly k^m unique co-occurrence types for neighborhoods of size m. This is the number of permutations with replacement. Each of these types of co-occurrence can be denoted by means of a symbol σ_i ($i=1, 2, \ldots, k^m$), and so each site can be uniquely associated with a specific symbol (that is, can be said to be of symbol i). In this way, we say that a location s is of σ_i-type if and only if $X_m(s) = \sigma_i$.

After the observations have been spatially embedded and their corresponding symbols identified, it is possible to calculate the observed relative frequency of each symbol as follows:

$$p_{\sigma_i} = \frac{n_{\sigma_i}}{S} \quad (3)$$

2 First published online on October 24, 2009.

where n_{σ_i} is the number of times symbol i appears among the symbolized locations, and S is the number of symbolized locations. The frequency of the symbols can be plotted to explore their distribution. Furthermore, the frequencies can be used to calculate a measure of symbolic entropy for a given embedding dimension m as follows:

$$h(m) = -\sum_j p_{\sigma_j} \ln\left(p_{\sigma_j}\right) \quad (4)$$

The entropy function is theoretically bounded between zero when only one of k^m symbols is observed in the empirical series (the map is highly organized), and $Q(m)$, the value of the entropy function calculated under the null hypothesis that the pattern of co-occurrences is random. The $Q(m)$ statistic is simply a likelihood ratio test of the empirical symbolic entropy and the expected entropy under the null. The expressions for $Q(m)$ are derived by Ruiz et al. (Ruiz et al., 2010). In the general case that outcomes are not equally probable, the statistic is expressed as:

$$Q(m) = 2S\left(\sum_{i=1}^{k^m} \frac{n_{\sigma_i}}{S} \sum_{j=1}^{k} p_{ij} \ln\left(q_j\right) - h(m)\right) \quad (5)$$

where ρ_{ij} is the number of times that class a_j appears in symbol σ_i and q_j is the probability of the jth outcome to occur under the null hypothesis of independence.

When implementing the statistic, the analyst must decide the size of the m-surrounding. This is an important consideration, because the number of symbols grows explosively as the number of outcomes k and/or as the size of the surrounding selected for analysis increase. A general guideline is that the number of symbolized observations S should be at least five times the number of symbols (Ruiz et al., 2010). While this is a practical constrain, it is possible to relax it by using equivalent symbols that trade-off some topological information for a more concise representation of the types of co-occurrence. This can be done, for instance, by considering only the number of outcomes of a certain class within an m-surrounding, but not their order of proximity and/or direction with respect to s_0 (Páez et al., 2011). The number of equivalent symbols is the number of combinations with replacement, in most cases a considerably smaller set than the standard symbols.

In terms of inference, the asymptotic results for $Q(m)$, as well as its small sample properties, are presented in Ruiz et al. (2010). When the number of symbolized observations S is small, the convergence of $Q(m)$ to its asymptotic distribution becomes dubious. In order to circumvent problems with approximating the distribution, it is possible to adopt a simulation (that is, bootstrapping) approach for inference. A distribution-free approach can be computationally expensive, but is still manageable for moderately sized samples and allows us to apply the statistic in situations where approximation to the theoretical distribution is difficult. The simulation procedure, for a fixed embedding dimension $m \geq 2$ with a number R of replications, is composed of the following steps:

1. Compute the value of the statistic $Q(m)$ for the original samples $\left\{X_s\right\}_{s\in S}$.

2. Re-label the set of coordinates by randomly drawing from the list of outcomes without replacement, to obtain the series $\left\{X_s^r\right\}_{s\in S}$ where r is the number of the replication.

3. Calculate the bootstrapped statistic $Q_b^r(m)$ for the simulated sample $\left\{X_s^r\right\}_{s\in S}$.

4. Repeat steps 2 and 3 R-1 times to obtain R realizations of the bootstrapped statistic $\left\{Q_b^r(m)\right\}_{r=1}^{R}$.

5. Compute the pseudo-p-value:

$$p_b = \frac{1}{R}\sum_{r=1}^{R} I\left(Q_b^r(m) > Q(m)\right)$$

where I (•) is the indicator function which assigns a value of 1 to a true statement and 0 otherwise.

6. Reject the null hypothesis if:

$$p_b < \alpha$$

for a nominal size α.

It is worth noting that $Q(m)$ is a global independence indicator. In the context of PC (and other spatial epidemiology applications) one might be interested in the empirical distribution of each symbol—that is, the frequency with which different diseases co-occur—and its relationship to the expected distribution under the null. Using a procedure similar to the one described above, it is possible to compute the 100(1-α) percent confidence interval for the relative frequency of a symbol σ_i, p_{σ_i}, by computing its bootstrapped realization $p_{\sigma_i}^r$ for each of the permuted series $\left\{X_s^r\right\}_{s\in S}$, for every symbol as follows:

$$I_\sigma(\alpha) = \left(\zeta_\alpha, \zeta_{1-\alpha}\right)$$

Where ζ_α is the percentile α of the distribution of $\left\{p_{\sigma_i}^r\right\}_{r=1}^{R}$. Accordingly, we will say that the null is rejected at the α level for symbol σ_i whenever the relative frequency of the symbol falls outside of the confidence interval, that is, when $p_{\sigma_i} \notin I_{\sigma_i}(\alpha)$. The interval of confidence can be used to determine, separately from the global significance of the statistic, whether a specific symbol is observed more or less frequently than expected by chance.

Example using simulated events

To illustrate the use of the statistics, we apply them to the simulated events shown in Figure 4.1. The results appear in Figure 4.2.

In the case of the spatial scan statistic, it can be seen there that the probability of the null hypothesis in Example 1 is 0.227, which would under typical rejection protocols (for example, $p \leq 0.10$ or $p \leq 0.05$) lead to the conclusion that no clustering is evident. In the case of Example 2, the null hypothesis would be rejected with a high level of confidence ($p = 0.001$). A large cluster is detected with an elliptical window that covers approximately half of the study region. An abundance of triangles there and a lack of circles and squares contribute to the significance of the cluster. Example 3 also displays clustering at a conventional 5% level of significance. The cluster in this case is located near the center of the region, and results from an absence of circles and a large number of triangles and squares relative to the expectation.

Next, we apply $Q(m)$ using surroundings of size $m = 3$. In the case of Example 1 no clusters were detected by the spatial scan statistic. In contrast, $Q(m)$ is significant ($p = 0.001$), indicating a non-random pattern of co-occurrence. Examination of the histogram of co-occurring types shows that the type of event represented by a triangle is found in triads in excess of what would be expected under the null ($p < 0.05$). Other types of co-occurrence are not significant or only marginally so at the 10% level. Significant co-occurrence in the absence of clustering, as in this example, may indicate localized but dispersed factors that affect the classes of events in the indicated symbol.

In Example 2 the most likely cluster of events covered a large sub-region of the study area. As a reflection of underlying geographical factors, the geographical extent may be indicative of a fairly ubiquitous factor, or the presence of localized but pervasive effects. In addition to the cluster, $Q(m)$ reveals that the pattern of co-occurrences is not random ($p < 0.001$), and displays an excess of homogeneous triads (that is, composed solely of three triangles, circles, or squares). Interestingly, two of these three types of co-occurrence are not part of the most likely cluster identified by the spatial scan statistic. Also evident are fewer cases than expected of two triangles co-occurring with one circle as well as mixed triads. This may suggest a pattern of repulsion between triangles and circles (for instance, one condition may pre-empt the other), or the presence of environmental factors with opposing effects.

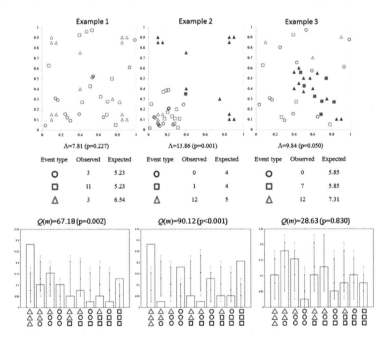

Figure 4.2   Multinomial spatial scan statistic and Q(m) applied to simulated examples

Note: For the scan statistic, results are for most likely cluster, and solid markers identify the element of significant clusters. For $Q(m)$ the histograms show the empirical frequency of co-occurring types, and intervals of confidence (90% and 95%).

The last case, in Example 3, was of a cluster composed of triangles and squares. Now, $Q(m)$ fails to detect a significant pattern of co-occurrence, and as seen in the histogram, all co-occurring types are within their confidence bands. This situation suggests that if the two events in the cluster are influenced by a common factor, the effect must be relatively uniform so that the cases emerge seemingly at random within the cluster—in other words, there is no evidence of pattern in the way events within the cluster are situated with respect to each other.

As these simulated examples demonstrate, the multinomial spatial scan statistic and $Q(m)$ are capable of providing complementary, and possibly non-overlapping information about a spatial pattern. The epidemiological implications are also distinct. To further illustrate the use of these two statistics, in the following sections, we apply them to an empirical case study of PC in Murcia, Spain.

Empirical Application

The Data

Information on cancer cases was originally obtained from the childhood population-based cancer registry in Murcia, Spain. The Environment and Pediatric Cancer Group compiled and geo-referenced the pediatric environmental histories (PEH) for all cases of children with cancer (younger than 15 years old) in the region, during the period 1998–2009 (Ortega-García et al., 2011). The registry includes over 99% of the diagnosed PC cases in Murcia. The cases are categorized according to the International Classification of Diseases for Oncology (Percy et al., 2000), and the International Classification of Childhood Cancer (Steliarova-Foucher et al., 2005).

The oncology PEH used in the study include a series of concise and basic questions that the pediatrician uses to identify environmental exposures in PC (Ferrís Tortajada et al., 2004; Ortega-García et al., 2011). The PEH document human carcinogens characterized by the International Agency for Research on Cancer (IARC) and by the U.S. National Toxicology Program (USNTP) (International Agency for Research on Cancer, National Toxicology Program). Completion of the PEH was conducted by a pediatrician with expertise on environmental health, oncology and experience interacting with PC patients and their families. Centralized referral and care of PC cases at Hospital Universitario Virgen de la Arrixaca (HUVA) since 1998 facilitated the access to the clinical histories of the patients. Following approval by HUVAs Research and Ethics Committee, families were identified from the base registry of the hospital and invited to participate in the study. Initial contact with families was by telephone. After informed consent was granted by participants (patients older than 12 years participated in the decision to consent), information was obtained in the following manner: 1) face-to-face interviews with one or both parents present, for all cases diagnosed since 2005; 2) telephone interviews with all patients, up to a maximum of seven calls, conducted between July 1, 2009, and July 1, 2010; and 3) the information was validated and complemented based on ambulatory clinical histories of the patients from their respective health centers/hospitals and regional databases OMI, Selene, and Civitas.

The resulting database includes detailed information concerning the address of the patient at the time of diagnostic (municipality, postal code, street name and number). Addresses were geo-referenced to support mapping using a Geographical Information System, and for spatial analysis (Ortega-García et al., 2011). Each record includes information about the gender of the patient, date of birth, diagnostic, and histopathological classification. While little is known about the sensitivity of $Q(m)$ to the accuracy of geocoding, in the present case, geocoding was highly precise based on addresses, and verified manually.

A total of 498 PC cases are registered in the region with a date of diagnostic between 1998 and 2009. For the purpose of this analysis we work with a subset of 298 cases, after excluding records as follows. We do not consider a number of specific cancer types that appear with very low frequency. Furthermore, we exclude patients from out of the region of Murcia who were seen at HUVA in order to complete their diagnostic, to initiate, or to continue treatment, and who have a temporary address or moved to Murcia to receive medical treatment. Likewise, we exclude patients from out of the region with an address elsewhere in Spain, who are seen at HUVA for a second opinion, diagnosis, and/or treatment. Some cases are not classifiable according to the ICCC-3. Finally, one case of hypophyseal stalk growth with an as yet indeterminate diagnosis was also excluded.

The 298 cases include five types of cancer with frequencies as follows: leukemia (Le: 108 cases), lymphoma (Ly: 32 cases), soft tissue (St: 38 cases), central nervous system (Cn: 91 cases), and sympathetic nervous system (Sn: 29 cases).

The Results

In this section we apply the multinomial spatial scan statistic and $Q(m)$ to the case of PC cases in Murcia.

In order to implement the spatial scan statistics, we select elliptical windows and limit the search to a maximum window size to 10% and subsequently to 50% of the total number of observations. For inference, the simulation is set to $R = 999$ replications. The results of the application are reported in Table 4.1. The table reports the results for the most likely cluster and the first secondary cluster. Clusters are selected for reporting based on their probabilities. The most likely cluster is the one with the lowest probability and thus the most likely to reject the null hypothesis. Secondary clusters always have greater probability, and are less likely to reject the null. In addition, the number of observed cases for each type of cancer and their expected values are reported. As seen in Table 4.1, there is no significant evidence of clustering beyond what would be expected by chance, as the lowest probability is 0.35 which fails to reject the null hypothesis under conventional levels of significance.

Table 4.1   Multinomial spatial scan statistic applied to pediatric cancer cases in Murcia

	Search: 10%				Search: 50%			
	MLC		SC		MLC		SC	
	$\Lambda=9.68$ ($p=0.570$)		$\Lambda=8.52$ ($p=0.850$)		$\Lambda=11.32$ ($p=0.350$)		$\Lambda=10.35$ ($p=0.528$)	
	Cases inside of cluster		Cases inside of cluster		Cases inside of cluster		Cases inside of cluster	
Types	Observed	Expected	Observed	Expected	Observed	Expected	Observed	Expected
Le	5	6.16	8	10.15	13	16.31	18	17.76
Cn	0	5.19	6	8.55	8	13.74	13	14.96
St	2	2.17	10	3.57	2	5.74	14	6.25
Ly	4	1.83	0	3.01	12	4.83	0	5.26
Sn	6	1.65	4	1.47	10	4.38	4	4.77

MLC= Most Likelihood Cluster; SC= Secondary Cluster.

In the case of $Q(m)$, the size of the m-surrounding is selected by the analyst based on practical and interpretive considerations (Páez et al., 2011; Ruiz et al., 2010). Presently, we select $m = 3$ to investigate triads of cases for co-occurrence. Selection of a larger neighborhood of four cases already implies 625 standard symbols and 70 equivalent symbols, and thus strains the limits of the method for our small sample. A smaller neighborhood of $m = 2$, while feasible, contains less spatial information. All bootstrapping procedures are conducted based on $R = 999$ replications.

Table 4.2 $Q(3)$ applied to pediatric cancer cases in Murcia

Classes	Leukemia (Le), lymphoma (Ly), soft tissue (St), Central nervous system (Cn), Sympathetic nervous system (Sn)
Number of symbolized locations	298
Number of standard symbols	125
Number of equivalent symbols	35
Frequency of categories	Le: 0.36, Ly: 0.31, St: 0.13, Cn: 0.11, Sn: 0.09
$Q(3)$	138.49
Pseudo-p-value	0.232

Table 4.2 presents the results of calculating $Q(m)$ and the pseudo-p-value for inference. Based on the pseudo-p-value, the statistic fails to reject the null hypothesis of a spatially random sequence of values globally for the qualitative variable. However, examination of the relative frequency of symbols and their confidence intervals (Figure 4.3) indicates that two patterns of co-occurrence are observed significantly more frequently than expected by chance alone (p <0.05): co-occurring leukemia triads, and triads consisting of co-located cases of lymphoma, soft tissue, and central nervous system tumors.

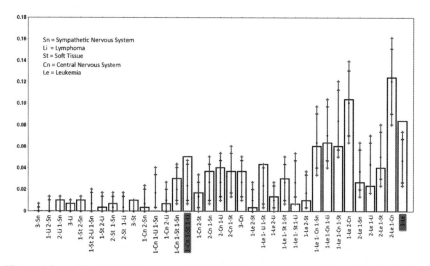

Figure 4.3   Empirical frequency of symbols (co-occurrence patterns) under k=5 and m=3, and bootstrapped confidence intervals (90% and 95%)

In order to further explore the co-occurrence of leukemia,[3] we reclassify the observations by labeling all cases other than leukemia as "Other" (Ot). The results of this variation in the analysis are presented in Table 4.3. It can be seen there that when leukemia is contrasted to other types of cancers, the statistic is now globally significant at p <0.10. The significance of the statistic is driven mainly by the co-occurrence of triads of leukemia cases, as indicated by the confidence intervals in Figure 4.4.

Table 4.3 $Q(3)$ **for reclassified types of pediatric cancer in Murcia, with a focus on leukemia**

Classes	Leukemia (Le), other (Ot)
Number of symbolized locations	298
Number of standard symbols	8
Number of equivalent symbols	4
Frequency of categories	Le: 0.36, Ot: 0.64
Q(3)	17.62
Pseudo-p-value	0.060

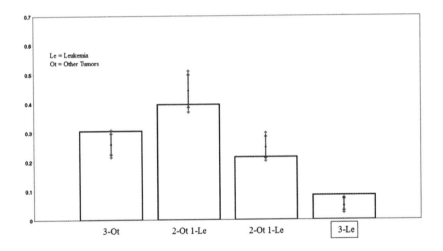

Figure 4.4   **Empirical frequency of symbols (co-occurrence patterns) under k=2 (leukemia and other) and m=3, and bootstrapped confidence intervals (90% and 95%).**

In order to contrast this result due to $Q(m)$ to the application of the spatial scan statistic, we apply the spatial scan statistic in its version for Bernoulli variables (Kulldorff and Nagarwalla, 1995; Kulldorff, 1997). The results are reported in Table 4.4. The most likely and secondary clusters are composed of small groups of leukemia cases (at most 6 cases),

---

3   The following analysis could be conducted as desired for other types of co-occurring cancers.

but none of these clusters are significant at conventional levels (the smallest p-value is 0.764) and thus we fail to reject the null hypothesis.

Table 4.4   Spatial scan statistic for reclassified types of pediatric cancer in Murcia, with a focus on leukemia

	MLC	SC
Cluster size	6	6
Leukemia cases in cluster	6	6
Expected cases	2.17	2.17
Λ	6.07	5.36
P	0.764	0.924

MLC= Most Likelihood Cluster; SC= Secondary Cluster.

Discussion

Significant co-occurrence in the absence of clustering suggests that any possible environmental factors may be local as opposed to regional in nature. This finding prompts a focused re-examination of the PEH. Information about the cases of co-occurring leukemia is summarized in Table 4.5.

Twenty-five cases of leukemia appear in 13 m-surroundings of size 3. There locations are indicated in Figure 4.5. None of the patients in these cases are consanguineal kin. The distance separating the cases tends to be relatively small, in some cases just a few meters, and on average a mere 1.36 km. The average time between the diagnostic of the cases on the other hand is approximately 4.31 years. This suggests that if a common environmental factor has been involved in some of these cases, it must be temporally persistent. Two particular localities in the region of Murcia are highlighted by the preceding analysis. In the municipality of Cieza, six individual cases are part of two triads of co-occurring leukemia cases. The average distance between elements of these triads is 256 m, well below the overall average for cases in all triads. The temporal separation between cases is in average 3.91 years, slightly below the overall average. In the municipality of San Javier, five individual cases are part of two triads. The average distance between cases in the two triads in this municipality is 591 m, and their average intervening time is 2.31 years.

Known RF associated with pediatric leukemia include (see Ries et al., 1999): sex, age, ionizing radiation in utero and postnatal (therapeutic use only), and genetic conditions such as Down syndrome, neurofibromatosis, Shwachman syndrome, Bloom syndrome, ataxia telangiectasia, Langerhans cell histiocytosis, Klinefelter syndrome, and nevoid basal cell syndrome. Factors for which evidence is suggestive but inconclusive include high birth weight, maternal history of fetal loss prior to the birth of the index child, older maternal age (>35 years at pregnancy), first born or only child, exposure to pesticides, and exposure to benzene. Factors for which evidence is inconsistent or limited are electromagnetic field exposure, smoking prior to and during pregnancy, father's occupation (parental exposure to motor vehicle exhaust, hydrocarbons, and paints), maternal alcohol consumption or use of recreational drugs during pregnancy, postnatal infections, postnatal use of chloramphenicol, vitamin K prophylaxis in newborns, and exposure to radon.

Table 4.5   **Further information regarding triads of co-occurring leukemia cases.**

Triad	Municipality	Time between cases in triad (yr)			Distance between cases in triad (km)		
		1–2	1–3	2–3	1–2	1–3	2–3
1	Alcantarilla	1.37	4.22	2.85	0.588	0.865	0.389
2	Murcia	1.67	6.66	4.99	0.123	0.652	0.53
3	Cieza	4.88	2.44	2.44	0.248	0.283	0.191
4	Beniel	5.81	7.33	1.52	0.594	0.963	0.391
5	San Javier	4.59	0.36	4.23	0.319	0.38	0.578
6	Mazarrón	7.79	9.44	1.64	1.414	5.964	5.534
7	Cieza	1.19	3.99	2.80	0.196	0.218	0.398
8	San Javier	0.39	1.93	2.32	0.031	1.136	1.132
9	Jumilla	1.70	8.71	7.02	0.413	0.455	0.852
10	Lorca	2.47	7.81	5.34	6.194	6.232	0.788
11	Bullas	4.40	0.47	4.87	0.442	0.871	0.484
12	Molina del Segura	8.26	8.42	0.17	2.194	2.362	0.611
13	Cartagena	10.59	10.79	0.19	1.248	3.411	3.25

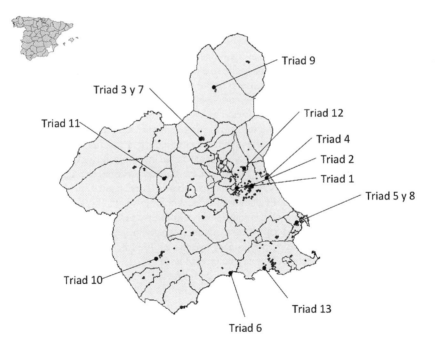

Figure 4.5   **Co-occurring leukemia triads in Murcia (dots are cases of pediatric cancer). Triad identifiers as in Table 4.5**

The PEH were carefully collected to include whenever present, any factors related to the above or other carcinogens characterized by IARC and USNTP. Considering the known or suspected RF, two co-occurring triads uncovered by the re-examination of PEH are of particular interest.

One instance of three co-occurring cases of leukemia is found in Cieza (co-occurrence 7 in Table 4.5; average time between cases is 2.66 yr, and average distance 0.27 km). This is composed of three females presenting ALL type-B. The PEH revealed three leading RF. One non-environmental factor in two of these cases was exposure to prenatal ionizing radiation. In addition, two potential environmental factors are observed. First, repeated viral infections were observed during periods between 24 and 3 months prior to the manifestation of the leukemia in the three cases. Contagious infections caused by agents with high infectivity and high pathogenicity, for example measles, frequently generate time clustering. One or more common infections during early infancy in genetically susceptible or environmentally conditioned individuals may play a role in the pathogenesis of childhood leukemia (Roman et al., 2007). And secondly, the three patients had their place of residence in proximity (<20 m) of power transformers or electrical power lines of high voltage (ELHV). ELHV are a powerful source of extremely low frequency electromagnetic fields (Ortega García et al., 2009). This is a factor acknowledged as a potential carcinogenic in children by the Working Group of the U.S. National Institute of Environmental Health Sciences, in the by the National Radiological Protection Board and the ICNIRP Standing Committee on Epidemiology.

Another set of co-occurring cases is found in the municipality of San Javier. Examination of the genetic and environmental RF in the PEH reveals that the only RF reported for two ALL type-B cases in triad 8 (Table 4.5; average time between cases is 1.55 years, and average distance 0.77 km) is a high level of residential exposure to emissions from burning of fossil fuels related to air traffic in a nearby (<1.4 km) civil and military airport. This airport saw a substantial increase in traffic since 2004 with the arrival of low-cost air carriers to the region. Aircraft engine emissions contain a number of known carcinogens, including the leukemogen benzene (Ries et al., 1999; Herndon et al., 2009). Research in the Netherlands, for instance, found a statistically significantly increase of ALL in children in the area around Schiphol airport, although it did not establish a conclusive link to pollution (Visser et al., 2005). The last case in this triad is an acute myeloblastic leukemia. In addition to proximity to the airport, the PEH for this patient showed exposure to intrauterine RF.

The results of the exploratory spatial data analysis do not conclusively link environmental factors to PC in Murcia. The individual risk assessment for PC by PEH is a complex process. It requires an understanding of carcinogenesis, and the resources and abilities to obtain and interpret the data. In this sense, the exploratory spatial analysis provides value, since it helps to discard a number of preliminary hypotheses (that is, various types of cancers neither cluster nor co-occur in Murcia), helps to validate other hypotheses (leukemia cases do not cluster, but tend to co-occur), and generates ideas for further research (the potential role of proximity to power lines and/or the airport).

Conclusions

Exploratory analysis of multinomial variables is a topic of interest in spatial epidemiology. Two recently developed methods, the multinomial spatial scan statistic and $Q(m)$, can be used to investigate different aspects of spatial pattern, namely the clustering and co-occurrence of various types of diseases. In this chapter, we discussed the differences between clustering and co-occurrence. Clustering is a commonly investigated property of spatially distributed events; however co-occurrence has seldom been explicitly identified as a separate characteristic of spatial pattern.

The distinction between clustering and co-occurrence is important. Comparison of the spatial scan statistic and $Q(m)$ using three simulated examples and an empirical application to PC in Murcia, shows that the two methods can be used to provide complementary—and at times non-overlapping—information regarding clustering and co-occurrence in an exploratory framework. This in turn can generate valuable insights into the possible origins and factors influencing spatial pattern. It is important to note, at this point, that analysis was conducted in a cross-sectional fashion—which carries implicit the assumption that detectable environmental effects must be long-lasting, and not due to ephemeral contextual conditions.

The findings of the empirical application, in addition to supporting our comparison of methods, are also interesting in their own right. Information about PC in Murcia has only recently become available thanks to a broader and ongoing project to develop and analyze PEH for children diagnosed with cancer (Ferrís Tortajada et al., 2004; Ortega-García et al., 2011). The analysis reported here is thus a first step towards a better understanding of the potential environmental factors that influence this disease in the region. In particular, the finding that PC cases do not form geographical clusters, but that leukemia tends to co-occur in triads significantly more frequently than expected, steers our attention away from broad regional factors and towards more localized effects. This result formed the basis for a focused re-examination of PEH, which revealed that the co-occurrence of childhood hematopoietic malignancies in Murcia may have been associated with several specific factors, such as viral infections, the presence of high voltage power lines, and proximity to an airport that saw increased traffic in the past decade. While the results, as is always the case in exploratory spatial data analysis, warrant further attention before they can be confirmed, it is intriguing that identification of these factors, supported by the spatial statistical analysis, points in similar direction as previous research on the occurrence of leukemia in minors. Further research, preferably in a multivariate framework, should be able to validate or refute the ideas generated by the exploratory analysis. As well, we should note that current plans to collect for the PEHs the address at the time of pregnancy should allow the examination of spatial patterns at different stages in the life of patients.

More generally, the examples and empirical analysis presented in this chapter suggest that the multivariate spatial scan statistic and $Q(m)$ should be used jointly in the exploratory analysis of spatially distributed events. In this chapter, the empirical application was to PC in Murcia, but the results of our comparison of methods clearly hold broader appeal in the exploratory spatial data analysis of any type of disease where environmental RF are of interest.

References

Abbaszadeh, F., Barker, K.T., Mcconville, C., Scott, R.H., and Rahman, N. 2010. A new familial cancer syndrome including predisposition to Wilms tumor and neuroblastoma. *Familial Cancer*, 9: 425–30.

Alfo, M., Nieddu, L., and Vicari, D. 2009. Finite mixture models for mapping spatially dependent disease counts. *Biometrical Journal*, 51: 84–97.

Amin, R., Bohnert, A., Holmes, L., Rajasekaran, A., and Assanasen, C. 2010. Epidemiologic Mapping of Florida Childhood Cancer Clusters. *Pediatric Blood & Cancer*, 54: 511–18.

Bailey, T.C. and Gatrell, A.C. 1995. *Interactive Spatial Data Analysis*. Essex: Addison Wesley Longman.

Bao, P.P., Zheng, Y., Gu, K., Wang, C.F., Wu, C.X., Jin, F., and Lu, W. 2010. Trends in Childhood Cancer Incidence and Mortality in Urban Shanghai, 1973–2005. *Pediatric Blood & Cancer*, 54: 1009–13.

Bell, B.S., Hoskins, R., Pickle, L., and Wartenberg, D. 2006. Current practices in spatial analysis of cancer data: Mapping health statistics to inform policymakers and the public. *International Journal of Health Geographics*, 5: 49.

Cassetti, T., La Rosa, F., Rossi, L., D'alo, D., and Stracci, F. 2008. Cancer incidence in men: A cluster analysis of spatial patterns. *Bmc Cancer*, 8.

Downing, A., Forman, D., Gilthorpe, M., Edwards, K., and Manda, S. 2008. Joint disease mapping using six cancers in the Yorkshire region of England. *International Journal of Health Geographics*, 7: 41.

Dreassi, E. 2007. Polytomous disease mapping to detect uncommon risk factors for related diseases. *Biometrical Journal*, 49: 520–29.

Elliott, P. and Wartenberg, D. 2004. Spatial epidemiology: Current approaches and future challenges. *Environmental Health Perspectives*, 112: 998–1006.

Ferrís Tortajada, J., Ortega García, J.A., Marco Macián, A., and García Castell, J. 2004. Medio ambiente y cáncer pediátrico. *Anales de Pediatría*, 61: 9.

Gatrell, A.C., Bailey, T.C., Diggle, P.J., and Rowlingson, B.S. 1996. Spatial point pattern analysis and its application in geographical epidemiology. *Transactions of the Institute of British Geographers*, 21: 256–74.

Gelfand, A.E., and Vounatsou, P. 2003. Proper multivariate conditional autoregressive models for spatial data analysis. *Biostatistics*, 4: 15.

Gilman, E.A. and Knox, E.G. 1998. Geographical distribution of birth places of children with cancer in the UK. *British Journal of Cancer*, 77: 842–9.

Goovaerts, P. 2010. Application of Geostatistics in Cancer Studies, in Atkinson, P.M. and Lloyd, C.D. (eds) *Geoenv Vii—Geostatistics for Environmental Applications*. Dordrecht: Springer.

Grupp, S.G., Greenberg, M.L., Ray, J.G., Busto, U., Lanctot, K.L., Nulman, I., and Koren, G. 2011. Pediatric cancer rates after universal folic acid flour fortification in Ontario. *Journal of Clinical Pharmacology*, 51: 60–65.

Haining, R., Wise, S., and Ma, J.S. 1998. Exploratory spatial data analysis in a geographic information system environment. *Journal of the Royal Statistical Society Series D-the Statistician*, 47: 457–69.

Held, L., Natario, I., Fenton, S.E., Rue, H., and Becker, N. 2005. Towards joint disease mapping. *Statistical Methods in Medical Research*, 14: 61–82.

Herndon, S.C., Wood, E.C., Northway, M.J., Miake-Lye, R., Thornhill, L., Beyersdorf, A., Anderson, B.E., Dowlin, R., Dodds, W., and Knighton, W.B. 2009. Aircraft hydrocarbon emissions at Oakland International Airport. *Environmental Science & Technology*, 43: 1730–36.

International Agency for Research on Cancer Overall Evaluations of Carcinogenicity to Humans. *IARC Monographs*.

Jin, X.P., Carlin, B.P., and Banerjee, S. 2005. Generalized hierarchical multivariate CAR models for areal data. *Biometrics*, 61: 950–61.

Jung, I., Kulldorff, M., and Richard, O.J. 2010. A spatial scan statistic for multinomial data. *Statistics in Medicine*, 29: 1910–18.

Kloog, I., Haim, A., and Portnov, B.A. 2009. Using kernel density function as an urban analysis tool: Investigating the association between nightlight exposure and the incidence of breast cancer in Haifa, Israel. *Computers Environment and Urban Systems*, 33: 55–63.

Knorr-Held, L. and Best, N.G. 2001. A shared component model for detecting joint and selective clustering of two diseases. *Journal of the Royal Statistical Society Series A-Statistics in Society*, 164: 73–85.

Knox, E.G. and Gilman, E.A. 1996. Spatial clustering of childhood cancers in Great Britain. *Journal of Epidemiology and Community Health*, 50: 313–19.

Kulldorff, M. 1997. A spatial scan statistic. *Communications in Statistics-Theory and Methods*, 26: 1481–96.

Kulldorff, M., Feuer, E.J., Miller, B.A., and Freedman, L.S. 1997. Breast cancer clusters in the northeast United States: A geographic analysis. *American Journal of Epidemiology*, 146: 161–70.

Kulldorff, M. and Nagarwalla, N. 1995. Spatial disease clusters—detection and inference. *Statistics in Medicine*, 14: 799–810.

Manda, S.O.M., Feltbower, R.G., and Gilthorpe, M.S. 2009. Investigating spatio-temporal similarities in the epidemiology of childhood leukemia and diabetes. *European Journal of Epidemiology*, 24: 743–52.

Mather, F.J., Chen, V.W., Morgan, L.H., Correa, C N., Shaffer, J.G., Srivastav, S.K., Rice, J.C., Blount, G., Swalm, C.M., Wu, X.C., and Scribner, R.A. 2006. Hierarchical modeling and other spatial analyses in prostate cancer incidence data. *American Journal of Preventive Medicine*, 30, S88-S100.

Mcnally, R., Basta, N., James, P., and Craft, A. 2009. Seasonal variation in birth and diagnoses of cancer in children and young people in Northern England, 1968–2005. *Epidemiology*, 20, S18-S18.

National Toxicology Program 12th Report on Carcinogens. Department of Health and Human Services.

Openshaw, S., Charlton, M., Wymer, C., and Craft, A.W. 1987. A Mark I geographical analysis machine for the automated analysis of point data sets. *International Journal of Geographical Information Systems*, 1: 335–58.

Ortega-García, J.A., López-Hernández, F.A., Sobrino-Najul, E., Febo, I., and Fuster-Soler, J.L. 2011. Medio ambiente y cáncer pediátrico en la Región de Murcia (España): Integrando la historia clínica medioambiental en un sistema de información geográfica. *Anales de Pediatría*, 74: 6.

Ortega García, J.A., Martin, M., Navarro-Camba, E., Garcia-Castell, J., Soldin, O.P., and Ferris-Tortajada, J. 2009. Pediatric Health Effects of Chronic Exposure to Extremely Low Frequency Electromagnetic Fields. *Pediatric Reviews*, 5: 234–40.

Páez, A., Ruiz, M., López, F., and Logan, J.R. 2012. Measuring Ethnic Clustering and Exposure with the Q statistic: An Exploratory Analysis of Irish, Germans, and Yankees in 1880 Newark. *Annals of the Association of American Geographers*, 102: 84–102.

Percy, C., Fritz, A., Jack, A., Shanmugarathan, S., Sobin, L., Parkin, D.M., and Whelan, S.2000. *International Classification of Diseases for Oncology (ICD-O)*, Geneve, Switzerland, World Health Organization.

Pickle, L., Szczur, M., Lewis, D., and Stinchcomb, D. 2006. The crossroads of GIS and health information: A workshop on developing a research agenda to improve cancer control. *International Journal of Health Geographics*, 5: 51.

Pollack, L.A., Gotway, C.A., Bates, J.H., Parikh-Patel, A., Richards, T.B., Seeff, L.C., Hodges, H., and Kassim, S. 2006. Use of the spatial scan statistic to identify geographic variations in late stage colorectal cancer in California (United States). *Cancer Causes & Control*, 17: 449–57.

Preston, J.R. 2004. Children as a sensitive subpopulation for the risk assessment process. *Toxicology and Applied Pharmacology*, 199: 10.

Rainey, J.J., Omenah, D., Sumba, P.O., Moormann, A.M., Rochford, R., and Wilson, M.L. 2007. Spatial clustering of endemic Burkitt's lymphoma in high-risk regions of Kenya. *International Journal of Cancer*, 120: 121–7.

Ries, L.A.G., Smith, M.A., Gurney, J.G., Linet, M., Tamra, T., Young, J.L., and Bunin, G.R. 1999. *Cancer Incidence and Survival among Children and Adolescents: United States SEER Program 1975–1995*, Bethesda, MD, National Cancer Institute, SEER Program.

Rigby, J.E. and Gatrell, A.C. 2000. Spatial patterns in breast cancer incidence in north-west Lancashire. *Area*, 32: 71–8.

Roman, E., Simpson, J., Ansell, P., Kinsey, S., Mitchell, C.D., Mckinney, P.A., Birch, J.M., Greaves, M., Eden, T., and United Kingdom Childhood Canc, S. 2007. Childhood acute lymphoblastic leukemia and infections in the first year of life: A report from the United Kingdom Childhood Cancer Study. *American Journal of Epidemiology*, 165: 496–504.

Ruiz, M., López, F., and Páez, A. 2010. Testing for spatial association of qualitative data using symbolic dynamics. *Journal of Geographical Systems*, 12: 281–309.

Schmiedel, S., Blettner, M., Kaatsch, P., and Schuz, J. 2010. Spatial clustering and space-time clusters of leukemia among children in Germany, 1987–2007. *European Journal of Epidemiology*, 25: 627–33.

Schmiedel, S., Jacquez, G.M., Blettner, M., and Schuz, J. 2011. Spatial clustering of leukemia and type 1 diabetes in children in Denmark. *Cancer Causes & Control*, 22: 849–57.

Shi, X. 2009. A geocomputational process for characterizing the spatial pattern of lung cancer incidence in New Hampshire. *Annals of the Association of American Geographers*, 99: 521–33.

Snow, J. 1855. *On the Mode of Communication of Cholera*. London: John Churchill.

Song, J.J., Ghosh, M., Miaou, S., and Mallick, B. 2006. Bayesian multivariate spatial models for roadway traffic crash mapping. *Journal of Multivariate Analysis*, 97: 28.

Steliarova-Foucher, E., Stiller, C., Lacour, B., and Kaatsch, P. 2005. International Classification of Childhood Cancer, third edition. *Cancer*, 103: 1457–67.

Tango, T. and Takahashi, K. 2005. A flexibly shaped spatial scan statistic for detecting clusters. *International Journal of Health Geographics*, 4: 11.

Tian, N., Wilson, J.G., and Zhan, F.B. 2010. Female breast cancer mortality clusters within racial groups in the United States. *Health & Place*, 16: 209–18.

Timander, L.M. and Mclafferty, S. 1998. Breast cancer in West Islip, NY: A spatial clustering analysis with covariates. *Social Science & Medicine*, 46: 1623–35.

Visser, O., Van Wijnen, J.H., and Van Leeuwen, F.E. 2005. Incidence of cancer in the area around Amsterdam Airport Schiphol in 1988–2003: A population-based ecological study. *Bmc Public Health*, 5.

Wheeler, D.C. 2007. A comparison of spatial clustering and cluster detection techniques for childhood leukemia incidence in Ohio, 1996–2003. *International Journal of Health Geographics*, 6: 16.

Yiannakoulias, N., Rosychuk, R., and Hodgson, J. 2007. Adaptations for finding irregularly shaped disease clusters. *International Journal of Health Geographics*, 6: 28.

Yin, N., Parker, D.F., Hu, S.S., and Kirsner, R.S. 2011. Geographic distribution of melanoma in Miami-Dade County, Florida. *Archives of Dermatology*, 147.

SECTION 2
Infectious Disease

Chapter 5
Spatio-Temporal Characteristics
of the Medieval Black Death

Brian H. Bossak and Mark R. Welford

Yersinia pestis, the Gram-negative bacterium responsible for the highly lethal flea-vectored disease known as bubonic plague, is generally considered to be responsible for the Medieval Black Death (MBD) of 1347–1351 (for example, Creighton, 1965; Shrewsbury, 2005; Biraben, 1975; Benedictow, 2004; Kelly, 2005). The MBD and subsequent outbreaks of disease referred to as 'plagues' in historical accounts, as well as in the later English 'Bills of Mortality,' ravaged medieval Europe during periodic epidemics until approximately 1801 (Gottfried, 1983). However, the primary wave (the first epidemic) of the MBD was by far the most devastating, destroying the burgeoning economy of the late Middle Ages and killing greater than 30% of the European population in a very short period. The true number of case fatalities and the actual mortality rate from this epidemic is still unknown and estimates are derived from the postulations of individual historians (Horrox, 1994; Benedictow, 2004; Kelly, 2005).

The paradigm of Black Death causation (bubonic plague) was adopted around the turn of the 20th century. First recorded in Europe at Messina, Sicily, in the October 1347, the transmission of MBD was so rapid and the disease was so lethal that it does not fit the typical velocity of diffusion in observed modern bubonic plague epidemics such as in India around the turn of the 20th century and in China (Manchuria) between 1910 and 1921 (Chermin, 1989; Kool, 2005; Nishiura et al., 2006). Plague transmission velocities averaged 13–19 km/year between 1866 and 1944 in China (Adjemian et al., 2007; Benedict, 1996), and in India between 1896 and 1906 where railroads affected disease diffusion across the country—but yet locally, plague failed to cross streets (Yu and Christakos, 2006). In contrast, MBD averaged velocities of 0.9 to 6 km *a day* (Benedictow, 2004; Christakos et al., 2005) with Noble (1974) estimating MBD transmission velocities of ~480 km/year. As a result MBD was transmitted from Messina to Gotland, Sweden in three years.

Figure 5.2 illustrates the velocity at which the MBD spread throughout Europe—within three years from entry it had reached Gotland, Sweden. The epidemic velocity of MBD, based on a detailed compilation of hundreds of individual MBD accounts, does not easily reconcile with known or estimated bubonic plague transmission velocity based on observations of genuine outbreaks. "Its [MBD] rapid propagation through Europe can be explained by the lengthy incubation period identified within burial registers at Penrith [England], which show an infectious period of 20–22 days, symptoms only for 5 days prior to death and an entire disease state taking 37 days from infection to death" (Welford and Bossak 2010a, 568, after Scott and Duncan, 2004).

Interestingly, MBD had an inverse seasonality of peak mortality from observed modern bubonic plague epidemics, for example India and Hong Kong in the 1900±20 yr time frame (Welford and Bossak, 2009). Modern bubonic plague outbreaks caused by *Y. pestis* have generally peaked in winter (colder months of the year), whereas peaks in

MBD infection and mortality occurred during the summer (warmest months of the year). This is contradictory not only to modern *Y. pestis*-variant plague, but also contrasts with some viruses such as influenza (Welford and Bossak 2010b). In the Middle Ages, most commerce took place during the warmer months of the year. Hence, peaks in infection and mortality during the MBD occur when the population would be outside working the fields, in the markets, or traveling, suggesting human-to-human transmission (Welford and Bossak, 2009) of the MBD. According to Welford and Bossak (2010b), inverse seasonality of infection and mortality seems logical in consideration of an approximately 32–37 day incubation period, as proposed by Scott and Duncan (2004). In this period, infectious (but not yet symptomatic) merchants and traders could travel far enough to spread the disease before becoming seriously ill, resulting in consonance between the peak time frame for trade and the peak of MBD in the late spring and summer (Bossak and Welford, 2009). This also might explain how the MBD could have traveled by ship without killing all onboard before making port.

MBD vs. Modern *Y. pestis* Plague: A Modern Scientific Debate

Adjemian et al. (2007) showed that plague transmission from animal-to-animal produces jagged traveling waves of plague highly sensitive to existing environmental and climatic conditions and landscape heterogeneities; in particular, mountains are barriers to plague transmission among animals. In contrast, the Pyrenees, the many large rivers that bisect the European Plain, and the high plateau of Spain, did not present an obstacle to the transmission of MBD. If MBD were in fact transmitted via rats and fleas, then a wave of MBD infection and mortality could have affected rural areas just as much—if not more than—urban areas (Christakos et al., 2005) due to the fact that rats and fleas do not necessarily travel via transportation networks (roads, canals, and by sea) and could have spread the disease to all cities and towns regardless of proximity to road networks. There is the possibility that pneumonic plague could have traveled along trade routes, but the chance that the infected humans could make it very far once infected were extremely slim as pneumonic plague is rapidly lethal (and has a much shorter incubation period than the historical data suggest). Thus, the purpose of our research was to use GIS and spatial analysis techniques to determine whether or not there was a correlation between MBD localities (cities and towns where records or documentation of MBD-related mortality have been found) and medieval transportation routes. If the MBD localities were randomly situated throughout Europe, with no association with transportation networks, then the results would be supportive of the current MBD paradigm (*Y. pestis*-related disease). However, if MBD mortalities were found to exist in a proximity cline along transportation networks (for example, more recorded MBD localities closer to roads, fewer localities recorded as distance from medieval roads increases), that could suggest a human-borne pathogen due to the fact that traders and pilgrims of the time traveled extensively back and forth between cities and towns via early road networks—many of which descended from older Roman Empire roads. While this study in historical medical geography could potentially provide evidence at odds with the established transmission mechanism for bubonic plague, it cannot provide enough evidence to determine the true MBD etiologic agent, nor can it provide enough evidence to rule out bubonic plague as a cause of MBD. Today we cannot assume modern and historic plague are genetically identical, in fact, several biovars have been identified in victims of MBD (Ayyaduri et al., 2010; Bos et al., 2011; Drancourt et al., 2004; Schueneman et al., 2011) and Justinian Plague (Drancourt et al., 2007; Wagner et al., 2014). Such variation could explain differences in modern and historic spatial transmission characteristics and velocities particularly among human-human pneumonic plague outbreaks but not bubonic outbreaks as humans remain dead-end and rat presence, densities and migration patterns still dictate very slow transmission velocities.

Data Collection and Methodology

Although there is a large body of literature on modeling space-time patterns of infectious diseases (a systematic review of 442 modeling studies by Lloyd-Smith et al. 2009 noted that most zoonotic modeling were dominated by pandemic influenza, SARS, and rabies) and identification of disease velocity metrics (Meentemeyer et al., 2011; Pioz et al., 2011) little work has been conducted on either historic or modern plague. Modeling MBD has limited utility given: (1) limited data (465 data points), and (2) the lack of independently gathered data to validate any MBD model (see Lloyd-Smith et al., 2009 arguments), but mapping traveling waves of MBD using GIS can still yield explanatory results. In order to utilize GIS and spatio-temporal analysis procedures, the first process is to identify as many localities where MBD was reported as possible and then geocode these locations into a GIS framework. Beginning with large-scale data mining using Google Scholar, Primary source data was obtained from digital copies of historical accounts, such as parish records (for example, Hatcher, 1977; Lock, 1992; Poos, 1991; Razi, 1980), burial registers (for example, Givry and St. Nizier in France; Gras, 1939 and Biraben, 1975), proxy records such as new consecrated bishops in the diocese of York (for example, Thompson, 1911), and documentary "plague tracts," composed during the time of the MBD (1347–1351), with secondary source data extracted from prior published accounts (for example, Creighton, 1965; Shrewsbury, 2005; Biraben, 1975; Benedictow, 2004; Kelly, 2005).

We collated thousands of point locations with MBD mortality data for the years 1347–1801, with 480 total point records dating to the time frame of the primary epidemic wave time from 1347–1351 (Figure 5.1). In many cases, records consisted of MBD mortality counts (465 records), while in other evidentiary sources, mortality percentages were described or in some cases, calculated from estimated mortality counts and population estimates (435 records). It is important to note that the primary wave of the MBD occurred more than 650 years ago, in multiple kingdoms scattered throughout Europe and Asia, and recorded in multiple languages and embellishments. Thus, all MBD data, whether case counts, case fatality rates, or mortality rates is estimated; no factual information on MBD mortality exists at modern levels of certainty and accuracy. Where allied to pre-existing MBD population estimates, MBD mortality estimates were calculated in the dataset's attribute table when mortality rates were the only MBD data available (Christakos et al., 2007). Many of the 457primary wave data points represent multiple mortality records collected for one location: for our analyses, the resulting mortalities were averaged by determining a mean value for each location. For example, large European population centers such as Paris or Florence were associated with multiple MBD mortality estimates; the mean values of these data were utilized to produce a mortality count estimate. The resulting GIS database contains 267 independent MBD localities, which represent the largest primary wave mortality database that we are aware of presently. Previously, Christakos et al. (2007) had located 197 places that had mortality data; Biraben's (1975) map contained 172 mortality points, Carpentier's (1962) map had 88 mortality locations, and Rasmussen (2000) possessed 184 mortality points. The end result of this data mining phase of our spatio-temporal research generated a geo-referenced GIS-based data set of mortality data for point locations during the MBD epidemic from 1347–1351 based on the many compendia of historical MBD accounts (see References for a complete list of documentary sources utilized).

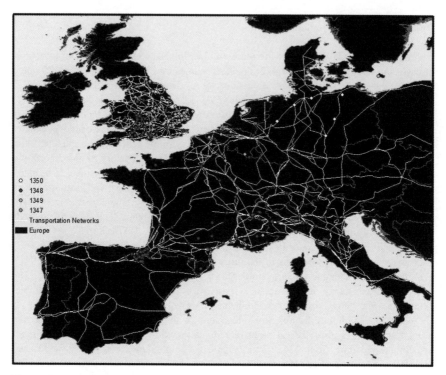

**Figure 5.1   Medieval Black Death Outbreaks (By Year) and Medieval
                   Transportation Networks**

The historical mortality data were then compiled in a geodatabase with attributes such as recorded deaths if available, the beginning date of the epidemic, the ending date of the epidemic, country name, and estimated mortality rate. Beginning and ending dates of the epidemic were based on stated dates in historical records, or where not present, based on estimates. The compiled location data were then imported into a geographic information system (ArcMap 9.3.3) and all locations geocoded in a point shapefile. Once the MBD locality shapefile was introduced into ArcMap, we proceeded to the next phase of the research process which was to digitally recreate the medieval transportation network estimated to be in use at the time of the primary wave of MBD.

In order to digitize historical road networks and transportation routes during the 1300's, we began by analyzing Peter Robin's (pilgrim.peterrobins.co.uk/itineraries/list.html) collection of pilgrimage routes throughout medieval Europe as well as an assortment of additional medieval atlases. Approximately half of the resulting medieval road network we digitized was developed from known pilgrimage routes, with the other networks located on medieval maps. Transportation nodes identified along the pilgrimage routes and in medieval atlas-derived roads were entered one by one into ArcMap 9.3 using the (x,y) function. Once all nodes were entered, they were connected using the "create new line function." The lines were then projected and the geographic coordinates of the transportation routes were validated using Google Earth. The MBD locality shapefiles and the transportation route shapefiles were created independently. The geocoded MBD localities were overlaid on top of the transportation network and then categorized into six groups and one subgroup based on their distance from transportation networks, with the subgroup consisting of port cities. The grouping distances from the transportation networks were: 8 kilometers (5 miles) or less, 16 kilometers (10 miles) or less, 24 kilometers (15 miles) or less, 32 kilometers (20 miles) or less, 40 kilometers (25 miles) or less, and greater than 40 kilometers. The MBD points were exclusively delegated into each of these groups by using the location query tool in ArcMap. In order to validate the proximities of the mortality-data points, the Buffer (analysis) tool in ArcMap 9.3.3 was used to create buffers at 8, 16, 24, 32, and 40 kilometer distances from the individual points. The buffer analysis supported the distance results from the location query, and the new shapefiles containing the distance based groups of localities were exported into Microsoft Excel. Basic statistical analyses were performed to analyze the presence of mortality clines.

In order to better understand exactly how far and fast this epidemic spread, a distance analysis was generated using the digitized medieval transportation network and known outbreaks of MBD. Messina, Sicily was used as the initialization point of the epidemic in 1347. From Messina, distance was measured in kilometers along the shortest route on the road network to each city that had an outbreak of MBD. Distances for each locality, separated by year, were placed into Excel files. Distances from the UK to Messina were measured along the shortest road connection from Messina to the nearest port city that had a medieval shipping route to a port in the UK. Once in the UK, distances were determined using the shortest road connection while shipping routes were created by finding all of the port cities and connecting them with simple route tools in ArcMap. The routes were calculated by taking known routes and after plotting them, using a measuring tool. We only measured known routes derived from Appendix 1 sources.

The epidemic map (Figure 5.2) was created by using a polygon feature creator on the editor toolbar. A polygon was drawn around the MBD localities in each year. Each polygon was saved as a shapefile, clipped to fit the country and overlaid. Differentially-colored "pockets" indicate areas with MBD in a different year than the surrounding areas.

Results and Discussion

We initially attempted inverse distance weighted and kriging-based models of MBD mortality in order to estimate transmission patterns; however, the spatially-concentrated nature of the data points generated a higher than acceptable perception of uncertainty in the resulting contours (both in RMSE values and analyst assessment), and we therefore opted to utilize the Buffer-based analysis methodology described earlier. Particularly when using data that is known to contain significant uncertainty, the use of kriging and IDW-based models could potentiate uncertainty and error over space/time. However, the kriging/IDW models were useful in order to validate the rapid spatial expansion of MBD during

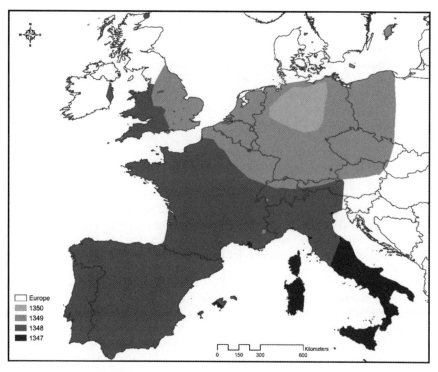

Figure 5.2   Geographic Expansion of the Medieval Black Death. Contours represent one-year intervals

the primary wave, as well as the directionality of transmission based on the point-based mortality data.

We found that 222 out of the 267 total MBD localities (83.2%) were within 8 km of a transportation network (including port cities; Table 5.1). 246 out of the 267 points, or 92.1%, were located within 16 km; 256 out of 267, or 95.9%, were located within 24 km; 260 out of 267, or 97.4%, were located within 32 km; and 263 out of 267, or 98.5%, were located within 40 km of a transportation network or port city (Table 5.1). Only four (4) recorded MBD localities (1.5% of all points) were located more than 40 km from a digitized transportation network (Table 5.1). These four cities may have been within a few miles of a coast, or possibly within 8 km of a road network not yet found; however, these points are currently classified as having no correlation with transportation networks. Nonetheless, the overall spatial assessment of 98.5% of locations within 40km of a medieval transportation route or port suggests a correlation between human trade networks and *recorded* MBD mortality. Figure 5.3 illustrates MBD point locations and buffers intersecting transportation networks in the UK.

Table 5.1 **Percentage of cities/towns affected by MBD within or outside of buffers (percentages rounded to nearest significant digit)**

Buffer radius in km	40 km	32 km	24 km	16 km	8 km	Percent outside of 40 km	Port Cities	Total
Percentage of locations inside buffer	98.5	97.4	95.9	92.1	83.1	1.5	12	
Number of cities	263	260	256	246	222	4	32	267

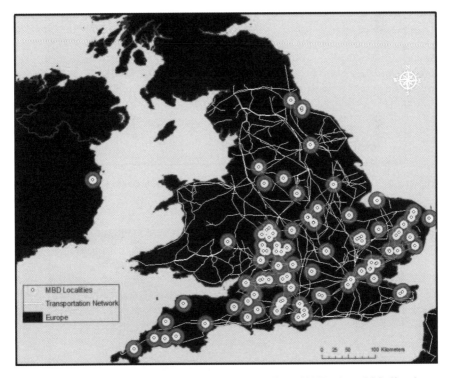

Figure 5.3 **Medieval Black Death Localities, Buffers (8/16 km), and Medieval Transportation Networks in the UK**

Figure 5.1 and Figures 5.4 and 5.5 illustrate that MBD expanded west and northwest in 1347 and 1348, and northeast in both 1349 and 1350. The distance traveled data indicate rapid expansion in 1347 and 1348, followed by a significant contraction in the area infected in 1349 and 1350. The average disease transmission distance was 1250 km (Euclidian, not network-derived) during 1348. In contrast, bubonic plague in the western U.S. has expanded at a maximum of 80km/year (Adjemian et al., 2007). The results indicate a very rapid, wave-like propagation of MBD through Europe, initially toward the west then arcing north and finally northeast and east. These data agree with the modeling by Christakos et al. (2007), although we have incorporated additional localities in our dataset. This wave-like propagation resembles human-to-human transmissible diseases such as the Spanish Flu (Olson et al., 2005; Chowell et al., 2008) rather than bubonic plague (Ari et al., 2010).

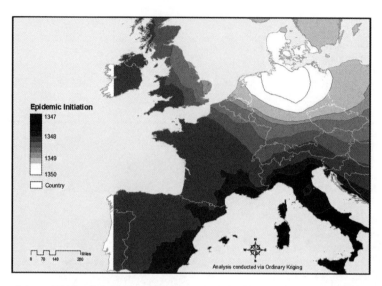

Figure 5.4   Epidemic initialization of the Medieval Black Death (ordinary kriging)

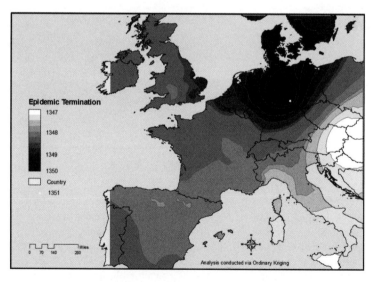

Figure 5.5   Epidemic termination of the Medieval Black Death (ordinary kriging)

Several factors could explain the initial rapid N/NW expansion from Messina. First, a high-density network based mostly on ancient Roman Empire roads (see Appendix 1) served northern Italy, the Rhone valley, and the northern slope of the Pyrenees at this time. Second, this road network connected many large urban centers and a large number of ports. For instance, Santiago de Compostela (Spain) became a pan-European place of pilgrimage beginning in the 10th century (Fletcher, 1984) and routes across northern Spain to Paris during the 1300s were well served with roads, smaller footpaths and sites for lodging (see Appendix 1 for sources). Moreover, once in the UK, MBD swept across England in less than a year. In contrast, in the mid-1300s, central and eastern Europe was served with a lower density road network and had limited access to ports (see Figure 5.2). Only the Rhine valley was a major corridor of trade and it was largely north-south, not west-east.

As these indications are considered, it is important to remember the words of Kool (2005): "humans do not readily transmit bubonic plague to other humans." Buboes do not generally appear on pneumonic plague victims as the focus of infection is different than with bubonic plague (lungs vs. cutaneous flea bite) and the disease is rapidly lethal, reducing the likelihood of widespread transmission, for example, along trade routes. Therefore, this spatio-temporal GIS-based research supports the *possibility* of human-to-human MBD transmission, which is at odds with the dominant paradigm consisting of a rat-flea transmission hypothesis in support of bubonic plague as the etiologic agent of the MBD.

Finally, alternative explanations must be considered. A recent paper by Ayyaduri et al. (2010) suggests that bubonic plague (biovar *orientalis*) may be transmitted by human body lice (*Pediculus humanus* sp.). Although this method of transmission may be theoretically possible, its efficacy and the likelihood of this transmission method being responsible for the spread of the MBD from southern Italy to the UK in a single year are still highly uncertain. Based on Boccaccio's observation that physical contact need not be present for transmission of the disease to occur among humans (1353), airborne or aerosol transmission may be more likely. Moreover, Boccaccio indicates that the MBD was a zoonosis as well, with a vivid description of livestock falling ill and dying in the street, reducing the likelihood that the human louse would have been the sole vector of the disease.

An additional non-biological possibility to explain transmission patterns of MBD is that historical accounts of the localities where MBD mortality were recorded may only be extant because such locations were near transportation networks; perhaps locations "farther off the beaten path" were less likely to have records of mortality still in existence or were lost to history permanently. Although this is certainly possible, we discount this scenario based on records from the United Kingdom, where church records in smaller towns such as Bury St. Edmunds (Gottfried, 1983) and villages such as Cleeve, Holy Cross, High Easter (Poos, 2004) and Walsham-de-Willows (Lock, 1992) were still recovered and collated by historians despite being located at some distance from the first-tier transportation nodes and routes.

Conclusion and Significance

It is still critically important to conduct research into the spatio-temporal aspects of the Medieval Black Death because there remains significant debate on the etiology of MBD, some researchers have suggested periodicity in human civilizational collapse-level epidemics (for example, Scott and Duncan, 2004) and there is therefore some question as to whether a similar disease(s) could strike again. Gottfried (1983) notes many historical 'plague' outbreaks, with the first true *Y. pestis*-like outbreak beginning in the 6th century, and an episodic recurrence every 10–24 years for 200 years, until the next great outbreak (MBD). Scott and Duncan (2004) proposed a roughly 700-year return frequency for virulent diseases similar to the MBD, associating the Plague of Athens and the Justinian Plague with the etiologic agent of the MBD. Noting that it has been approximately 700 years since the MBD, if their hypothesis is correct, then the world could be due for another great lethal epidemic in the near future. Gottfried (1983) also states that the geographical location of Europe could have been a huge factor in why "plague" outbreaks occurred so often between 1347 and 1801. Since Europe is located adjacent to Asia, which is a known "inveterate foci" or permanent disease reservoir, it is highly likely that many diseases could make their way to Europe with little interference during that period. Recent research has also suggested that a massive bottleneck in the European population must have occurred around 700 years ago in order for an allele deletion (CCR5-Δ32, which helps to prevent against HIV-1 and smallpox infection—both viral diseases), to occur in as much as 10% of the European population (Cantor, 2002). That time frame correlates with the MBD, providing yet another piece of evidence to consider while continuing investigations into the MBD.

Recent biological evidence dating to the time of the MBD has been found in purported MBD grave sites; the teeth of some the buried remains contain molecular biomarkers for the presence of *Y. pestis* (biovars *orientalis* and *medievalis*) in extracted dental pulp (Drancourt et al., 1998; Drancourt et al., 2007). These forms of plague could have been of the pneumonic form and therefore we must keep in mind that although some evidence may point away from the bubonic and septicemic forms of plague, we cannot eliminate the pneumonic form as the cause. Haensch et al. (2010) also found evidence of *Y. pestis* in gravesites of purported MBD victims—however, the two biovars (strains) of *Y. pestis* detected do not match the three biovars that are currently in existence. Both Schueneman et al. (2011) and Bos et al. (2011) identified ancestral strains of *Y. pestis* in Black Death victims, but neither group could consistently identify genetic differences between detected strains and modern *Y. pestis* that would explain differences in transmission or lethality. This elicits one possibility that the true etiologic agent of MBD may have been an ancestral strain of *Y. pestis* that, according to historical documentation, spread with the infectiousness of influenza (Welford and Bossak 2010b) but possessed a mortality rate similar to the Ebola virus. A significant and unaddressed issue with these studies is that they have proceeded through deductive science; falsification of alternate hypotheses (for example, a viral agent or co-morbidity) has yet to be published in recent literature, and thus questions regarding the etiology of MBD remain valid. In all respects, investigations of the geography, epidemiology, biology, and history of the MBD continue to generate additional questions for further investigation. GIS modeling and subsequent spatial analysis provide tools within a set of methodological approaches to elicit additional evidence pertaining to the MBD and other historical lethal epidemics.

References

Adjemian, J.Z., Foley, P., Gage, K.L., and J.E. Foley. 2007. Initiation and spread of traveling waves of plague, Yersinia pestis, in the western United States. *The American Journal of Tropical Medicine and Hygiene*, 76(2): 365–75.

Amori, G. and Cristaldi, M. 1999. Rattus rattus, in *The Atlas of European Mammals*, edited by A.J. Mitchell-Jones, G. Amori, W. Bogdanowicz, B. Kryštufek, P.J.H. Reijnders, F. Spitzenberger, M. Stubbe, J.B.M. Thissen, V. Vohralík, and J. Zima. London: Academic Press, 278–9.

Ari, T.B., Gershunov, A., Tristan, R., Cazelles, B., Gage, K., and N.C. Stenseth. 2010. Interannual variability of human plague occurrence in the Western United States explained by Tropical and North Pacific Ocean climate variability. *The American Journal of Tropical Medicine and Hygiene*, 83(3): 624–32.

Ayyadurai, S., Sebbane, F., Raoult, D., and M. Drancourt. 2010. Body lice, Yersinia pestis Orientalis, and Black Death. *Emerging Infectious Diseases*, 16: 892–3.

Bos, K.I., Schuenemann, V.J., Golding, G.B., Burbano, H.A., Waglechner, N., Coombes, B.K., McPhee, J.B., DeWitte, S.N., Meyer, M., Schmedes, S., Wood, J., Earn, D.J.D., Herring, D.A., Bauer, P., Poinar, H.N., and J. Krause. 2011. A draft genome of *Yersinia pestis* from victims of the Black Death. *Nature*, 478: 506–10.

Begier, E.M., Asiki, G., Anywaine, Z., Yockey, B., Schriefer, M.E., Aleti, P., Ogen-Odoi, A., Staples, J.E., Sexton, C., Bearden, S.W., and J.L. Koll. 2006. Pneumonic Plague Cluster, Uganda, 2004. *Emerging Infectious Diseases*, 12(3): 460–67.

Benedictow, O.J. 2004. *The Black Death 1346–1353: The Complete History*. Woodbridge: Boydell Press.

Biraben, J-N. 1975. *Les hommes et al peste en France et dans les pays europeens et mediterraneens* 1 & 2, Mouton, Paris, France.

Bossak, B.H. and M.R. Welford. 2009. Did medieval trade activity and a viral etiology control the spatial extent and seasonal distribution of Black Death Mortality? *Medical Hypotheses*, 72: 749–52.

Byrne, J.P. 2004. *The Black Death*. Greenwood (Guides to Historic Events of the Medieval World), UK.

Cantor, N.F. 2002. *In The Wake of The Plague: The Black Death and The World it Made*. New York: Free Press.

Carmichael, A.G. 1986. *Plague and the Poor in Renaissance Florence (Cambridge Studies in the History of Medicine)*. Cambridge: Cambridge University Press.

Carpentier, É.C. 1962. Autour de la Peste Noire: Famines et épidémies dans l'histoire du XIVe siècle. *Annales: E.S.C.* 17: 1062–1092.

Chernin, E. 1989. Richard Pearson Strong and the Manchurian Epidemic of Pneumonic Plague 1910–11. *Journal of the History of Medicine and Applied Sciences*, 44(3): 296–319.

Choa, G.H. 1994. The Lowson Diary: A record of the early phase of the Hong Kong Bubonic Plague 1894. Talk given to the Hong Kong Royal Asiatic Society on Oct 5th 1994.

Chowell, G., Bettencourt, L.M.A., Johnson, N., Alonso, W.J., and C. Viboud. 2008. The 1918–1919 influenza pandemic in England and Wales: Spatial patterns in transmissibility and mortality impact. *Proc. R. Soc. B*, 275: 501–9.

Christakos, G., Olea, R.A., and H.L. Yu. 2007. Recent results on the spatiotemporal modelling and comparative analysis of Black Death and bubonic plague epidemics. *Public Health*, 121: 700–20.

Christakos, G., Olea, R.A., Serre, M.L., Yu, H.L., and L-L. Wang. 2005. *Interdisciplinary Public Health Reasoning and Epidemic Modelling: The Case of Black Death*. New York: Springer Verlag.

Cohn, S.K. Jr. 2002. *The Black Death Transformed: Disease and Culture in Early Renaissance Europe*. London: Hodder Arnold.

Cohn, S.K. and L.T. Weaver. 2006. The Black Death and AIDS: CCR5-D32 in genetics and history, *Quart J Med*, 99: 497–503.

Comba, R. 1977. Vicende demografiche in Piemonte nell'ultimo medioevo. *Bollettino storico-bibliografico subalpino*, 75: 39–125.

Creighton, C.A. 1965. *History of Epidemics in Britain,* Cambridge University Press 1st ed. 1894; 2nd ed. Cambridge: Frank Cass.

Curson, P. and K. McCraken. 1989. *Plague in Sydney: The Anatomy of an Epidemic. Kensington, N.S.W.* New South Wales University Press, Australia.

Davis, D.E. 1986. The scarcity of rats and the Black Death: An ecological history. *Journal of Interdisciplinary History*, 16: 455–70.

Del Panta, L. 1980. *Le epidemie nella storia demografica italiana (secoli XIV-XIX).* Lescher editore. Torino.

Drancourt, M., Aboudharam, G., Signoli, M., Dutour, O., and D. Raoult. 1998. Detection of 400-year-old Yersinia pestis DNA in human dental pulp: An approach to the diagnosis of ancient septicemia. *Proceeding of the National Academy of Science* 95: 12637–40.

Drancourt, M., Signoli, M., Dang, L.V., Bizot, B., Roux, V., Tzortzis, S., and D. Raoult. 2007. Yersinia pestis Orientalis in Remains of Ancient Plague Patients. *Emerging Infectious Diseases*, 13(2): 332.

Duncan, C.J. and S. Scott. 2005. What caused the Black Death? *Postgraduate Medical Journal*, 81: 315–20.

Fiumi, E. 1968. *Demografia, Movimento urbanistico e classi sociali in Prato dall'eta comunale ai tempi moderni.* Florence, Italy.

Fletcher, R.A. 1984. *Saint James's Catapult: The Life and Times of Diego Gelmirez of Santiago de Compostela.* Oxford: Clarendon Press.

Gani, R. and S. Leach. 2004. Epidemiologic determinants for modeling pneumonic plague outbreaks. *Emerging Infectious Diseases*, 10: 608–14.

Gottfried, R.S. 1983. *The Black Death: Natural and Human Disaster in Medieval Europe.* Free Press, NYC.

Gras, P. 1939. Le registre paroissial de Givry (1334–1357) et la peste noire en Bourgogne. *Biblio. de l'école des chartes*, 100: 295–308.

Haensch, S., Bianucci, R., Signoli, M., Rajerison, M., and M. Schultz. 2010. Distinct clones of Yersinia pestis caused the Black Death. *PLoS Pathogens*, 6(10). e1001134.

Horrox, R. 1994. *The Black Death.* Manchester: Manchester University Press.

Indian Plague Commission Report 1898–99. 1899. Bombay: Times of India. Available at: http://www.nls.uk/indiapapers/find/vol/index.cfm?vbid = 3734941. Accessed July 31, 2009.

Karlsson, G. 1996. Plague without rats: The case of fifteenth century Iceland. *Journal of Medieval History*, 22: 263–84.

Kelly, J. 2005. *The Great Mortality: An Intimate History of the Black Death, the Most Devastating Plague of all Time.* New York: HarperCollins.

Kool, J.L. 2005. Risk of person-to-person transmission of pneumonic plague. *Clinical Infectious Diseases*, 40: 1166–72.

Lock, R. 1992. The Black Death in Walsham-le-Willows. *Proceedings of the Suffolk Institute of Archaeology and History*, 37(4): 316–37.

Nishiura, H. 2006. Epidemiology of a primary pneumonic plague in Kantoshu, Manchuria, from 1910–1911: Statistical analysis of individual records collected by the Japanese empire. *International J. Epidemiology*, 35: 1059–65.

Nishiura, H., Schwehm, M., Kakehashi, M., and M. Eichner. 2006. Transmission potential of primary pneumonic plague: Time-inhomogeneous evaluation based on historical documents of the transmission network. *Journal of Epidemiological Community Health*, 60: 640–45.

Noble, J.V. 1974. Geographic and temporal development of plague. *Nature*, 250: 726–8.

Olson, D.R., Simonsen, L., Edelson, P.J., and S.S. Morse. 2005. Epidemiological evidence of an early wave of the 1918 influenza pandemic in New York City. *PNAS*, 102(31): 11059–63.

Perry, R.D. and J.D. Fetherston. 1997. Yersinia pestis-etiologic agent of plague. *Clinical Microbiology Review*, 10(1): 35–66.

Poos, L.R. 1991. *A Rural Society after the Black Death: Essex 1350–1525*. Cambridge: Cambridge University Press.

Rasmussen, P.R. 2000. The Black Death in Western Europe. http://scholiast.org/history/blackdeath/procsrc.html. Accessed March 1, 2013.

Ratsitorahina, M., Chanteau, S., Rahalison, L., Ratsifasoamanana, L., and P. Boisier. 2000. Epidemiological and diagnostic aspects of the outbreak of pneumonic plague in Madagascar. *Lancet*, 355: 111–13.

Rielly, K. 2010. The black rat, in *Extinctions and Invasions. A Social History of British Fauna* ed. T. O'Connor and N. Sykes. Windgather Press: Oxford, 134–45.

Rotelli, C. 1973. *Una campagns medieval. Storia agrarian del Piemonte fra il 1250 e il 1450*. Turin: Giulio Einaudi editore.

Schuenemann, V.J., Bos, K., DeWitte, S., Schmedes, S., Jamieson, J., Mittnik, A., Forrest, S., Coombes, B.K., Wood, J.W., Earn, D.J.D., White, W., Krause, J., and H.N. Poinar. 2011. Targeted enrichment of ancient pathogens yielding the pPCP1 plasmid of *Yersinia pestis* from victims of the Black Death. *PNAS Early Edition*, 1105107108: 1–7.

Scott, S. and C.J. Duncan. 2001. *Biology of Plagues: Evidence from Historical Populations*. Cambridge: Cambridge University Press.

Scott, S. and C.J. Duncan. 2004. *Return of the Black Death*. Chichester: Wiley.

Seal, S.C. 1969. Epidemiological studies of plague in India: The present position. *Bulletin of the World Health Organization*, 23: 283–92.

Shrewsbury, J.F.D. 2005. *A History of Bubonic Plague in the British Isles*. Cambridge: Cambridge University Press.

Sloane, B. 2011. *The Black Death in London*. London: The History Press.

Twigg, G. 1984. *The Black Death: A Biological Appraisal*. London: Batsford Academic & Educational.

Welford, M.R. and B.H. Bossak. 2010a. Revisiting the Medieval Black Death of 1347–1351: Spatiotemporal dynamics suggestive of an alternate causation. *Geography Compass*, 4: 561–75.

Welford, M.R. and B.H. Bossak. 2010b. Body Lice, Yersinia pestis Orientalis, and Black Death. *Emerging Infectious Diseases*, 16(10): 1650–51.

Welford, M.R. and B.H. Bossak. 2009. Validation of inverse seasonal peak mortality in medieval plagues, including the Black Death, in comparison to modern Yersinia pestis-variant diseases. *PLoS ONE*, 4(12). e8401.

Wood, J.W., Ferrell, R.J., and S.N. DeWitte-Aviña. 2003. The temporal dynamics of the fourteenth-century Black Death: New evidence from English ecclesiastical records *Human Biology*, 75(4): 427–48.

Yu, H.L. and G. Christakos. 2006. Spatiotemporal modeling and mapping of the Bubonic Plague Epidemic in India. *International Journal of Health Geographics*, 5(12). [Online: http://www.ij-healthgeographics.com/].

Chapter 6
Space-Time Visualization of Dengue Fever Outbreaks

Eric Delmelle, Meijuan Jia, Coline Dony, Irene Casas, and Wenwu Tang

The outbreak and communicability of infectious diseases across the world are driven by an array of complex interrelated factors and processes, such as rapid urbanization, human mobility, changes in public health policy, and global climate change (Cooley et al., 2008; Eastin et al., 2014; Hu et al., 2012; Kanobana et al., 2013; Wu et al., 2009). Infectious diseases such as malaria or dengue fever, pose a critical threat to vulnerable human populations such that timely responses are necessary to reduce the burden caused by the diseases. Dengue fever for instance is known to vary through time and space, due to a number of factors including human host, virus, mosquito acting as disease-vector and environment (Mammen et al., 2008). To implement appropriate control measures, public health organizations and policy makers must rely on accurate and timely predictions of disease for monitoring and analyzing them under critical space-time conditions. A better understanding on the space-time signature of infectious diseases such as their rate of transmission and tendency to cluster should help epidemiologists and public health officials better allocate prevention measures.

Geographical Information Systems (GIS) and associated spatial analysis methods (Haining, 1990; Goodchild et al., 1992; Anselin, 1999; Fischer and Getis, 2009) have received considerable attention in the field of spatial epidemiology (Cromley and McLafferty, 2011). GIS can effectively organize individual or aggregated data, and map the magnitude and expansion of these diseases over time. Ultimately, results from GIS and spatial analysis have the potential to reduce disease burden by generating new information for the public and health agencies, leading to improvement on development of preventive measures and strategic allocation of control resources.

In spatial epidemiology, the development of computational techniques for the identification and the visualization of space-time clusters in large databases is a challenge (Delmelle et al., 2013; Goovaerts and Goovaerts, this volume; Delmelle et al., 2014). Our chapter contributes to the field of space-time clustering, specifically the use of space-time kernel density estimation (STKDE) for infectious diseases and the visualization of disease dynamics in a 3D environment. The results provide a sense of where outbreaks of dengue fever linger around for a longer time (persistence). These patterns might generate clues on why dengue outbreaks tend to remain around particular areas. Together with information and expertise from local authorities, intervention programs may be conducted about to address particular (unique) settings.

The chapter is organized as follows: Section 2 discusses the importance of spatial and space-time analytical methods, and existing visualization frameworks for epidemiology. Section 3 discusses the contribution of visual analytics for space-time data and introduces the extension of the kernel density estimation (KDE) in time (STKDE). STKDE is particularly demanding, from a computational perspective, and to address this issue, we integrate

STKDE with a parallel computing environment, in Section 4. We illustrate our approach on a set of dengue fever cases in the city of Cali, Colombia (2010). A static 3D visualization is used to illustrate the spatio-temporal characteristics of this infectious disease. Directions for future research are presented in the last section.

Space-Time Visual Analytics

Representing the multi-dimensionality of geographical space and its phenomena is a complex task. Over time, cartographers have proposed a wide variety of techniques; mostly two-dimensional, static maps. In spatial epidemiology, disease maps are generally presented in one of two forms (Cromley and McLafferty, 2011): at an aggregated level, divided by the population at risk, resulting in a choropleth map, or at the individual level as a scatter map. As scatter maps become very cluttered when incorporating a large number of events, an alternative consists of producing a kernel distribution map that generates a smooth probability raster that facilitates the discovery of clusters (Bailey and Gatrell, 1995).

As part of the multi-dimensionality of spatial data, temporal coordinates are often present, for instance the time stamp associated with a particular event. Since the development of the space-time cube concept (Hägerstrand, 1970), several analytical and visualization strategies have been developed to integrate the temporal dimension in spatial analysis (Kwan, 2004; Miller, 2005; Jacquez, Greiling, and Kaufmann, 2005). The availability of personal computers, advances in geographical information systems technology, and increasingly efficient graphic processing power (Peterson, 1995) has allowed the development of new tools and techniques to integrate the temporal dimension in the form of cartographic movies (Tobler, 1970), animations in two and three dimensions (Moellering, 1980; Harrower and Fabrikant, 2008; Brunsdon, Corcoran, and Higgs, 2007) and interactive maps or geo-browsers (Harrower, 2004; Dykes, MacEachren, and Kraak, 2005; Hruby, Miranda, and Riedl, 2009). Graphic user interfaces have been tailored to address particular needs (Jacquez, Greiling, and Kaufmann, 2005; Rogerson and Yamada, 2008; Delmelle et al., 2011). The latter techniques are particularly well-suited to analyze infectious diseases exhibiting cyclic patterns—outbreaks can surge and resurface at particular space-time locations. Thakur and Hanson (2010) developed a pictorial representation based on the space-time cube framework, providing, in a single display, an overview and details of a large number of time-varying information. Their approach is discrete using aggregated county data.

In this chapter, we follow the continuous approach set forth by Andrienko et al. (2010) and Demšar and Virrantaus (2010) by extending the space-time prism framework to monitor and visualize disease dynamics. Recently, Fang and Lu (2011) proposed a 3D space-time cube, which integrates air pollution scenarios. The base of this cube describes the spatial variation of air pollution across a traditional 2D space while the height is used as the temporal dimension. As time progresses, the change of air pollution for each location within the base 2D space is continuously updated. Time periods are represented as layers, which are separated from each other by an offset. However, 3D layers may form cluttered maps, even with a small number of layers.

In certain cases, animations may work better than three-dimensional mapping and lead to a more accurate detection of space-time clusters. However, several variables prevent full control over these animations, as, among others, frame duration and speed of transition must be calibrated (Harrower and Fabrikant, 2008). In spite of the advances, 3D visualization remains a cartographic challenge, and not enough is known about its effectiveness for mapping the temporal dimension along the third axis (Fabrikant, 2005; Andrienko et al., 2010). In this chapter, we use a 3D space-time cube framework to map the stability and strength of spatial clusters over time.

Methodology

Kernel Density Estimation (KDE)

Kernel Density Estimation (KDE) (Bailey and Gatrell, 1995; Delmelle, 2009) is a spatial analytical technique, which maps the spatial distribution of point events across an area. KDE produces a raster surface reflecting the spatial variation of the probability that an event will occur. One of the benefits of KDE is its ability to delimit the spatial extent of disease occurrences (Delmelle et al., 2011). For infectious diseases, these maps are used in conjunction with prevention and control programs to guide disease-vector control and/or surveillance activities (Eisen and Eisen, 2011; Eisen and Lozano-Fuentes, 2009; Delmelle et al. 2014b). KDE is computed at each grid cell, which receives a higher weight if it has a larger number of observations in its surrounding. Let $s(x,y)$ represents a location in an area R where the kernel density estimation needs to be estimated, and $s_1, \ldots s_n$ the locations of n observed events (Silverman, 1986):

$$\hat{f}(x,y) = \frac{1}{nh_s^2} \sum_i I(d_i < h_s) k_s \left(\frac{x - x_i}{h_s}, \frac{y - y_i}{h_s} \right) \quad (1)$$

Where I is an indicator function, the term h_s is the search radius, or spatial bandwidth, which governs the amount of smoothing, and k_s is a standardized weighting function, known as the kernel that determines the shape of the weighting function. Here d_i is the distance between location s and event s_i, which is constrained by $d_i < h_s$, such that only the points which fall within the chosen bandwidth contribute to the estimation of the kernel density at s. The number of points participating in the kernel density estimation affects the computation time linearly, while smaller cell sizes (finer grid) result in an exponential increase in the overall computation time.

Point events generally have a time stamp associated with them $s(x,y,t)$. One critical challenge is time integration in KDE, which may help to address the questions of (1) whether the distribution of events varies over time, and (2) whether the occurrence of the underlying process is repeated in certain areas and exhibit a cyclical pattern, in turn indicating a spatio-temporal process. Two approaches exist to integrate the temporal dimension in the KDE. In the first approach, we query data for different temporal intervals -time period 1, time period 2 ... time period t, and the spatial kernel density is repeated sequentially for those intervals. We call this approach sequential KDE. Usually, the resulting kernel density images are arranged in an animation framework or side-by-side as a multiple. In the second approach, the kernel density estimation method is explicitly extended in time STKDE by adding a temporal bandwidth (Nakaya and Yano, 2010; Demšar and Virrantaus, 2010), and visualized in a 3D environment. Computationally, the space-time kernel density estimation (STKDE) is much more challenging and cartographically more complex, but it has the advantage of summarizing the spatio-temporal clustering pattern in one image. The two approaches are explained in detail in the following paragraphs.

In the sequential Kernel Density Estimation (sKDE) the KDE procedure can be repeated over several time intervals (granularity of one day, one week, one month), producing different KDE layers. These layers are then arranged in a mosaic framework side-by-side. This so-called "multiple" arrangement has the disadvantage that users must move from one image to another and reconstruct the movement of the clusters. An alternative proposed earlier is the use of animation (Dorling, 1992; Harrower and Fabrikant, 2008; Brunsdon, Corcoran, and Higgs, 2007). The sequential Kernel Density Estimation does not explicitly take time into account; rather, each map is a snapshot of a particular situation at a specific time interval.

According to Nakaya and Yano (2010) and Demšar and Virrantaus (2010), a mosaic of two-dimensional kernel densities does not facilitate simultaneous visualization of the geographical extent and duration of point patterns and spatial clusters, unless they are adequately arranged in an animation framework, or stacked on top of each other in a 3D environment. An alternative is the extension of the kernel density estimation in time (STKDE). The STKDE approach requires mapping a volume of probabilities. Its main advantage lies in seeing all probabilities and clusters at once, rather than the need to recreate a mental image through animation. We apply an extension of Silverman's spatial kernel density which incorporates the temporal dimension, as described in Nakaya and Yano (2010). The space-time density $\hat{f}(x,y,t)$ at s is thus estimated by:

$$\hat{f}(x,y,t) = \frac{1}{nh_s^2 h_t} \sum\nolimits_i I(d_i < h_s, t_i < h_t) k_s \left(\frac{x - x_i}{h_s}, \frac{y - y_i}{h_s} \right) k_t \left(\frac{t - t_i}{h_t} \right) \quad (2)$$

where d_i and t_i are the Euclidean (or network) distance and temporal differences between location s and event s_i. $I(d_i < h_s, t_i < h_t)$ is an indicator function taking value 1 if $d_i < h_s$ and $t_i < h_t$. h_s is the search radius (or bandwidth), governing the amount of smoothing. h_t is the temporal search window. k_s is a standardized weighting function, known as the kernel that determines the shape of the weighting function. k_t is a standardized temporal weighting function. Generally, a space-time count statistic (K-function) can be used to estimate the optimal global bandwidths (h_s^2, h_t) in space and time, although different approaches have been suggested.

STKDE Implementation

We used the procedure presented in Figure 6.1 to compute the STKDE. First, based on the extent of the study region and a desired voxel[1] size, a 3D grid is generated. This process is also known as three-dimensional discretization. Second, for each voxel centroid, we compute the space-time kernel density $f(z0)$ using appropriate bandwidths (h_s^2, h_t) and Equation 2 to identify the disease events that will participate in the computation. An additional constraint is added that only considers nearby points. Voxels that fall outside of the study region or where the density value is equal to zero ($f(z0)=0$) are excluded. Both added constraints reduce computational time. The STKDE algorithm is coded in the Python

---

1   A voxel can be conceived as the extension of a grid cell in a third dimension. It is a volumetric element.

environment. Finally, density results are visualized in a three-dimensional visualization environment (Voxler, Golden, CO) where the appearance of the cluster can be adjusted according to the distribution of density values. To detect micro-patterns, a fine grid is generally preferred to the cost of a heavier computation and longer rendering time. Results are visualized in a 3D environment, allowing a better understanding of complex disease dynamics in an interactive manner.

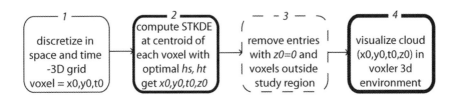

Figure 6.1 **STKDE computing framework. Boxes in bold (2) and (4) denote computing intensive procedures, while the dashed box (3) is optional**

A Parallel Computing Approach

The estimation of space-time kernel densities in point pattern data consumes considerable amount of computing resources. This method requires handling each voxel in the space-time cube at a time. Consequently, when the number of voxels or the number of points in the spatiotemporal point pattern is excessively large, a significant amount of computation may be required to complete the derivation of space-time kernel densities. Thus, we prefer using a parallel computing approach (Armstrong, 2000; Wang and Armstrong, 2009; Tang and Bennett, 2011; Tang, 2013) to cope with the intense computational efforts related to the derivation of spatiotemporal kernel density. In this study, we follow the strategy proposed by Wilkinson and Allen (2004) for the efficient estimation of space-time kernel densities. The parallel computing resources that we used include 54 nodes with 432 computing cores. A dual Intel Xeon X5570 processor with 2.93 GHz of clock rate was used, with each processor having 24 GB of memory.

Results

Dengue Fever and Geospatial Datasets

Dengue fever, which is an arboviral disease endemic to tropical and subtropical areas is among the most dangerous infectious diseases. It has a global presence and poses a threat to more than 2.5 million people. Dengue fever is a problem to communities and health entities, necessitating the control and the prevention of the virus (Méndez et al., 2006; WHO, 2009). Dengue fever is transmitted to humans by a mosquito of the genus *Aedes* (Monath, 1988). The *Aedes aegypti* mosquito lives and reproduces in warm climates where temperatures range between 18° and 25°C (Wu et al., 2009).

Colombia is one of the many countries where dengue fever is endemic and constitutes a serious health problem. The population living in areas at risk of contracting the disease totals 26 million individuals. These areas are characterized by elevations below 1,800 meters (Colombianos, 2011). The city of Cali, which is the focus of this study, is considered an endemic dengue fever zone. It is located in the valley of the Cauca River 1,000 meters above sea level and has an average daily temperature of 23°C. Annual precipitation in the driest zones reaches 900 mm and 1,800 mm in the rainiest. The city's annual precipitation average is 1000 mm (Cali, 2008). According to the health municipality of the city of Cali (Cali, 2010) there have been three severe dengue outbreaks between 1990 and 2009: 1995, 2002, and 2005 (Cali, 2010). During 2009 and the first quarter of 2010 more than 7,000 cases of dengue fever where reported in the city (Cali, 2010). By the end of January of that year a total of 990 dengue fever cases were registered with the number increasing to 3,540 by week 10 (mid-March). At this point the signs of an epidemic were evident. By August of 2010 a total of 9,310 cases of dengue fever had been reported. In this chapter, we map space-time clusters of individuals infected by dengue fever within the city of Cali for the year 2010, a year that corresponds to a strong outbreak of the disease.

The dengue fever dataset corresponds to cases of dengue reported in the "Sistema de Vigilancia en Salud Pública" (SIVIGILA, English: Public health surveillance system) for the city of Cali (provided by the health municipality of the city of Cali) for the year 2010. An entire description of the geospatial dataset, and geocoding procedure is given in (Delmelle, Casas et al., 2013). Individuals report dengue fever symptoms on a daily basis (unit: Julian date) at a local hospital. Hospital records include patient information (sex, age, race, address, neighborhood, and occupation), date of diagnosis, epidemiological week, and day symptoms started. There were 11,760 cases reported in 2010. The data was geocoded to the intersection level to guarantee patient privacy (Kwan, Casas, and Schmitz, 2004). Eighty-one percent of the total cases were perfectly matched. The 19% that could not be matched constitute non-existent addresses or addresses with errors that could not be fixed. From the 9,555 that were successfully geocoded, as described in (Delmelle, Casas et al., 2013), the majority of the cases occurred during the first 4–5 months of the year, after which the municipality invested time and efforts to further eradicate mosquito breeding habitats.

We now illustrate the results from the STKDE algorithm on all patients exhibiting symptoms of dengue fever in Cali, Colombia, in 2010. We only visualize intensities during the first seven months of the year. We discuss the benefits of the parallel approach, and the sensitivity of the STKDE.

Parallel Computing Results

We use a space-time K-function, a scan statistic (see HELP in Delmelle et al. (2011)), to identify the geographical and temporal scale at which clusters of dengue are the strongest (determined as $h_s = 1500$ m and $h_t = 10$ days). We implement the STKDE algorithm with those space-time parameters. We also estimate the robustness and sensitivity of the STKDE algorithm against different cell sizes and bandwidths, among others.

Figure 6.2a reports the running time to conduct the STKDE for the entire city of Cali on a single CPU, and hereafter referred to as 'sequential.' The STKDE is estimated on a 3D grid of resolution 100m × 100m × 1 day. The running time significantly decreases when using larger voxel sizes, since STKDE is estimated on a much smaller set of 3D gridded data points. The same reduction in running time is observed in Figure 6.2b which summarizes the maximum computational effort when using a parallel computing approach. Although the shapes of the curves in Figure 6.2a and Figure 6.2b are similar, the time range is much smaller in the latter. As the cell size increases, the running time to compute the STKDE

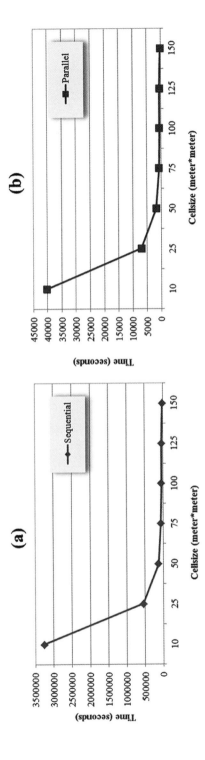

Figure 6.2   Running time for the STKDE on one CPU (a) 96 CPUs (b) based on different cell sizes. Note the values along the Y-axis are different

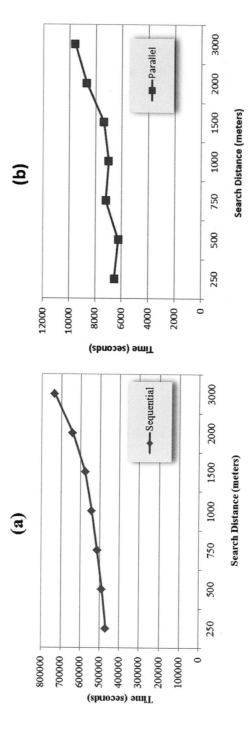

Figure 6.3 Sensitivity of the STKDE running with variation in spatial bandwidth, on one CPU (a) 96 CPUs (b) Note the values along the Y-axis are different

decreases, in an exponential fashion. Figure 6.3a reports the 'sequential' time for different spatial bandwidths. As expected, the running time increases in a linear fashion when using larger bandwidths, since a greater number of points are included in the computation. The same pattern is observed in Figure 6.3b, which illustrates the maximum computational effort when using a parallel computing approach.

3D Visualization

The kernel density volume is extracted by computing the kernel density estimation of each voxel in our 3D area. It is then visualized by color-coding each voxel according to its kernel density value. The transparency level is also adjusted to concentrate the focus on those regions with higher density values (Figure 6.4). The KDE values range from 0 to 1 (on the map). Voxels having a KDE of 0.3 or less are colored in dark and lighter blue shades and voxels with a value between 0.3 and 0.5 are colored yellow. Voxels between 0.5 and 0.7 are colored green and voxels with higher KDE values are colored in red or purple tones. The highest KDE values are usually found in a couple of cores within the voxel cloud (not at the edges). Since those higher densities are the phenomenon we are interested to capture and thus visualize, different transparency levels are used depending on the KDE value of each voxel. In this case, voxels with lower KDE value (<0.25) were brought to a high transparency level, allowing the visualization of underlying voxels, with higher KDE values. For values of 0.3 to 1 the transparency was lowered in a linear fashion.

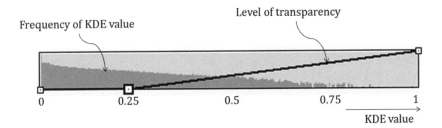

Figure 6.4   Transparency scheme adopted in Voxler for visualizing STKDE values ranging from 0 to 1 (on the map)

We reinforce the importance of clusters by using so-called 'egg shells.' These are created by computing an isosurface around voxels having a value of 0.7. When visualizing these isosurfaces, interestingly they create an envelope capturing areas with higher KDE values. Also, these isosurfaces allow us to visualize the shape of the outbreak in space and time. This isosurface can be visualized in two ways, one is a smooth surface, and the other one is a TIN-like surface. Both surface types have the same shape and encapsulate the same space-time areas, but are represented in a different way.

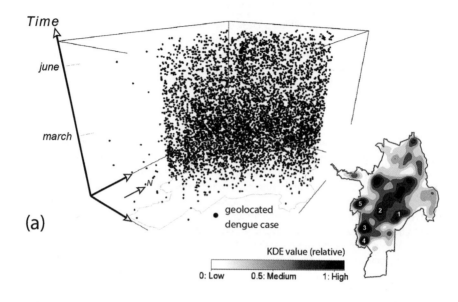

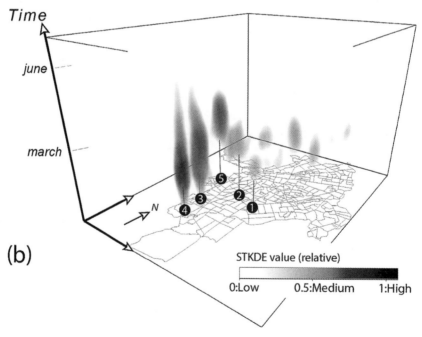

Figure 6.5   Space-time distribution of dengue fever cases for the
              first seven month of the year 2010 in Cali, Columbia

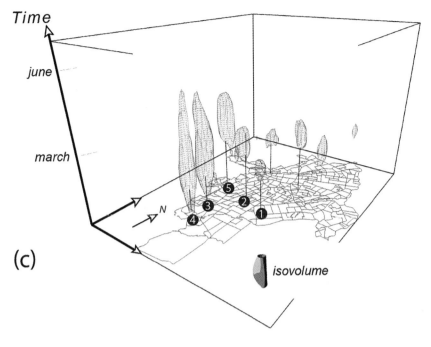

Figure 6.5 (*concluded*)

3D Visualization Results

We map the space-time stamp of dengue cases in Figure 6.5a and the space-time kernel density in Figures 6.5b with isovolumes and 6.5c with triangulated irregular networks using $h_s = 1500$ m and $h_t = 10$ days at a discretization level of 125m*125m*1 day (finer discretization level may provide more details, but at a cost of a larger storage). In Figures 6.5b and 6.5c, the observed pattern clearly shows cyclic reoccurrence of the cluster of the disease, although this particular infectious disease may follow a pyramid like-pattern: dengue fever may saturate in the population over time and as a consequence exhibits skinnier clusters. Several space-time clusters are constant, with high intensity. These correspond to areas in the central part of the city (clusters 1 and 2), the west side close to the foothills (cluster 5), while clusters (3) and (4) are in proximity in the south-west part of the city (see Figure 6.5b). All clusters are more intense in the beginning of the year since the highest number of cases occurred between January and March of 2010. Among the five clusters, clusters (1) and (2) cover a larger spatial extent, but short in time. Cluster (1) is concentrated throughout a relatively shorter time period compared to the other four clusters. This is an area with old urban neighborhoods composed of households with large families, which increases the risk of people being exposed to the disease. On the other hand, policy measures might work more effectively in those areas because of the pre-existing bond between people and their willingness to work together to eradicate the epidemic, which might be the reason for the quick eradication of the disease in that area. Cluster (5) has a more focused spatial extent in the foothills to the west of the city. It corresponds to a neighborhood of low stratification where individuals are more likely to maintain water containers serving as breading sites for the mosquito to reproduce. This cluster prevails over time, more so than cluster (1) and (2). Clusters (3) and (4) are located in Commune 18 (a commune is a group of neighborhoods). This commune is characterized by the presence of a military base with a high concentration of individuals living and interacting in very close proximity due to the confined space. This explains the shape of the cluster, small extent and long period. This particular commune has been identified by the secretary of health municipality of the city of Cali as being at high risk of contracting dengue fever.

Discussion and Future Research

Dengue fever is a vector-borne infectious disease, which can take dramatic proportions when conditions (for example, population, disease-vector, climate, behavior) are optimal. Mobility of humans is known to affect the spread of the disease. However, less research has been conducted on visualizing clusters of infectious diseases, in particular, in a spatiotemporal dimension. In this chapter, we apply a spatial and temporal extension of the kernel density estimation algorithm to map space-time clusters of dengue fever in Cali, Colombia in 2010. We used a parallelized computation framework which, significantly reduces the running time and thus allows the problem solving to be more manageable. We illustrate and extract complex disease clusters using a 3D environment. We can observe different cluster shapes, which might correspond to different neighborhood and population setting. Elongated temporal clusters denote that the disease tend to remain contained over a geographic area for a longer time (for instance, a military base that tend to be hermetic, meaning the disease does not spread further form the base borders). Larger yet less intense clusters which are less intense generally occur in areas with a high degree of human movement. Such visualizations are very valuable to better understand the resurgence cycle of infectious diseases and necessary for public health managers to decide on the allocation of appropriate resources. The findings from our work can help better understand the dynamics of a quickly spreading disease and may help build different strategies depending on distinct socio-environmental settings within the area at stake.

We see the followings for future research. First, the extension of the kernel density estimation in time can be used to estimate the speed at which the disease tend to spread, and whether it tends to have a directional preference. Second, computational time could further be reduced by using space-time adaptive (and directional) windows. Third, it is necessary to simulate points datasets (Monte Carlo simulations) to extract the significance of the clusters. Of importance is the accuracy of the geocoded data itself (Delmelle et al., 2014a, DeLuca and Kanaroglou, this volume): if the accuracy is very low, authorities might want to improve their surveillance strategies in the future. Finally, more research is needed to evaluate the effectiveness of 3D versus animated cartography when mapping disease cluster dynamics.

References

Andrienko, G., N. Andrienko, U. Demšar, D. Dransch, J. Dykes, S.I. Fabrikant, M. Jern, and M-J. Kraak, H. Schumann, and C. Tominski. 2010. Space, time and visual analytics. *International Journal of Geographical Information Science*, 24(10): 1577–1600.

Anselin, L. 1999. Interactive techniques and exploratory spatial data analysis, in *Geographical Information Systems*, edited by P. Longley, M. Goodchild, D. Maguire, and D. Rhind. New York: Wiley.

Armstrong, M.P. 2000. Geography and computational science. *Annals of the Association of American Geographers*, 90(1): 146–56.

Bailey, T. and Q. Gatrell. 1995. *Interactive Spatial Data Analysis*. Edinburgh Gate: Pearson Education Limited.

Brunsdon, C., J. Corcoran, and G. Higgs. 2007. Visualizing space and time in crime patterns: A comparison of methods. *Computers, Environment and Urban Systems*, 31: 52–75.

Cali, A. d. S. d. 2008. *Cali en cifras*. Cali: Alcaldia de Cali.

Cali, S. d. S. P. M. d. 2010. *Historia del dengue en Cali. Endemia o una continua epidemia*. Cali: Secretaria de Salud Publica Municipal de Cali.

Colombianos, M.G. *Guia de atención al dengue* [last accessed December 2014].Available from http://www.saludpereira.gov.co/documentos/Guias_medicos/Urgencias/33-DENGUE.pdf

Cooley, P., L. Ganapathi, G. Ghneim, S. Holmberg, W. Wheaton, and C.R. Hollingsworth. 2008. Using influenza-like illness data to reconstruct an influenza outbreak. *Mathematical and Computer Modelling*, 48(5): 929–39.

Cromley, E. and S. McLafferty. 2011. *GIS and Public Health*. New York: Guilford Press.

Delmelle, E. 2009. Point Pattern Analysis, in *International Encyclopedia of Human Geography*, edited by R. Kitchin and N. Thrift. Oxford: Elsevier, 204–11.

Delmelle, E., I. Casas, J. Rojas, and A. Varelo. 2013. Modeling spatio-temporal patterns of dengue fever in Cali, Colombia. *International Journal of Applied Geospatial Research*, 4(4): 58–75.

Delmelle, E., E. Delmelle, I. Casas, and T. Barto. 2011. H.E.L.P: A GIS-based health exploratory analysis tool for practitioners. *Applied Spatial Analysis and Policy*, 4(2): 113–37.

Delmelle, E., C. Kim, N. Xiao, and W. Chen. 2013. Methods for space-time analysis and modeling: An overview. *International Journal of Applied Geospatial Research*, 4(4): 1–18.

Delmelle, E., Dony, C., Casas, I., Jia, M., and W. Tang. 2014a. Visualizing the impact of space-time uncertainties on dengue fever patterns. *International Journal of Geographical Information Science*, 28(5): 1107–27.

Delmelle, E. M., Zhu, H., Tang, W., and Casas, I. 2014b. A web-based geospatial toolkit for the monitoring of dengue fever. *Applied Geography*, 52: 144–52.

Deluca, P. and P. Kanaroglou (this volume). An Assessment of Online Geocoding Services for Health Research in a Mid-Sized Canadian City, in *Spatial Analysis in Health Geography*, edited by P. Kanaroglou, E. Delmelle, and A. Páez. Farnham: Ashgate.

Demšar, U. and K. Virrantaus. 2010. Space–time density of trajectories: Exploring spatio-temporal patterns in movement data. *International Journal of Geographical Information Science*, 24(10): 1527–42.

Dorling, D. 1992. Stretching space and splicing time: From cartographic animation to interactive visualisation. *Cartography and Geographic Information Systems*, 19: 215–27.

Dykes, J., A.M. MacEachren, and M-J. Kraak. 2005. *Exploring Geovisualization*. San Diego, CA: Elsevier.

Eastin, M., Delmelle, E.M., Casas, I., Wexler, J., and C. Self. 2014. Intra- and Interseasonal Autoregressive Prediction of Dengue Outbreaks Using Local Weather and Regional

Climate for a Tropical Environment in Colombia. *American Journal of Tropical Medicine and Hygiene*, 91(3): 598–610.

Eisen, L., and R. Eisen. 2011. Using geographic information systems and decision support systems for the prediction, prevention, and control of vector-borne diseases. *Annual Review of Entomology*, 56(1): 41–61.

Eisen, L., and S. Lozano-Fuentes. 2009. Use of mapping and spatial and space-time modeling approaches in operational control of aedes aegypti and dengue. *PLoS Negl Trop Dis*, 3(4):e411.

Fabrikant, S.I. 2005. Towards an understanding of geovisualization with dynamic displays: Issues and prospects. Paper read at Proceedings, American Association for Artificial Intelligence (AAAI) 2005 Spring Symposium.

Fang, T.B. and Y. Lu. 2011. Constructing a near real-time space-time cube to depict urban ambient air pollution scenario. *Transactions in GIS*, 15(5): 635–49.

Fischer, M. and A. Getis. 2009. *Handbook of Applied Spatial Analysis: Software Tools, Methods and Applications*. New York: Springer.

Goodchild, M.F., R. Haining, S. Wise, G. Arbia, L. Anselin, E. Bossard, C. Brunsdon, P. Diggle, R. Flowerdew, M. Green, D. Griffith, L. Hepple, T. Krug, R. Martin, and S. Openshaw. 1992. Integrating GIS and spatial data analysis—Problems and possibilities. *International Journal of Geographical Information Systems*, 6(5): 407–23.

Goovaerts, P. and M. Goovaerts (this volume). Space-time analysis of late-stage breast cancer incidence in Michigan, in *Spatial Analysis in Health Geography*, edited by P. Kanaroglou, E. Delmelle, and A. Páez. Farnham: Ashgate.

Hägerstrand, T. 1970. What about people in regional science? *Papers in Regional Science*, 24(1): 7–24.

Haining, R. 1990. *Spatial Data Analysis in the Social and Environmental Sciences*. Cambridge: Cambridge University Press.

Harrower, M. 2004. A look at the history and future of animated maps. *Cartographica: The International Journal for Geographic Information and Geovisualization*, 39(3): 33–42.

Harrower, M. and S. Fabrikant. 2008. The role of map animation for geographic visualization, in *Geographic Visualization:Concepts, Tools and Application*, edited by M. Dodge, M. McDerby, and M. Turner. San Francisco, CA: John Wiley & Sons, 49–65.

Hruby, F., R. Miranda, and A. Riedl. 2009. Bad Globes & Better Globes–multilingual categorization of cartographic concepts exemplified by "map" and "globe" in English, German and Spanish. Paper read at Proceedings of the 24th International Cartography Conference (ICC, 2009), Santiago, Chile 2009.

Hu, W., A. Clements, G. Williams, S. Tong, and K. Mengersen. 2012. Spatial patterns and socioecological drivers of dengue fever transmission in Queensland, Australia. *Environmental Health Perspectives*, 120(2): 260.

Jacquez, G., D. Greiling, and A. Kaufmann. 2005. Design and implementation of a space-time intelligence system for disease surveillance. *Journal of Geographical Systems*, 7(1): 7–23.

Kanobana, K., B. Devleesschauwer, K. Polman, and N. Speybroeck. 2013. An agent-based model of exposure to human toxocariasis: A multi-country validation. *Parasitology*, 140: 986–98.

Kwan, M-P. 2004. GIS methods in time-geographic research: Geocomputation and geovisualization of human activity patterns. *Geografiska Annaler B*, 86: 205–18.

Kwan, M-P., I. Casas, and B.C. Schmitz. 2004. Protection of geoprivacy and accuracy of spatial information: How effective are geographical masks? *Cartographica*, 39(2): 15–28.

Mammen, M.P., C. Pimgate, C.J.M. Koenraadt, A.L. Rothman, J. Aldstadt, A. Nisalak, R.G. Jarman, J.W. Jones, A. Srikiatkhachorn, C.A. Ypil-Butac, A. Getis, S. Thammapalo, A.C. Morrison, D.H. Libraty, S. Green, and T.W. Scott. 2008. Spatial and temporal clustering of dengue virus transmission in Thai villages. *PLoS Medicine*, 5: 1605–16.

Méndez, F., M. Barreto, J.F. Arias, G. Rengifo, J. Muñoz, M.E. Burbano, and B. Parra. 2006. Human and mosquito infections by dengue viruses during and after epidemics in a dengue-endemic region of Colombia. *The American Journal of Tropical Medicine and Hygiene*, 74: 678–83.

Miller, H. 2005. A measurement theory for time geography. *Geographical Analysis*, 37: 17–45.

Moellering, H. 1980. The real-time animation of three-dimensional maps. *The American Cartographer*, 7(1): 67–75.

Monath, T. ed. 1988. *The Arboviruses: Epidemiology and Ecology*. Boca Raton, FL: CRC Press.

Nakaya, T. and K. Yano. 2010. Visualising crime clusters in a space-time cube: An exploratory data-analysis approach using space-time kernel density estimation and scan statistics. *Transactions in GIS*, 14(3): 223–39.

Peterson, M.P. 1995. Interactive and animated cartography. *Faculty Books and Monographs*. Book 50. Englewood Cliffs, NJ: Prentice Hall.

Rogerson, P. and I. Yamada. 2008. *Statistical Detection and Surveillance of Geographic Clusters*. Boca Raton, FL: CRC Press.

Silverman, B.W. 1986. *Density Estimation for Statistics and Data Analysis*. London: Chapman & Hall.

Tang, W. 2013. Parallel construction of large circular cartograms using graphics processing units. *International Journal of Geographical Information Science*, 27(11): 2182–206.

Tang, W. and D.A. Bennett. 2011. Parallel agent-based modeling of spatial opinion diffusion accelerated using graphics processing units. *Ecological Modelling*, 222: 3605–15.

Thakur, S. and A.J. Hanson. 2010. A 3D visualization of multiple time series on maps, in *Information Visualisation (IV), 2010 14th International Conference* (336–43). IEEE

Tobler, W.R. 1970. A computer movie simulating urban growth in the Detroit region. *Economic Geography*, 46: 234–40.

Wang, S. and M.P. Armstrong. 2009. A theoretical approach to the use of cyberinfrastructure in geographical analysis. *International Journal of Geographical Information Science*, 23(2): 169–93.

WHO 2009. *Dengue and severe dengue—Fact sheet no. 117*. Available from http://www. who.int/mediacentre/factsheets/fs117/en/ [Last accessed December 2014].

Wilkinson, B. and Allen, M., 2004. *Parallel Programming: Techniques and Applications Using Networked Workstations and Parallel Computers (Second Edition)*. Upper Saddle River, NJ USA: Pearson Prentice Hall.

Wu, P-C., J-G. Lay, H-R. Guo, C-Y. Lin, S-C. Lung, and H-J. Su. 2009. Higher temperature and urbanization affect the spatial patterns of dengue fever transmission in subtropical Taiwan. *The Science of the Total Environment*, 407: 2224–33.

Chapter 7

Disease at the Molecular Scale: Methods for Exploring Spatial Patterns of Pathogen Genetics

Margaret Carrel

Despite confidence in the mid-20th century that modern medicine had conquered infectious disease, morbidity and mortality due to infectious diseases remains a serious threat to global population health in the 21st century (Burnet and White, 1962). Diarrheal diseases, HIV/AIDS, tuberculosis and other infectious diseases rank in the top 10 causes of mortality (World Health Organization (WHO) 2011). The continued significance of infectious diseases is due to three phenomena: the emergence of novel infectious diseases, the re-emergence of "old" infectious diseases in places where they had previously been nearly or totally eradicated, and the persistence of infectious disease despite decades of international public health efforts, particularly in low and middle income countries.

The emergence and re-emergence of infectious diseases is influenced both by changing population-environment interactions and by changing pathogen genetics (Mayer, 2000). Population growth, urbanization, agricultural intensification and extensification, rapid and long-distance travel, dietary shifts and other factors are all linked to increased opportunities for infectious disease transmission. Human behaviors, such as decreased compliance with vaccination campaigns and widespread antimicrobial usage in both people and livestock, combined with these increased opportunities for transmission, present numerous occasions for mutations in the genetic code of pathogens.

The changing nature of human interactions with environments, and the selective pressure this places on pathogens, from viruses to bacteria to protozoa, lends urgency and importance to our efforts to understand where, when, and why pathogens evolve. Spatial analysis should not be limited to patterns of disease incidence. Understanding the landscapes that produce drug resistance, or landscapes that encourage evolutionary development of characteristics such as efficient human-to-human transmission, are important for the success of public health efforts, providing information on how to prolong the effectiveness of pharmaceuticals and decrease chances of major pandemics. Combining molecular data on pathogens with disease ecology perspectives on human–environment interactions and spatial analytic methods, spatial epidemiologists can begin to consider disease at the molecular scale, and answer questions about what spaces and places produce or prevent pathogenic evolution.

Background

The disease ecology branch of medical geography is concerned with linking humans and their environments, exploring interactions between the two that drive patterns of disease (Meade and Emch, 2010). Such interactions are in constant flux and negotiation, humans alter yet are also constrained by their environments (Dubos, 1987; Mayer, 2000). This 'dynamic equilibrium' is subject to disturbances, such as population growth or climate shifts, which can cause new diseases or new disease patterns to emerge (Hunter, 1974; Dubos, 1965). Disease in a person cannot be separated from the environment where that person exists, a holistic approach is necessary.

Pathogens are not static elements responding passively to humans and environments, they are adaptive evolutionary organisms, changing to increase their chances of survival and reproduction (Ewald, 2000). Pathogens act in their own self-interest, changing in response to increased contact with hosts, increased opportunities for transmission, or increased contact with medical treatments. For instance, the antibiotic paradox explains why the success and popularity of antibiotics drives the emergence of antibiotic resistance: that modern medicine kills the weak bacteria and leaves the strong to thrive (Levy, 2002). Resistance to penicillin and methicillin as treatment for *Staphylococcus aureus* was documented within a decade after its first usage (National Institute of Allergy & Infectious Diseases (NIAID) 2008). Among malaria protozoa, resistance to antimalarial drugs such as chloroquine was observed within two decades of their first use (Peters, 1970). Similarly, treatment of human immunodeficiency virus (HIV) with antiretroviral therapy has spurred the emergence of drug-resistant HIV (Clavel and Hance, 2004). Vaccination can also influence the molecular character of viruses, selecting for those strains which have genetic code which render vaccines ineffective, or allowing previously out-competed strains of virus to emerge as newly dominant (Martcheva, Bolker, and Holt, 2008).

A disease ecology perspective posits that human–environment interactions are dynamic and that both the person and the place where disease occurs have to be considered. This view lends itself to studying pathogenic evolution, the adaptive responses of pathogens to pressures from hosts and environments. By adding a third element to the disease ecology framework, to a *pathogen* in a *person* in a *place*, we can begin to answer questions surrounding what landscapes are conducive to the molecular evolution of human pathogens.

Landscape Genetics

The ambition of disease ecologists, to understand both patterns and underlying processes of disease, is echoed in the emergent field of landscape genetics. Foundational to landscape genetics is the principle that patterns of species genetics can indicate underlying influences of environmental features on processes such as migration, extinction and disease response (Manel et al., 2003; Holderegger and Wagner 2006; Balkenhol et al., 2009; Storfer et al., 2007; Storfer et al., 2010). In other words, processes within the landscape have effects that are signaled in the molecular characteristics of individuals or populations. Landscape genetics studies combine spatially referenced genetic data with spatial methods, primarily from landscape ecology, to answer questions of pattern and process (Storfer et al., 2007).

To date, landscape genetic studies have been confined mainly to the fields of ecology and biology, focusing on plants and animals and insects (Mock et al., 2007; Sork et al., 1999; Spear et al., 2005; Dionne et al., 2008). More recently there has been recognition that the goals and methods of landscape genetics are as applicable to understanding disease in humans as disease in other species (Archie, Luikart, and Ezenwa 2009; Carrel and Emch, 2013). The combination of landscape genetics and disease ecology offers great potential to understand patterns and processes of the evolution of infectious diseases in humans across space and time.

Measuring Pathogenic Evolution

There are multiple methods to quantify the molecular character of human pathogens. Most rely on the decreased time necessary for polymerase chain reaction (PCR) and genotyping of genetic sequences, as well as the decreased cost of these steps and increased computational speeds. There are two broad categories of methods used in landscape genetics for measuring pathogen genetics: to calculate genetic distance or genetic relatedness, or to test for specific genes that confer drug resistance.

Phylogenetics is used to assess the evolutionary relationships between pathogen isolates, such as viruses or bacteria. An evolutionary tree is created, showing branching of different lineages from a single ancestor (Figure 7.1) (Page and Holmes, 2009). Attaching a spatial location to those isolates, phylogeography, can indicate places where several branches are in circulation, or places where a single evolutionary line predominates (Avise et al., 1987; Wu et al., 2010). In addition to illustrating the evolutionary relationships between isolates, phylogenetic trees enable the measurement of genetic distance. Genetic distance between isolates or between an isolate and the ancestral is measured via the length of the phylogenetic tree branches connecting pathogen pairs, or connecting an observed isolate and the ancestor. Increased genetic distance indicates increased amounts of genetic change between isolates, while small genetic distances indicate a high degree of genetic relatedness.

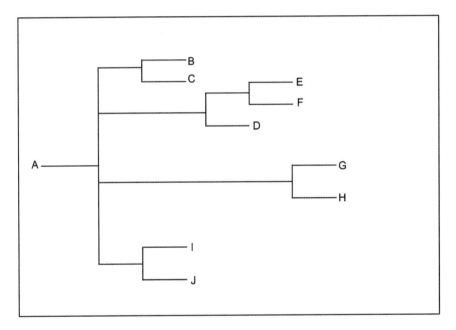

Figure 7.1   Example of a phylogenetic tree indicating the relationships between viruses (or bacteria) B-J, from an ancestral or root virus (A). Viruses G and H have the greatest degree of genetic change from virus A, while B and C are closely related to each other and to A

Single nucleotide polymorphisms (SNPs) are instances where a single nucleotide (adenine (A), guanine (G), cytosine (C), thymine (T)) in a genome differs between members of a species, creating two alleles. By measuring these SNPs in the sequences of pathogens and generating a measure of allele frequencies, genetic relatedness between two pathogens, two malarial parasites for instance, can be assessed. The fixation index (FST) is one such measure, assessing how closely two individual parasites are as compared to the total population of parasites (Holsinger and Weir, 2009; Wright, 1965). Slatkin's RST is another measure of genetic differentiation, based on a different presumption of how frequently mutations occur (Slatkin, 1995).

To determine if a pathogen is drug-resistant or drug sensitive, either genotypic or phenotypic methods may be used. In phenotypic methods, the pathogen is exposed to the drug and a response or lack of response is measured, a bacterial colony on an agar plate exposed to an antibioitc, for instance. In genotypic methods, known mutations in the genetic code of pathogens are assessed. For HIV, there is a database of drug resistance mutations used in surveillance of drug-resistant HIV (Bennett et al., 2009). There are also known mutations in the genes of malaria plasmodium that can be tested via PCR and genotyping (Wongsrichanalai et al., 2002). The outcome of such tests is a binary indicator of drug resistance in observed pathogens.

All of these measures allow spatial epidemiologists and disease ecologists to treat infectious disease as a continuous rather than dichotomous variable. Alternately, disease can be broken down into more detailed dichotomous variables, reflecting not simply presence/absence of disease but rather presence/absence of drug resistance or drug sensitivity. These more specific measures of disease in a landscape can then be used to explore patterns and processes of evolution of infectious diseases.

Methods for Exploring Spatial Variation in Disease Genetics

There are a multitude of spatial methodologies available to understand both pattern and process in genetic variation across a landscape. A selection of such methods is presented below, with examples from the literature of how they have been applied to the study of landscape genetics of pathogens.

Pattern

Mapping and interpolation
The most basic methods for determining patterns of pathogen genetics in a landscape are mapping and interpolation. Point or choropleth maps of where drug resistance does and does not occur, or where pathogen genetic change is elevated, can indicate spatial variation in genetic outcomes. Such maps can range from a global scale (as seen in a world map of methicillin-resistant *Staphylococcus aureus* prevalence by country) to the national scale (as seen in a drug resistance map of malarial parasites across the island of Madagascar) to a local scale (Grundmann et al., 2006; Andriantsoanirina et al., 2009). Interpolation between observed locations can indicate how genetic characteristics of pathogens might vary across unobserved locations, and suggest spaces where genetic change or amounts of drug resistance are either high or low. Inverse distance weighting interpolation methods have been used to map malaria parasitemia prevalence in the Democratic Republic of Congo, and kriging interpolation methods have been used to map spatial heterogeneity

in population genetics across China (Taylor et al., 2011; Xue et al., 2005). Alternately, Bayesian geostatistical methods for interpolation of genetic patterns are based upon probability distributions and are well-suited for use in areas where the distribution of data points is relatively sparse (Patil et al., 2011).

Cluster analysis
Cluster analysis is used to explore whether phenomena are grouped in non-random patterns. In the case of understanding patterns of pathogen genetics, cluster methods are useful to assess whether observations grouped according to genetic characteristics also reflect the spatial arrangement of sample locations, that is, are genetically similar pathogens located in the same geographic locations. K-means clustering and model-based clustering techniques are both appropriate to cluster observations based on one or many genetic measures. K-means clustering finds "natural" clusters in datasets by partitioning observations into groups around centroids such that within-cluster sum of squares are minimized (Hartigan and Wong, 1979). Unlike other cluster detection methods, k-means clustering can detect clusters in multiple types of space, including geographic, temporal and genetic dimensions. The technique is similar to principal component analysis, in that it reveals the internal structure of the dataset by minimizing data variability. An initial number of seeds (or clusters) have to be indicated for k-means clustering algorithms, however, which introduces uncertainty about whether the true number of genetic clusters has been established.

In model-based clustering, as conducted using the *mclust* package in R, problems associated with heuristic decisions about the number of clusters to search for, which clustering methodology to use and how non-normally distributed data should be handled are overcome using a principled statistical approach (Fraley and Raftery, 2010; R Development Core Team, 2011). In model-based clustering, the various genetic distance measures for pathogens (that is, phylogenetic tree distance for influenza viruses) are used to partition the observations such that within-cluster likeness and between-cluster difference is maximized. The model cluster algorithm investigates a variety of shape and size constraints for the clusters, including equal and unequal volume, equal and unequal shape, spherical, diagonal or ellipsoidal orientation, etc., and returns indications of how well each of these models fit the dataset across different numbers of clusters. The number of clusters that pathogen genetic observations will be divided into, the shape of the clusters and the optimum cluster assignment is determined by the lowest Bayesian Information Criterion (BIC). The BIC is calculated given the log-likelihood, the dimensionality of the data (the number of genetic measures used), and the number of mixture components (observations), and varies greatly according to the model type and number of clusters. Once pathogen isolates are assigned to clusters based on their genetic characteristics, those cluster assignments can be mapped in a GIS according to geographic location of parasite isolation. This allows for the assessment of whether the cluster assignments generated in the model clustering algorithm reflect the spatial locations where pathogens were isolated, or if there are other patterns inherent in the genetic characteristics of the pathogen under study. Similar methods have been used by to explore patterns of H5N1 evolution in Vietnam, determining that clustering viruses according to genetic distance on eight viral gene segments reflected time more than space as a driver of pathogenic evolution (Carrel et al., 2011a).

Mantel tests

Mantel tests, originally developed to explore clustering of cancer, have been adapted by ecologists to study the correlation between environmental variables and organism distribution, and can be further used to study relationships between genetic differences between observed organisms and their attendant spatial locations (Mantel, 1967). Mantel tests make use of distance matrices, testing for levels of correlation between matrices and estimating the probability that those correlations could have occurred by chance. They are highly useful for uncovering spatial patterns because they explicitly account for the non-independence of observations found in spatial data. Mantel tests overcome the lack of independence by randomly shuffling the values in one of the matrices multiple times and calculating correlations between the shuffled and original matrices. The probability distribution of the test statistic (the Mantel r) is generated by this random permutation process and used as a basis for assigning a probabilistic interpretation of the true correlation statistic between the observed response and predictor matrices. In this way, the relationship between the geographic space between pathogen isolates and their genetic differences can be assessed. Genetic measures of distance or relatedness between pairs of viruses, bacteria or parasites, such as phylogenetic tree distance, RST, and FST, can all be used.

Mantel tests measure whether genetic and geographic distances are related, but they indicate little about the form of this relationship. In a Mantel correlogram, the geographic distance between observations is divided into lags, and Mantel statistics (including significance) are calculated for case pairs which fall within each lag (Figure 7.2). Correlograms thus display the degree of likeness or difference among malarial parasites at specific geographic distances. Mantel correlograms thus allow one to assess whether the degree of genetic dissimilarity among pathogens corresponds to the scale of geographic distance between the pathogens, that is, whether an isolation-by-distance pattern is observed. Isolation-by-distance is a basic and important measure of pattern in landscape genetics, indicating that as species move further apart in geographic space they share less genetic characteristics (Wright, 1943). Mantel tests indicate isolation-by-distance among malarial parasites in Southeast Asia (Anderson et al., 2005). Mantel tests and Mantel correlograms have also indicated an isolation-by-distance pattern among H5N1 avian influenza in Vietnam, with a spike in genetic dissimilarity of viruses at approximately the distance between Vietnam's two main population centers, Hanoi and Ho Chi Minh City (Carrel et al., 2010). Mantel tests and Mantel correlograms can be explored in the *ecodist* package in R.

Process

Environmentally-informed Mantel tests

Mantel tests can also be used to explore processes underlying patterns of genetic differentiation. An adaptation of traditional Mantel tests is to use them to explore how species move through geographic space. For example, if an organism is hypothesized to disperse along river networks, two different geographic distance matrices could be generated to test this assumption. If a stronger correlation exists between the genetic distance matrix and a geographic distance matrix measured along river networks than between genetics and Euclidean (as-the-crow-flies) distance matrices, this suggests that the species does disperse along the river. Mantel tests can also be multivariate in nature, using multiple predictor distance matrices that describe variation in the environment. In addition to testing the relationship between genetic distances of pathogens and distance as measured along a

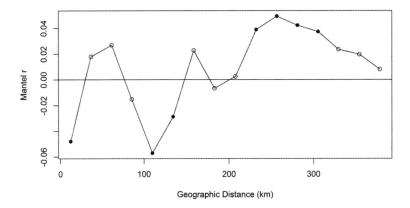

Figure 7.2   **A Mantel spatial correlogram indicating genetic similarity or dissimilarity at various distance thresholds. Points above the zero line indicate similiarity, points below indicate dissimilarity. Filled circles indicate statistical significance. In this case, an isolation-by-distance pattern is not observed. Instead, samples that are further apart geographically exhibit statistically significant similarity, while those at low geographic distances are genetically dissimilar**

river network, road network matrices, population density matrices and altitudinal difference matrices can all be generated to test theories about the impacts on genetic characteristics of varying spatial, population and environmental distances between observed pathogen locations. In this multiple matrix environment, the effect of each predictor matrix is measured while controlling for the effects of other predictor matrices (a partial-Mantel test), allowing a clear understanding of the varying correlation between each predictor variable and malarial genetic outcomes. Such tests have been used to determine the effects of environmental factors such as slope and elevation on the genetics of *Escherica coli* bacteria after accounting for their spatial distribution (Bergholz, Noar and Buckley, 2011).

Barrier analysis
Genetic diversity relies either on selection pressures in local environments or random chance and on limited gene flow among populations (Barbujani and Sokal, 1990). Identifying boundaries to gene flow, common in ecology and in population genetics, is useful to generate hypotheses about underlying processes that drive disease incidence and pathogenic evolution (Jacquez, Maruca and Fortin, 2000; Jacquez, 2010). Barriers are located in spatial zones of rapid genetic change, and indicate the presence of either a sharp environmental transition and/or a low rate of genetic exchange (Barbujani, Oden, and Sokal, 1989). Barriers can exist either for host (humans or other animals) or vector (mosquitoes) movement, limiting both disease incidence and the exchange of genes among pathogens.

Monmonier's algorithm is one of the original methods used for delineation of barriers to gene flow. Monmonier proposed that when examining genetic difference across geographic space, barriers to genetic exchange exist where the steepest gradient between linked (that is, neighboring) locations (Monmonier, 1973; Manni, Guerard, and Heyer, 2004). Within barrier analysis, sample points may be associated with either populations or individuals. The geographic locations (in latitude/longitude) of observations are associated with one or more matrices of genetic distances between observations (again, a variety of genetic measures may be used). Using freely available software such as BARRIER or the *adegenet* package in R or, a spatial network is created among the points by means of Voronoi tessellations and Delaunay triangulations, which is then associated with the genetic distance network between the points (Figure 7.3) (Manni, Guerard 2004, Manni, Guerard and Heyer, 2004). Barriers are calculated for each genetic distance matrix (that is, one per gene in influenza viruses or one per microsatellite for malarial parasites) associated with the spatial network, and then can be overlaid to gain a sense of how barriers are similar or different between them. The presence or absence of spatial barriers to gene flow at multiple points in the pathogen's genetic structure can then be assessed and maps of barriers can be generated.

A second method for detecting barriers or boundaries is Womble's method (or wombling). Barrier analysis of H5N1 influenza genetics in Vietnam indicated spatially and temporally transitory barriers to gene flow, suggesting that influenza viruses move across the Vietnamese landscape unimpeded by population or environmental discontinuity

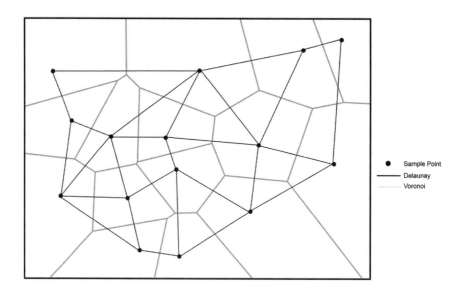

Figure 7.3   Sample points (black circles) are used to create a triangular network
                  of neighbors (using Delaunay triangulation, black lines) and Voronoi
                  tessellations (gray lines). In Monmonier's algorithm the genetic
                  distance or other measure of genetic relatedness is associated with
                  each connection between points on the triangular network. Barriers
                  are then drawn between points, along the Voronoi tessellation lines,
                  that have the highest degree of genetic distance. Individual barrier
                  segments, dividing points, are linked across the network to either
                  form a closed loop or end at the edge of the study area

(Carrel et al., 2011b). Womble proposed that when examining genetic cline across geographic space, barriers to genetic exchange exist where the cline exhibits steep gradients (Womble, 1951). Bayesian wombling for spatial point processes is an adaptation of Womble's method. While traditional wombling allows the detection of barriers to genetic exchange between neighboring areas, it does not provide any direct association between barriers and landscape variables (Lu and Carlin, 2005). A Bayesian spatially varying coefficient (SVC) model allows for the assessment of how underlying population and environmental variables influence barriers to genetic exchange, and have been adapted from areal processes to point processes (Liang, Banerjee, and Carlin 2009, Wheeler and Waller, 2008). Using an SVC model, one can measure whether barriers to pathogen genetic exchange found in a wombling analytic framework are correlated with physical features, such as mountains, climate features, such as temperature changes, or population variables, such as low population density. Mountains and rivers, for instance, have been shown to act as barriers to the spread of rabies along the Eastern coast of the United States (Wheeler and Waller, 2008). Under a Bayesian curvilinear approach, boundaries that follow curvilinear features such as rivers or roads can be detected (Banerjee and Gelfand, 2006). Understanding what types of landscape features act as barriers to molecular exchange can provide an understanding of not just the ecology of disease in a place, but the ecology of disease evolution.

Geographically weighted regression
Geographically weighted regression (GWR) can be used to test for the strength of relationships between landscape variables and genetic variability. GWR does not assume that the relationship between genetics and landscape features is constant across the study area, but rather leaves open the possibility of non-stationarity. Traditional regression models assume that the relationships being measured are equal at all points in space, that they are stationary. This assumption, however, contradicts the goals of geographic, spatially-explicit studies that attempt to understand where, when, and why an outcome differs across a study area. Geographically weighted regression (GWR) models can be used to explore relationships that exhibit spatial non-stationarity, varying across space (Brunsdon, Fotheringham and Charlton, 1998; Fotheringham, Charlton and Brunsdon, 1998). In GWR, the global regression model is modified to allow local parameters to be estimated, and is written as:

$$y_i = \beta_0(u_i, v_i) + \sum_k \beta_k(u_i, v_i) x_{ik} + \varepsilon_i$$

where (u_i, v_i) denote the coordinates of the ith datapoint in space, and $\beta_k(u_i, v_i)$ is the realization of the continuous function $\beta_k(u, v)$ at point i (Fotheringham, Charlton and Brunsdon, 1998; Fotheringham, Charlton and Brunsdon 1998; Fotheringham, Brunsdon and Charlton, 2002). GWR4.0 software is freely available online.

Mixed models may also be used in GWR, based upon knowledge that some coefficients do not vary in space to the degree that others do, and the global coefficients are modeled additively with the local coefficients for each point in space (Brunsdon, Fotheringham and Charlton, 1999; Mei, He, and Fang, 2004). Mixed-models in GWR allow some variables to have constant effects on genetic outcomes, while others are allowed to vary in strength from place to place. In the case of a viral genetic dataset, for instance, this could allow for the exploration of whether the relationship between viral evolution is differentially affected by

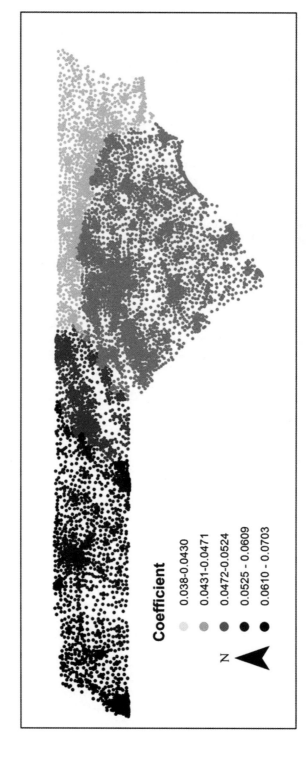

Coefficient

- 0.038-0.0430
- 0.0431-0.0471
- 0.0472-0.0524
- 0.0525 - 0.0609
- 0.0610 - 0.0703

N

Figure 7.4   In this hypothetical example, the influence of an explanatory variable on genetic distance data varies acros the dataset, with the strongest influence observed in the west and declining association in the east

population density in one area versus another while holding constant the effect of climate (that is, humidity).

To assess whether there is significant spatial variation in the local parameter estimates, the data is randomly rearranged in space and the GWR procedure is repeated. Then, the standard deviation distribution obtained from the random permutations is compared to the standard deviation of the original localized parameter estimates. As well as producing local parameter estimates, GWR techniques provide localized versions of standard regression diagnostics, including R2 and other goodness-of-fit measures (Fotheringham, Charlton and Brunsdon, 1998). GWR model output will show not only which population and environmental variables in a landscape are significantly related to pathogenic genetic change, but also how these relationships vary in space (that is, do the drivers of evolution vary from east to west across a region) (Figure 7.4). The local parameter estimates from GWR can be displayed graphically using ArcGIS software, indicating areas of high and low effects. The local R2 values can also be mapped, to indicate where modeled variables do a good job of predicting the outcome variable, and places where some untested explanatory variable is needed. GWR has been used to explore variation in climate drivers of drug-resistant tuberculosis at a global scale (Liu et al., 2011).

It should be noted that GWR has limitations which may be relevant to the study of infectious disease evolution. In particular, GWR is sensitive to the number of observations, and small sample sizes (for example, under 400) can cause collinearity and mis-estimation of local coefficients (Páez, Farber, and Wheeler 2011; Wheeler and Páez 2010; Wheeler and Tiefelsdorf, 2005). GWR findings are also potentially biased by bandwidth definitions, which are user defined, that indicate how much influence neighboring observations have on local observations (Griffith 2008, Jetz, Rahbek and Lichstein, 2005). Despite these limitations, however, GWR remains an interesting tool for exploring spatial variation in population and environmental drivers of pathogenic change.

Ecological niche modeling
The Russian medical geographer Pavlovsky described landscape epidemiology as the study of natural nidi of disease (Pavlovsky, 1966). A natural nidi is the combination of environmental features, such as climate, vegetation, host, agent, etc., that are necessary to the continuous circulation of a disease. Pavlovsky believed that by ascertaining the natural nidus of a disease, other areas that could play host to the disease could be predicted. Ecological Niche Modeling (ENM) is premised on the same theory: the features of an environment currently associated with an organism can be used to predict where else that organism will be found or could exist. ENM is a set of methodologies designed to explore the combinations of environmental and population variables that provide an appropriate habitat for species occurrence. In the case of malaria, for example, the presence of a certain species of mosquito can act as a predictor of disease range. Knowledge about preferred or viable environments is vital to studies of disease, particularly zoonotic diseases that are conditioned by host ranges. Genetic Algorithm for Rule-Set Prediction (GARP) software, for instance, has been used to analyze the predicted range of dengue fever, Chagas disease and monkeypox (Levine et al., 2007; Peterson et al., 2005; Peterson et al., 2002). MaxEnt software (freely available) has been used to predict malaria based on suitability of *Anopheles* mosquito habitat (Kulkarni, Desrochers and Kerr 2010, Phillips, Anderson and Schapire, 2006). Extending the use of ENM to explore landscapes with the combinations of population and environmental characteristics which comprise the fundamental niche

of observed pathogenic mutations would allow for the prediction of places where drug resistance or high levels of genetic evolution could be found.

ENM operates on a presence/absence dichotomous outcome variable. The presence/absence of drug resistance or drug sensitivity could comprise one outcome variable, the presence/absence of high levels of genetic change (initially measured as a continuous variable) could also be dichotomized. Thus, it is possible to predict not only the environmental niches suitable to disease incidence, but also those niches which host drug-resistant strains or those niches which encourage rapid evolution of pathogens. ENM methods all consist of algorithmic iterations, using known presence locations spatially associated with population and environmental layers (such as landcover, elevation, climate, population density) to predict other areas of presence (Stockwell, 2006; Stockwell, 1999; Phillips, Anderson, and Schapire, 2006). Helping to inform the predictor layers included in ENM could be those found in a barrier analysis to be associated with boundaries to gene flow or in a partial Mantel test to be associated with genetic outcomes. Rule sets regarding environments where known occurrences took place are developed through evolutionary refinement, with a goal of maximizing significance and predictive accuracy without overfitting. Significance is assessed with a χ^2 test of the difference in the probability of the predicted value (presence/absence) before and after the rule is applied (Stockwell, 1999). The output of ENM is a probability surface, indicating areas where presence of drug-resistant disease or strongly evolving disease, is highly expected given underlying environmental and population variables. This information indicates not only where the event is expected, but also the necessary precursors to presence.

Conclusion

Traditionally, studies of the ecology of infectious disease treated the diseases themselves as fixed outcomes, either present or absent in an individual or population. In reality, a disease outcome represents a spectrum of pathogenic adaptation, comprised of a regularly changing set of mutations designed to overcome host immune response, evade medical treatments and circumvent environmental limitations. This is especially the case of rapidly mutable RNA viruses such HIV or influenza. Additionally, there is increasing concern over the development of drug-resistant pathogens, such as chloroquine and mefloquine resistant malaria, extensively drug-resistant tuberculosis (XTB) and Methicillin-resistant *Staphylococcus aureus* (MRSA). Going forward, combining disease ecology and landscape genetics can enhance the contributions that medical geographers and spatial epidemiologists make to the study of infectious diseases. Linking the two provides a framework for understanding that the spatial patterns of disease within a landscape can reveal the processes, that is, interactions between people and their environment, that allow the disease to evolve. Adding this genetic component to more traditional disease ecology studies enables an examination of both the etiology and ecology of pathogenic evolution. Not only does the disease exist in a person and a place, it has a particular genetic signature that can indicate pathogenic response to environmental pressures.

Using the spatial methods outlined above to explore pattern and process of disease at the molecular scale is useful both for diseases of known and unknown etiologies. For diseases of known etiologies, a genetically-informed disease ecology perspective provides insights into how variation in population-environment factors that are hypothesized to increase or decrease transmission is linked to patterns of disease genetics, such as spaces of increased genetic distance. For diseases of unknown etiology, molecular scale disease ecology studies, can provide insight into what variables in an individuals' social or natural environment cause or correlate with pathogenic evolution, such as the development of drug resistance. By considering infectious disease from an evolutionary perspective, spatial analysis can help to answer questions such as: Is pathogenic evolution spatially structured? Where is drug resistance taking place, and why there? Where are spaces that pathogens are evolving quickly and are more likely to develop efficient human-to-human transmission capabilities? Answering these questions will be increasingly important in the coming decades, as infectious diseases persist as a major cause of morbidity and mortality globally, as human mobility increases, as vector range changes due to global climate change, and as modern medical treatments lose efficacy.

References

Anderson, T.J., Nair, S., Sudimack, D., Williams, J.T., Mayxay, M., Newton, P.N., Guthmann, J., Smithuis, F.M., Hien, T.T., and Ingrid V.F. van den Broek. 2005. Geographical distribution of selected and putatively neutral SNPs in Southeast Asian malaria parasites. *Molecular Biology and Evolution*, 22(12): 2362–74.

Andriantsoanirina, V., Ratsimbasoa, A., Bouchier, C., Jahevitra, M., Rabearimanana, S., Radrianjafy, R., Andrianaranjaka, V., Randriantsoa, T., Rason, M.A., and M. Tichit. 2009. Plasmodium falciparum drug resistance in Madagascar: Facing the spread of unusual pfdhfr and pfmdr-1 haplotypes and the decrease of dihydroartemisinin susceptibility. *Antimicrobial Agents and Chemotherapy*, 53(11): 4588–97.

Archie, E.A., Luikart, G., and V.O. Ezenwa. 2009. Infecting epidemiology with genetics: A new frontier in disease ecology. *Trends in Ecology and Evolution*, 24(1): 21–30.

Avise, J.C., Arnold, J., Ball, R.M., Bermingham, E., Lamb, T., Neigel, J.E., Reeb, C.A., and N.C. Saunders. 1987. Intraspecific phylogeography: The mitochondrial DNA bridge between population genetics and systematics. *Annual Review of Ecology and Systematics*, 18: 489–522.

Balkenhol, N., Gugerli, F., Cushman, S.A., Waits, L.P., Coulon, A., Arntzen, J.W., Holderegger, R., and H.H. Wagner. 2009. Identifying future research needs in landscape genetics: Where to from here? *Landscape Ecology*, 24(4): 455–63.

Banerjee, S., and A.E. Gelfand. 2006. Bayesian wombling: Curvilinear gradient assessment under spatial process models. *Journal of the American Statistical Association*, 101(476): 1487–1501.

Barbujani, G., Oden, N.L., and R.R. Sokal. 1989. Detecting regions of abrupt change in maps of biological variables. *Systematic Biology*, 38(4): 376.

Barbujani, G. and R.R. Sokal. 1990. Zones of sharp genetic change in Europe are also linguistic boundaries. *Proceedings of the National Academy of Sciences of the United States of America*, 87(5): 1816.

Bennett, D.E., Camacho, R.J., Otelea, D., Kuritzkes, D.R., Fleury, H., Kiuchi, M., Heneine, W., Kantor, R., Jordan, M.R., and J.M. Schapiro. 2009. Drug resistance mutations for surveillance of transmitted HIV-1 drug-resistance: 2009 update. *PloS one*, 4(3):e4724.

Bergholz, P.W., Noar, J.D., and D.H. Buckley. 2011. Environmental patterns are imposed on the population structure of Escherichia coli after fecal deposition. *Applied and Environmental Microbiology*, 77(1): 211–19.

Brunsdon, C., Fotheringham, A.S., and M. Charlton. 1999. Some notes on parametric significance tests for geographically weighted regression. *Journal of Regional Science*, 39(3): 497–524.

Brunsdon, C., Fotheringham, S., and M. Charlton. 1998. Geographically weighted regression-modelling spatial non-stationarity. *The Statistician*, 47(3): 431–43.

Burnet, M. and D. White. 1962. *Natural History of Infectious Disease*. Cambridge, MA: Cambridge University Press, 275.

Carrel, M. and M.E. Emch. 2013. Genetics: A New Landscape for Medical Geography. *Annals of the Association of American Geographers*, 103(6): 1452–67.

Carrel, M., Wan, X.F., Nguyen, T., and M. Emch. 2011a. Genetic Variation of Highly Pathogenic H5N1 Avian Influenza Viruses in Vietnam Shows Both Species-Specific and Spatiotemporal Associations. *Avian Diseases*, 55(4): 659–66.

Carrel, M., Wan, X.F., Nguyen, T., and M. Emch. 2011b. Highly Pathogenic H5N1 Avian Influenza Viruses Exhibit Few Barriers to Gene Flow in Vietnam. *EcoHealth*: 1–10.

Carrel, M.A., Emch, M., Jobe, R.T., Moody, A., and X.F. Wan. 2010. Spatiotemporal structure of molecular evolution of H5N1 highly pathogenic avian influenza viruses in Vietnam. *PloS* one, 5(1):e8631.

Clavel, F., and A.J. Hance. 2004. HIV drug resistance. *New England Journal of Medicine*, 350(10): 1023–35.

Dionne, M., Caron, F., DODSON, J.J., and L. Bernatchez. 2008. Landscape genetics and hierarchical genetic structure in Atlantic salmon: The interaction of gene flow and local adaptation. *Molecular Ecology*, 17(10): 2382–96.

Dubos, R.J. 1987. *Mirage of Health: Utopias, Progress and Biological Change*. New Brunswick: Rutgers University Press.

Dubos, R.J. 1965. *Man Adapting*, Yale University Press, New Haven.

Ewald, P.W. 2000. *Plague Time: How Stealth Infections Cause Cancers, Heart Disease, and Other Deadly Ailments*. Free Press, New York.

Fotheringham, A.S., Charlton, M.E., and C. Brunsdon. 1998. Geographically weighted regression: A natural evolution of the expansion method for spatial data analysis. *Environment and Planning A*, 30: 1905–27.

Fotheringham, A.S., Brunsdon, C., and M. Charlton. 2002. *Geographically Weighted Regression: The Analysis of Spatially Varying Relationships*. Wiley, Chichester.

Fraley, C. and A. Raftery. 2010. *mclust: Model-Based Clustering / Normal Mixture Modeling*.

Griffith, D.A. 2008. Spatial-filtering-based contributions to a critique of geographically weighted regression (GWR). *Environment and Planning* A, 40(11): 2751.

Grundmann, H., Aires-de-Sousa, M., Boyce, J., and E. Tiemersma. 2006. Emergence and resurgence of meticillin-resistant Staphylococcus aureus as a public-health threat. *The Lancet* 368(9538): 874–85.

Hartigan, J.A. and M.A. Wong. 1979. Algorithm AS 136: A K-Means Clustering Algorithm. *Applied Statistics*, 28(1): 100–108.

Holderegger, R. and H.H. Wagner. 2006. A brief guide to landscape genetics. *Landscape Ecology*, 21(6): 793–6.

Holsinger, K.E. and B.S. Weir. 2009. Genetics in geographically structured populations: Defining, estimating and interpreting FST. *Nature Reviews Genetics*, 10(9): 639–50.

Hunter, J.M. 1974. The challenge of medical geography, in *The Geography of Health and Disease*, edited by J.M. Hunter, Department of Geography, University of North Carolina at Chapel Hill, Chapel Hill: 1–31.

Jacquez, G.M. 2010. Geographic boundary analysis in spatial and spatio-temporal epidemiology: Perspective and prospects. *Spatial and Spatio-temporal Epidemiology*, 1(4): 207–18.

Jacquez, G.M., Maruca, S., and M.J. Fortin. 2000. From fields to objects: A review of geographic boundary analysis. *Journal of Geographical Systems*, 2(3): 221–41.

Jetz, W., Rahbek, C., and J.W. Lichstein. 2005. Local and global approaches to spatial data analysis in ecology. *Global Ecology and Biogeography*, 14(1): 97–8.

Kulkarni, M.A., Desrochers, R.E., and J.T. Kerr. 2010. High resolution niche models of malaria vectors in northern Tanzania: A new capacity to predict malaria risk? *PLoS One*, 5(2):e9396.

Levine, R.S., Peterson, A.T., Yorita, K.L., Carroll, D., Damon, I.K., and M.G. Reynolds. 2007. Ecological niche and geographic distribution of human monkeypox in Africa. *PLoS One*, 2(1): e176.

Levy, S.B. 2002. *The Antibiotic Paradox: How the Misuse of Antibiotics Destroys their Curative Power.* Cambridge, MA: Perseus Publishing.

Liang, S., Banerjee, S., and B.P. Carlin. 2009. Bayesian wombling for spatial point processes. *Biometrics*, 65(4): 1243–53.

Liu, Y., Jiang, S., Wang, R., Li, X., Yuan, Z., Wang, L., and F. Xue. 2011. Spatial epidemiology and spatial ecology study of worldwide drug-resistant tuberculosis. *International Journal of Health Geographics*, 10: 50.

Lu, H. and B.P. Carlin. 2005. Bayesian areal wombling for geographical boundary analysis. *Geographical Analysis*, 37(3): 265–85.

Manel, S., Schwartz, M.K., Luikart, G., and P. Taberlet. 2003. Landscape genetics: Combining landscape ecology and population genetics. *Trends in Ecology & Evolution*, 18(4): 189–97.

Manni, F. and E. Guerard. 2004. *Barrier v.2.2 user's manual. Population genetics team, Musée de l'Homme, Paris.*

Manni, F., Guerard, E., and E. Heyer. 2004. Geographic patterns of (genetic, morphologic, linguistic) variation: How barriers can be detected by using Monmonier's algorithm. *Human Biology; An International Record of Research*, 76(2): 173–90.

Mantel, N. 1967. The detection of disease clustering and a generalized regression approach. *Cancer Research*, 27(2) Part 1: 209–220.

Martcheva, M., Bolker, B.M., and R.D. Holt. 2008. Vaccine-induced pathogen strain replacement: What are the mechanisms? *Journal of the Royal Society Interface*, 5(18): 3–13.

Mayer, J.D. 2000. Geography, ecology and emerging infectious diseases. *Social Science & Medicine*, 50(7–8): 937–52.

Meade, M.S. and M. Emch. 2010. *Medical Geography* 3rd edn. New York: Guilford Press,

Mei, C., He, S. and K. Fang. 2004. A Note on the Mixed Geographically Weighted Regression Model*. *Journal of Regional Science*, 44(1): 143–57.

Mock, K.E., Bentz, B., O'neill, E., Chong, J., Orwin, J., and M. Pfrender. 2007. Landscape-scale genetic variation in a forest outbreak species, the mountain pine beetle (Dendroctonus ponderosae). *Molecular Ecology*, 16(3): 553–68.

Monmonier, M.S. 1973. Maximum-Difference Barriers: An Alternative Numerical Realization Method. *Geographical Analysis*, 5(3): 245–61.

National Institute of Allergy and Infectious Diseases (NIAID) 2008. March 4, 2008-last update, *Methicillin-Resistant Staphylococcus aureus: History* [Homepage of NIAID], [online]. Available at: http://www.niaid.nih.gov/topics/antimicrobialresistance/ examples/mrsa/pages/history.aspx [April 12, 2013].

Páez, A., Farber, S., and D. Wheeler. 2011. A simulation-based study of geographically weighted regression as a method for investigating spatially varying relationships. *Environment and Planning-Part* A, 43(12): 2992.

Page, R.D. and E.C. Holmes. 2009. *Molecular Evolution: A Phylogenetic Approach.* Oxford: Wiley-Blackwell.

Patil, A.P., Gething, P.W., Piel, F.B., and S.I. Hay. 2011. Bayesian geostatistics in health cartography: The perspective of malaria. *Trends in Parasitology*, 27(6): 246–53.

Pavlovsky, E.N. 1966. Natural Nidality of Transmissible Diseases with special reference to the Landscape Epidemiology of Zooanthroponoses. *Natural Nidality of Transmissible Diseases with special reference to the Landscape Epidemiology of Zooanthroponoses.*

Peters, W. 1970. *Chemotherapy and Drug Resistance in Malaria.* London: Academic Press.

Peterson, A.T., Martínez-Campos, C., Nakazawa, Y., and E. Martínez-Meyer. 2005. Time-specific ecological niche modeling predicts spatial dynamics of vector insects and human dengue cases. *Transactions of the Royal Society of Tropical Medicine and Hygiene*, 99(9): 647–55.

Peterson, A.T., Sánchez-Cordero, V., Beard, C.B., and J.M. Ramsey. 2002. Ecologic niche modeling and potential reservoirs for Chagas disease, Mexico. *Emerging Infectious Diseases*, 8: 662–7.

Phillips, S.J., Anderson, R.P., and R.E. Schapire. 2006. Maximum entropy modeling of species geographic distributions. *Ecological Modelling*, 190(3): 231–59.

R Development Core Team 2011. *R: A Language and Environment for Statistical Computing.*, R Foundation for Statistical Computing, Vienna, Austria.

Slatkin, M. 1995. A measure of population subdivision based on microsatellite allele frequencies. *Genetics*, 139(1): 457.

Sork, V.L., Nason, J., Campbell, D.R., and J.F. Fernandez. 1999. Landscape approaches to historical and contemporary gene flow in plants. *Trends in Ecology & Evolution*, 14(6): 219–24.

Spear, S.F., Peterson, C.R., Matocq, M.D., and A. Storfer. 2005. Landscape genetics of the blotched tiger salamander (Ambystoma tigrinum melanostictum). *Molecular Ecology*, 14(8): 2553–64.

Stockwell, D. 1999. The GARP modelling system: Problems and solutions to automated spatial prediction. *International Journal of Geographical Information Science*, 13(2): 143–58.

Stockwell, D.R. 2006. Improving ecological niche models by data mining large environmental datasets for surrogate models. *Ecological Modelling*, 192(1): 188–96.

Storfer, A., Murphy, M.A., Spear, S.F., Holderegger, R., and L.P. Waits. 2010. Landscape genetics: Where are we now? *Molecular Ecology*, 19(17): 3496–514.

Storfer, A., Murphy, M.A., Evans, J.S., Goldberg, C.S., Robinson, S., Spear, S.F., Dezzani, R., Delmelle, E., Vierling, L., and L.P. Waits. 2007. Putting the "landscape" in landscape genetics. *Heredity*, 98(3): 128–42.

Taylor, S.M., Messina, J.P., Hand, C.C., Juliano, J.J., Muwonga, J., Tshefu, A.K., Atua, B., Emch, M., and S.R. Meshnick. 2011. Molecular malaria epidemiology: Mapping and burden estimates for the Democratic Republic of the Congo, 2007. *PloS one*, 6(1):e16420.

Wheeler, D.C. and L.A. Waller. 2008. Mountains, valleys, and rivers: The transmission of raccoon rabies over a heterogeneous landscape. *Journal of Agricultural, Biological, and Environmental Statistics*, 13(4): 388–406.

Wheeler, D.C. and A. Páez. 2010. Geographically weighted regression, in *Handbook of Applied Spatial Analysis*. Heidelberg: Springer: 461–86.

Wheeler, D. and M. Tiefelsdorf. 2005. Multicollinearity and correlation among local regression coefficients in geographically weighted regression. *Journal of Geographical Systems*, 7(2): 161–87.

Womble, W.H. 1951. Differential Systematics. *Science*, 114(2961): 315–22.

Wongsrichanalai, C., Pickard, A.L., Wernsdorfer, W.H., and S.R. Meshnick. 2002. Epidemiology of drug-resistant malaria. *The Lancet Infectious Diseases*, 2(4): 209–18.

World Health Organization (WHO) 2011. June-last update, *The Top 10 Causes of Death*. [Homepage of World Health Organization], [online]. Available at: http://www.who.int/mediacentre/factsheets/fs310/en/index.html [March 19, 2013].

Wright, S. 1965. The interpretation of population structure by F-statistics with special regard to systems of mating. *Evolution*, 19(3): 395–420.

Wright, S. 1943. Isolation by Distance. *Genetics*, 28: 114–38.

Wu, L., Lees, D.C., Yen, S., and Y. Hsu. 2010. The complete mitochondrial genome of the near-threatened swallowtail, Agehana maraho (Lepidoptera: Papilionidae): Evaluating sequence variability and suitable markers for conservation genetic studies. *Entomological News*, 121(3): 267–80.

Xue, F., Wang, J., Hu, P., and G. Li. 2005. The "Kriging" model of spatial genetic structure in human population genetics. *Yi Chuan Xue Bao*, 32: 219–33.

SECTION 3
Chronic Disease

Chapter 8
Modeling Spatial Variation in Disease Risk in Epidemiologic Studies

David C. Wheeler and Umaporn Siangphoe

Many diseases have risk factors that are distributed unevenly in the environment. Examples for cancers include bladder cancer and arsenic (Silverman et al., 2006), leukemia and benzene (Linet et al., 2006), and lung cancer and radon (Spitz et al., 2006). It is reasonable, therefore, to expect spatial pattern in disease risk, which may be explained by the uneven distribution of risk factors that are known or unknown. When risk factors are unknown, studying spatial-temporal patterns in disease events may reveal clues about disease etiology. There is a long history of research analyzing geographic patterns in cancer incidence and mortality with the objective of discovering environmental determinants of disease (Devesa, Grauman and Fraumeni, 1998; Devesa et al., 1999; Fraumeni and Blot, 1977). Examples of risk factors revealed by analytic epidemiologic studies that followed upon observations of geographic patterns of cancer include chronic use of snuff as a risk factor for oral and pharyngeal cancer among women in the southern United States (Winn et al., 1981) and exposure to asbestos from shipyards as a risk factor for lung cancer among men along the southeastern United States seaboard (Blot et al., 1979). Other examples of investigating spatial patterns in cancer risk include studies on childhood leukemia (Alexander, 1993; Bithell and Vincent, 2000; Wheeler, 2007) and bladder cancer (Jacquez et al., 2006).

Many of the studies in the literature analyzing geographic patterns of disease have been ecological studies, using readily available data on disease and the population at risk aggregated to areal units, such as counties. Unfortunately, ecological studies face a number of inherent analytic challenges (Beale et al., 2010) that limit their role in etiologic research. These challenges include exposure misclassification, spatial inaccuracy of data, and ecological bias (Elliott and Savitz, 2008; Wakefield and Elliott, 1999). In addition, analyses in ecological studies are usually based on administrative geographic boundaries that are not inherently meaningful for studying disease. Moreover, these studies lack information on risk factors at the individual level and information about residential and occupational locations over time. Furthermore, available environmental exposure data of interest typically will have been collected on different spatial scales. As more spatial and temporal information is being collected for individuals through epidemiologic studies and disease registries, attention is focusing on modeling and explaining risk at the individual level (Richardson et al., 2013).

In spatial analyses of cancer risk, the residence at diagnosis is typically used as a surrogate for unknown environmental exposures, defined broadly to include lifestyle factors as well as pollutants. However, due to the long latency, or lag time, between exposure to a relevant risk factor and diagnosis of cancer, and to population mobility, it is reasonable to believe that residential locations many years before cancer diagnosis are more relevant for cancer risk. Researchers in geography (Bentham, 1988; Han et al., 2004; Sabel et al., 2009) and public health (Jacquez et al., 2005; Paulu, Aschengrau, and Ozonoff, 2002; Vieira et al., 2005) have recognized the importance of population mobility when studying disease patterns. Ignoring migration when studying health outcomes with long latencies can lead to exposure misclassification, biased risk estimates, and diminished study power (Tong, 2000).

When analyzing disease risk in epidemiologic studies which include data on residential histories, one may use historic residential locations for individuals to study spatial risk over time. Residential histories can be informative for both the location and the timing of potential environmental exposures associated with residential locations, which is especially useful when the average latency for a cancer with suspected environmental causes is unknown. In such an analysis, one can adjust for known or suspected risk factors that are typically collected in epidemiologic studies and then examine the unexplained risk for spatial-temporal patterns (Kelsall and Diggle, 1998). Researchers have conducted this type of research using generalized additive models (GAMs) and residential histories for several cancers (Vieira et al., 2005; Webster et al., 2006; Wheeler et al., 2011). Researchers have either used all the residential locations available for each subject to estimate one risk surface without modeling lag times (Vieira et al., 2005; Webster et al., 2006), or estimate a spatial risk surface for one time period or lag time in a model for several time periods of interest, yielding a different risk surface for each model (Vieira et al., 2008; Wheeler et al., 2011; Wheeler, Ward, and Waller, 2012). Other work has modeled spatial risk surfaces at several different times together in one model using residential histories (Wheeler, Ward, and Waller, 2012). Such an approach can be thought of as providing an estimate of unmeasured lifecourse environmental exposures, which is in concordance with the increasing popular vision in epidemiology of the "exposome" that seeks to characterize the totality of environmental exposures for disease risk (Rappaport and Smith, 2010; Wild, 2005). The rationale is that relevant cumulative environmental exposures could occur over multiple residential locations across time and models should attempt to encompass such lifecourse environmental exposures. There is a recent growth in interest in the public health literature on modeling early-life and lifecourse exposures (Murray et al., 2011; Murray et al., 2012; Strand et al., 2010; Richardson et al., 2013).

The discussion in this chapter summarizes our work on modeling spatial variation in disease risk in epidemiologic studies. We use an analysis approach based on generalized additive models with bivariate smoothing functions to model residual spatial variation in disease risk after adjusting for known risk factors and potential confounders. We consider modeling residual spatial variation in risk at different times if residential histories are collected in an epidemiologic study or a disease registry. Using residential history data, we also explore modeling jointly several spatial risk components for different exposure times to allow for a more cumulative measure of spatial risk in a study area. We specify multi-component models with penalized thin plate regression splines, indicators for residential change, and weights for residential duration. We present a simulation study to evaluate the performance of different bivariate smoothing functions used in GAMs. We also present results from analyses of a population-based case-control study of non-Hodgkin lymphoma (NHL) incidence in Los Angeles County that includes residential histories.

Methods

Statistical Models

The main focus in this research is modeling the spatial variation in disease risk after adjusting for confounders and known risk factors. There are several important references for approaches to model the spatial variation in risk based on spatial point pattern theory (Diggle, 2003; Kelsall and Diggle, 1995; Kelsall and Diggle, 1998). Our discussion focuses on a generalized additive model approach to describe spatial variation in risk given the ease in adjusting for individual level covariates with this approach. Consider $i=1,\ldots,n$ subjects located within a region A, each with known residential location s_{it} at a particular lag time t and a binary label Y_i for disease status. Defining the spatial risk function as $r(s)$, equal to the probability that a person at location s will be a case, we can define the general spatial log-odds function as $l(s)=\log[r(s)/\{1-r(s)\}]$. We can model the log-odds of disease through either a crude generalized additive model as

$$\log[P(Y_i=1)/P(Y_i=0)]=\alpha+l_t(s_{it}), \quad (1)$$

or with an adjusted model

$$\log[P(Y_i=1)/P(Y_i=0)]=\alpha+X_i\beta+l_t(s_{it}), \quad (2)$$

which adjusts for covariates X_i parametrically with coefficients β and an intercept α. Note that the covariates could include known environmental exposures. Both the crude and adjusted models include a nonparametric term for the spatial log-odds using residential locations at time t. The spatial log-odds is a bivariate function of spatial coordinates and models residual spatial variation in risk. The residential locations at only one time are used in this model specification, but models with multiple spatial components are considered later.

In a typical study, the time would be the time of diagnosis of disease or time of study enrollment if the study is a case-control study. If spatial information for other times is available, through residential histories for example, then the spatial risk component at specific time lags can be evaluated. In this case, t could be the lag time of interest or a particular calendar year, and would typically be the same for all subjects. The residual odds ratio surface may be calculated through the spatial log-odds estimated at points in space and the mean of the spatial log-odds by $\exp[l(s)-\bar{l}]$, where the mean is taken over all the points in the study area. The spatial component estimated by the GAM can be visualized by mapping the log-odds or odds ratios, or by mapping it using a logarithmic base 2 scale where each unit increase corresponds to a doubling in risk (Diggle, 2003).

Two convenient forms of the spatial log-odds function are a locally weighted scatterplot smoother (LOESS; Cleveland and Devlin, 1988) and a thin plate regression spline (TPRS) smoother (Wood, 2006). Both types of smoothers have been used previously in GAMs to model spatial variation in risk, although LOESS has been used more (Vieira et al., 2005; Webster et al., 2006; Wheeler et al., 2011; Wheeler, Ward and Waller, 2012) than TPRS (Wheeler, Ward, and Waller, 2012). In a previous simulation study, a GAM with a bivariate LOESS smoother was found to have greater power and sensitivity to detect areas of elevated risk when compared with the popular local spatial scan statistic (Young et al., 2010). GAMs with a TPRS smoother have not previously been evaluated in a simulation study. There is a spatial span parameter to estimate in LOESS and several smoothing parameters to estimate in a thin plate regression spline. The span parameter in LOESS is typically found by minimizing the Akaike Information Criterion (AIC; Akaike, 1973) and the smoothing parameters in the thin plate regression spline are estimated by minimizing the unbiased risk estimator (UBRE; Craven and Wahba, 1979), which is effectively a linear transformation of the AIC (Wood, 2006). The LOESS smoother is available in a GAM in the R

(R Development Core Team, 2010) package gam (Hastie, 2013) and the thin plate regression spline smoother is available in a GAM in the R package mgcv (Wood, 2014).

The null hypothesis of primary interest in this type of analysis is that the risk is constant over space, that is, $r(s)=r$. One can evaluate the significance of the overall spatial variation in risk at a particular time for the LOESS or thin plate regression spline smoothers using a p-value from an analysis of deviance of nested models, where the test is approximately chi-square distributed (Hastie and Tibshirani, 1990). Two models that differ only in the inclusion of the spatial component are nested. If the spatial component added to the model significantly lowers the model deviance, then the spatial term is statistically significant. An issue, however, is that the overall p-values from analysis of deviance in GAMs have been found to be biased (Young et al., 2011). For the thin plate regression spline, another option is to use the p-value from an approximate chi-square test of the smoothing parameters being equal to zero, although the p-values are typically underestimated with penalized splines when the smoothing parameters are estimated (Wood, 2006).

A better approach to evaluate the significance of the spatial component in the GAM for both types of the smoothing function is to use a Monte Carlo permutation test. This test conditions on the number of cases and the locations of subjects at one time, performs randomization of the case labels among the locations, fits the model to the randomized data, and calculates the change in deviance with the spatial component in the model (Waller and Gotway, 2004). Typically, 999 randomizations are used to build the permutation distribution for the test. Random labeling of subjects as cases or controls is equivalent to the null hypothesis of constant risk over space (Diggle, 2003). Thus, the random labeling process is akin to repeatedly generating data from the null hypothesis. The p-value for the overall spatial term is calculated by dividing the rank for the observed data of the change in deviance among the permutation distribution by the number of randomizations. Statistical significance of the local log-odds or odds ratio can also be evaluated with the Monte Carlo permutation procedure. The local log-odds or odds ratio, calculated either at each data point or predicted for each cell of a grid overlaid on the study area, are regarded as statistically significant if they are outside the 2.5% and 97.5% ranked values from the local permutation distribution. When multiple locations for subjects are available over a range of time, this randomization procedure may be used to determine the temporality of the spatial component that is most associated with risk of disease (Wheeler et al., 2011; Wheeler, Ward, and Waller, 2012).

There are different versions of the Monte Carlo permutation test, most notably the unconditional and conditional permutation tests (Young et al., 2011). In an unconditional permutation test, the smoothing function parameters are estimated for each Monte Carlo randomization of the data, whereas in a conditional permutation test the smoothing function parameters estimated from the observed data are applied in the smoothing of the randomized data. In this way, the smoothing function for the randomized data is conditional on the smoothing function for the observed data. In a simulation study, the unconditional permutation test was found to be unbiased, but the conditional permutation test has been found to have inflated type I error (Young et al., 2011). The cost for less bias in this case is more computation time, as the unconditional permutation test requires considerably more computational time than does the conditional permutation test. In the R implementations of the smoothers, the unconditional permutation test is generally faster computationally with a thin plate regression spline than with a LOESS smoother.

Multiple Spatial Risk Components

In studies where several residential or occupational locations are collected for study subjects over a period, one may consider including several spatial components in a GAM to represent unmeasured spatial risk at several different times. The sum of the estimated spatial components would represent the cumulative unmeasured spatial risk over a set of time windows in the lifecourse. The rationale for a model with multiple spatial components is that relevant cumulative environmental exposures could occur over multiple residential locations across time and models should attempt to encompass lifecourse environmental exposures. There are several possible ways to specify a GAM with multiple spatial risk components for an epidemiologic study with residential histories. The simplest approach is to add a smoothing function for each time lag of interest in the model for log-odds of disease

$$\log[P(Y_i=1)/P(Y_i=0)]=\alpha+X_i\beta+l_1(s_{i1})+l_2(s_{i2})+\ldots+l_k(s_{ik}), \quad (3)$$

where l_2 is the bivariate smoothing function of residential locations at time 1, l_2 is the smoother for time 2, and l_k is the smoother for the kth time at diagnosis. The times could be expressed in calendar years or in years before diagnosis of disease. One issue with this model specification is potential collinearity effects. Given that many study subjects may not move over the time span of interest of a study, there could be potentially strong correlation between spatial log-odds surfaces at different times. We can address this issue by using for the smoothing function a thin plate regression spline with a ridge-type shrinkage penalty, which can shrink some terms to zero when components are strongly correlated (Wood, 2006). The ridge penalty effectively adds a small multiple of the identity matrix to the penalty matrix of the smooth.

Alternatively, we can specify a model that includes a spatial component for some base time of interest and then augments this spatial component with additional components based on residential change at select time windows. The base spatial component would have the residential locations for all subjects, and the additional components would have the residences that changed for subjects since the time for the previous spatial component in the model. We use an indicator variable z_{it} for each subject to indicate a change in spatial location at a fixed time. Specifically, z_{it} equals 1 if a subject i changed residence from the previous lag time and 0 otherwise. This is similar to a variable coefficient model (Hastie and Tibshirani, 1993), where smooths are multiplied by a known covariate. In this case, the smooth for time t is only applied to subjects who experienced a change in location between time t and the previously specified time in the model. Using the indicator variables, the log-odds model for the residential change model is

$$\log[P(Y_i=1)/P(Y_i=0)]=\alpha+X_i\beta+l_1(s_{i1})+l_2(s_{i2})z_{i2}+\ldots+l_k(s_{ik})z_{ik}, \quad (4)$$

where z_{i2} indicates a change in location from the base time to time 2 for subject i and z_{ik} indicates a location change between time $k-1$ and time k for subject i.

Another approach to modeling spatial variation in risk with several spatial components is to weight each spatial location by residential duration. Extending the residential change model, we can apply weights w_{it}, which equal the duration spent by each subject at the residential location at time lag t divided by the total time in the study area. The duration-weighted residential change model is

$$\log[P(Y_i=1)/P(Y_i=0)]=\alpha+X_i\beta+l_1(s_{i1})\,w_{i1}+l_2(s_{i2})z_{i2}w_{i2}+...+l_k(s_{ik})z_{ik}w_{ik},\quad(5)$$

where w_{i1} is the proportion of time spent at the base time residence for the ith subject, w_{i2} is the proportion of time spent at the location reported for time 2, and w_{ik} is the time spent at the location reported at time k. The weights are only applied to subjects for a spatial component after the baseline if the subject reported a change in location at the time of that spatial component. This particular model specification uses more of the information contained in an epidemiologic study with residential histories.

The significance of individual or groups of spline-smoothed lag functions can be evaluated with the approximately chi-square p-value from an analysis of deviance of nested models. For individual functions one may also use the approximately chi-square p-value from a test of the smoothing parameters equal to zero. However, confidence intervals for the linear predictor and model component functions from simulation studies show that it is easier to estimate the overall smoothing correctly than the smoothing parameters for individual model components (Wood, 2006). In addition, these approximate chi-square p-values are known to be biased (Wood, 2006; Young et al., 2011). In simulations, the p-values are often lower than the true values. Therefore, a permutation-based p-value is preferred to evaluate the overall and local significance of individual and groups of spatial log-odds components.

Different strategies are possible for the Monte Carlo randomization procedure to evaluate the significance of the overall spatial risk when several spatial components are included in the model. The Monte Carlo procedure for one spatial risk component conditions on the locations of subjects at one time and then randomizes case labels among these locations. The values for any covariates in the model are assigned concurrently with the case labels so that disease status and covariate values are paired as in the observed data. In this way, only the locations are randomized for subjects in the simulated data. For evaluating several spatial components, one possible approach is to condition on the known locations at each time and the migration rate at that time separately for each time point when assigning case labels to locations. An approach more consistent with the Monte Carlo routine for one spatial component is to assign the case label to the set of locations for each subject. This keeps the long-term path through space consistent for each subject and conditions on location and population mobility simultaneously. We use the second approach when evaluating the overall and local significance of the total spatial component of risk. The goodness-of-fit of models with different model specifications for spatial risk can be compared using the AIC.

Simulation Studies

We evaluated the ability of some of the modeling approaches described above to identify disease risk that varies spatially in two simulation studies. The motivation for this was to determine the absolute and relative accuracy of the models over a set of varied conditions of simulated data with known risks. The first study compared the LOESS smoother with a conditional permutation test with the thin plate regression spline with and without a shrinkage penalty. This study included the null hypothesis of constant disease risk and two realistic scenarios of alternative hypotheses to the null hypothesis. In a hypothetical square study region we defined areas of true elevated risk and set the levels of risk using the odds ratio of disease in these areas. We varied the shape, location, and level of risk of the true elevated risk areas in the scenarios. More specifically, we defined the true risk area as either

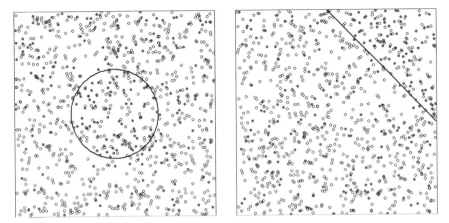

Figure 8.1   **Two simulation study scenarios and one realization from the alternative hypothesis for the first simulation study. The true risk area is either a circle or triangle with an odds ratio of 3.5 and a probability of disease of 0.2 outside the elevated risk area. Case (solid) and control (unfilled) locations are shown as circles for one realization.**

a circle in the middle of the study area (scenario 1) or a triangle in the upper right corner of the study area (scenario 2), and set the odds ratio inside the elevated risk area to increase from 1.0 to 3.5 in increments of 0.5. A rationale for setting the true risk area in the corner of the study area was to explore the ability of the models to detect risk on a boundary. A triangle is an appropriate shape for defining a risk area in the corner of a square study area, and a circle is a commonly used shape in disease cluster detection. The probability of disease outside the elevated risk area was set to be 0.2. The sample size was set to 1,000, and subject locations were generated randomly uniformly over the study area. For each risk scenario, we drew 500 realizations from the data-generating process and then estimated the model parameters for each realization. An example of one data realization is depicted for scenarios 1 and 2 in Figure 8.1.

In the second simulation study, we extended scenario 1 to multiple times by allowing for population mobility. To do so, we set the odds ratio = 3.5 in the elevated risk area at time 1, randomly generated locations for subjects at time 1, and then allowed subjects to move randomly from their locations at subsequent times 2, 3, and 4 at systematic levels of probability of 0.00, 0.25, 0.50, 0.75, 1.00 at each time. For this simulation study, we used a sample size of 1000 and 100 data realizations. We compared the performance of the simple cumulative spatial component model (equation 3) with a thin plate regression spline with and without a shrinkage penalty. A motivation for this comparison was to determine if using the shrinkage penalty would affect performance at this level of population mobility. An example of one data realization for this simulation study is depicted in Figure 8.2.

To evaluate the performance of the models in the simulation study, we used several metrics that assess either overall error or spatial accuracy. To investigate the overall performance of the models to evaluate the null hypothesis of constant risk we calculated the type I error and the power to reject the null hypothesis. We also calculated the approximate chi-square p-value and Monte Carlo p-value for the test of the overall null hypothesis. We calculated the performance measures by combining the values across the data realizations.

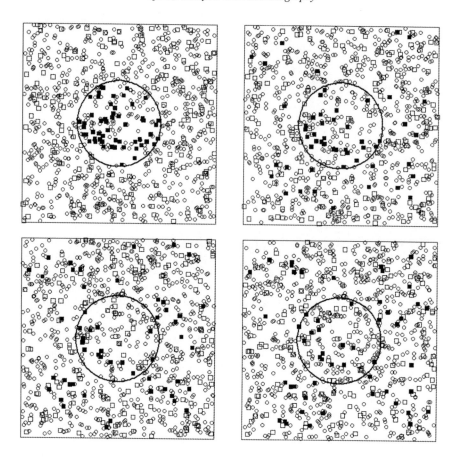

Figure 8.2   **One realization from the alternative hypothesis for the risk scenario in the second simulation study. Data were generated with a true elevated risk area at time 1 with an odds ratio of 3.5, a probability of disease of 0.2 outside the circle, and a 0.5 probability of moving at three later times: time 2 (upper right), 3 (lower left) and 4 (lower right). Case (unfilled square) and control (unfilled circle) locations are shown as circles. Cases located in the true risk area at time 1 are shaded black**

Due to the finding of inflated type I error with the conditional permutation test by Young et al. (2011), we used an alpha = 0.025 for this test. The overall measures have been used in other model assessment studies in the context of modeling disease risk at single time points (Waller, Hill, and Rudd, 2006; Young et al., 2010; Young et al., 2011). To evaluate the performance of the models to detect local areas of significant risk, we calculated the detection rate of the area of elevated risk and quantified the spatial accuracy of detection of the high-risk area using the measures of sensitivity and false positive rate. We defined detection as the proportion of datasets where the true area of significant risk is identified. Sensitivity was the proportion of the true significant risk area detected. The false positive rate was the proportion of true non-significant risk areas detected as a significant risk area. Similar performance measures have been used previously in disease cluster detection studies (Aamodt, Samuelsen, and Skrondal, 2006; Ozonoff et al., 2007; Waller, Hill, and Rudd, 2006). The rationale for using these performance metrics in this series of experiments was that it is important to know how frequently a model will falsely detect an area of elevated risk as well as how frequently it will correctly detect a true area of elevated risk. If a model performs well, we can confidently apply it in epidemiologic studies of disease.

Results

Simulation Studies

Results for the first simulation study show that the GAM models with individual spatial risk components performed well when true risk was highly elevated inside either the circular area centered in the study area or the triangular area in a corner of the study area (Table 8.1). The type I error rate was approximately the nominal level for all three spatial smoothing functions in both scenarios. The power to detect significant spatial variation in risk was very high with the highest true risk level (OR = 3.5), and consistently high across the methods for OR = 3.0. Power was generally higher for LOESS than for the TPRS smoothers. Within methods, LOESS had higher power in scenario 2 than in scenario 1, while TPRS had higher power in scenario 1 than in scenario 2. The models were better at detecting the true elevated risk area than detecting overall spatial variation in risk, which is not surprising given that the design of the simulation study was based on defining one area of true elevated risk. Probability of detection exceeded 0.9 for all models when OR was 2.5 or higher, and was still relatively high for OR = 2.0. The probability of detection was overall higher for LOESS than for the TPRS smoothers. In scenario 1, the sensitivity to detect the area of the true elevated risk was higher for TPRS than for LOESS for OR of 2.0 or greater. In scenario 2, TPRS had a larger sensitivity than did LOESS for OR of 2.5 or greater. The false positive rate was higher for TPRS than for LOESS with all elevated risks in scenario 1. However, the false positive rate for TPRS was lower than for LOESS in scenario 2 for lower risks (2.5 and below for TPRS without shrinkage and 2.0 for TPRS with shrinkage). The p-values behaved as expected, with very small p-values for highest true risk and increasing p-values with decreasing true risk. The approximate chi-square p-values were consistently lower than the Monte Carlo p-values. Comparing methods, the Monte Carlo p-values were consistently lower for LOESS than for TPRS smoothers.

Results from the second simulation study show that the power to detect spatial variation in risk varied according to the level of population mobility when using several risk components (Table 8.2). Power was very high (0.97) for both TPRS smoothers when there was no population mobility. Power decreased substantially with population mobility between 0.25–0.75 and then increased with complete population mobility. The overall trend of increasing power with increasing population mobility is possibly due to having less correlated spatial components in the model. Sensitivity was also very high when there was no population mobility, but then decreased systematically with increasing population mobility. The Monte Carlo p-values all suggested evidence for the alternative hypothesis, although the smallest p-values were associated with no population mobility. Results show that the models had less success at detecting overall spatial variation in risk compared with identifying the elevated risk area. Overall, results were very similar for the TPRS smoothers with and shrinkage.

Table 8.1	**Results from the first simulation study**

Measures	OR	Single circular cluster			Single triangular cluster		
		CPT	TPRS	TPRS-S	CPT	TPRS	TPRS-S
Type I error rate	1.0	0.044	0.048	0.048	0.044	0.052	0.054
Power	3.5	0.980	0.960	0.960	0.998	0.944	0.944
	3.0	0.920	0.848	0.848	0.98	0.836	0.834
	2.5	0.754	0.66	0.662	0.900	0.604	0.602
	2.0	0.430	0.336	0.340	0.644	0.324	0.326
	1.5	0.154	0.118	0.116	0.220	0.08	0.080
Detection	3.5	0.994	1.000	1.000	1.000	1.000	1.000
	3.0	0.994	0.994	0.994	0.994	0.994	0.99
	2.5	0.946	0.920	0.920	0.966	0.970	0.966
	2.0	0.786	0.738	0.734	0.818	0.808	0.782
	1.5	0.372	0.300	0.296	0.432	0.366	0.350
	1.0	0.072	0.070	0.070	0.058	0.052	0.064
Sensitivity[a]	3.5	0.861	0.973	0.973	0.937	0.974	0.978
	3.0	0.832	0.940	0.941	0.922	0.945	0.950
	2.5	0.741	0.815	0.817	0.863	0.880	0.884
	2.0	0.538	0.565	0.565	0.676	0.665	0.657
	1.5	0.215	0.196	0.197	0.333	0.264	0.251
	1.0	0.060	0.050	0.051	0.059	0.051	0.054
False positive rate[a]	3.5	0.043	0.117	0.125	0.063	0.077	0.083
	3.0	0.045	0.106	0.114	0.064	0.068	0.074
	2.5	0.042	0.082	0.090	0.059	0.056	0.061
	2.0	0.038	0.059	0.065	0.048	0.044	0.046
	1.5	0.025	0.032	0.035	0.030	0.024	0.024
	1.0	0.056	0.046	0.047	0.059	0.049	0.049
Chi-square p-value[a]	3.5	0.001	0.001	0.001	0.000	0.000	0.000
	3.0	0.006	0.006	0.005	0.002	0.002	0.002
	2.5	0.028	0.036	0.019	0.009	0.008	0.007
	2.0	0.088	0.097	0.043	0.050	0.048	0.024
	1.5	0.236	0.256	0.073	0.197	0.197	0.053
	1.0	0.328	0.351	0.065	0.326	0.349	0.064
Monte Carlo p-value[a]	3.5	0.002	0.008	0.008	0.000	0.009	0.010
	3.0	0.010	0.024	0.023	0.003	0.025	0.025
	2.5	0.044	0.074	0.074	0.014	0.058	0.059
	2.0	0.128	0.167	0.172	0.070	0.143	0.143
	1.5	0.319	0.367	0.369	0.260	0.343	0.344
	1.0	0.429	0.486	0.489	0.426	0.488	0.488

Note: Three smoothers in generalized additive models: LOESS with conditional permutation test (CPT), thin plate regression spline without shrinkage (TPRS) and with shrinkage (TPRS-S).
[a] Average values were used.

Table 8.2   Results from the second simulation study allowing for population mobility

	TPRS					TPRS-S				
Population mobility	0.00	0.25	0.50	0.75	1.00	0.00	0.25	0.50	0.75	1.00
Power	0.970	0.670	0.690	0.690	0.750	0.970	0.670	0.670	0.680	0.740
Sensitivity[a]	0.967	0.898	0.867	0.848	0.814	0.967	0.888	0.866	0.854	0.814
Chi-square p-value[a]	0.001	0.004	0.001	0.000	0.001	0.001	0.004	0.000	0.000	0.004
Monte Carlo p-value[a]	0.008	0.055	0.047	0.038	0.040	0.008	0.057	0.046	0.039	0.048

Note: Two smoothers in GAMs: Thin plate regression spline without shrinkage (TPRS) and with shrinkage (TPRS-S).

[a] Average values were used.

Data Analysis

As an example of modeling the spatial variation in disease risk at different times, we present some results from an analysis of non-Hodgkin lymphoma (NHL) risk in a case-control study. The National Cancer Institute (NCI)-Surveillance, Epidemiology and End Results (SEER) NHL study is a case-control study of 1,321 cases aged 20–74 years that were diagnosed between July 1, 1998 and June 30, 2000 in four SEER cancer registries, including Detroit, Iowa, Seattle, and Los Angeles County. The study has been described previously (Chatterjee et al., 2004; Morton et al., 2008; Wheeler et al., 2011). Briefly, population controls (1,057) were selected from residents of the SEER areas using random digit dialing (<65 years of age) or Medicare eligibility files (65 and over) and were frequency matched to cases by age (within 5-year groups), sex, race, and SEER area. Among eligible subjects contacted for an interview, 76% of cases and 52% of controls participated in the study. Cases and controls with a history of NHL or known HIV infection were not included in the study. The goal of the NCI-SEER NHL study was to investigate potential environmental and genetic risk factors for NHL.

Computer-assisted personal interviews were conducted during a visit to each subject's home to obtain lifetime residential and occupational histories, medical history, and other information including date of birth, gender, race, education, and pest treatments including home treatment for termites before 1988 (a surrogate for the insecticide chlordane). Written informed consent was obtained during the home visit and human subjects review boards approved the study at the NCI and at all participating institutions. Historic addresses were collected in a residential history section of an interviewer-administered questionnaire. Participants were mailed a residential calendar in advance of the interview and were requested to provide the complete address of every home in which they lived from birth to the current year, listing the years they moved in and out (De Roos et al., 2010). Interviewers reviewed the residential calendar with respondents and probed to obtain missing address information. Residential addresses were matched to geographic address databases to yield geographic coordinates that were used in this analysis.

We have restricted the focus of this analysis to the Los Angeles County study area, which was previously found to have the most significant unexplained spatial risk of NHL (Wheeler et al., 2011). To maximize the power to detect local variation in spatial risk of NHL that could be related to environmental exposures within the study center, we limited the

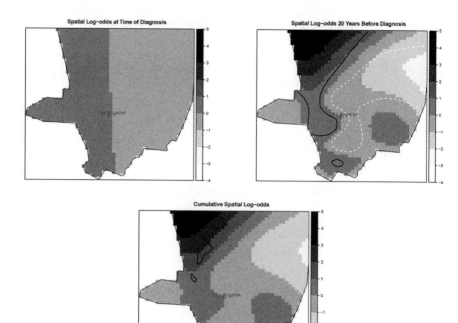

Figure 8.3   Spatial log-odds estimated at time of diagnosis and at 20 years before diagnosis for models with individual spatial risk components and cumulative spatial log-odds estimated for a model with individual spatial risk components at time of diagnosis and 5, 10, 15, and 20 years before diagnosis. Statistically significant local areas of elevated (lowered) risk are outlined in black solid (white dashed) lines

analysis to subjects residing in the study area for at least 20 years prior to study enrollment. There were 190 cases and 161 controls for analysis in the Los Angeles study area using this selection criterion. In previous analyses, we investigated the significance of unexplained spatial risk among the potentially etiologically relevant latencies for NHL of 5, 10, 15, and 20 years before diagnosis, as well as at time of diagnosis. Using a GAM with a LOESS function and a conditional permutation test (Wheeler et al., 2011) and a GAM with a thin plate regression spline without a shrinkage penalty (Wheeler, Ward, and Waller, 2012), we found that the most significant time lag for Los Angeles was 20 years. In the analysis presented here, we estimated models with individual (equation 2) or cumulative spatial risk components (equation 3) using a thin plate regression spline with a shrinkage penalty. The models with individual spatial risk components were based on a time lag of 20 years and, as a comparison, time at diagnosis. The cumulative spatial risk component was based on time lags of 5, 10, 15, and 20 years and time of diagnosis. In all models, we adjusted for age at study enrollment, gender, race, education level, and a surrogate for exposure to the insecticide chlordane. We evaluated the statistical significance of the spatial risk components using Monte Carlo permutation testing.

Monte Carlo permutation testing found no significant spatial variation in NHL risk at time of diagnosis, but significant spatial variation in risk at a lag of 20 years. The Monte Carlo p-value for the overall spatial variation in risk for the model with one spatial risk component for time of diagnosis was 0.792. The corresponding p-value for the model with one spatial component at a time lag of 20 years was 0.026. The Monte Carlo p-value for the cumulative spatial risk was 0.170. The estimated spatial log-odds components in the individual spatial component models exhibit very different patterns (Figure 8.3). No areas of statistically significant risk were identified at the time of diagnosis. In contrast, several statistically significant risk areas were detected at a time lag of 20 years. The most prominent area of concern is the elevated risk area west of Los Angeles that extends into the northwest portion of the study area. This area contained 34 cases and 10 controls at a lag of 20 years. The cumulative spatial log-odds pattern resembles the spatial log-odds pattern at a lag of 20 years (Figure 8.3). However, the cumulative spatial risk detects a much smaller significant area of elevated risk, composed of several small areas of elevated risk northwest of Los Angeles. These significantly elevated areas do spatially overlap with the larger significantly elevated area detected in the model with a spatial risk component at lag 20 years.

Discussion and Conclusions

In the analysis of non-Hodgkin risk, we found statistically significant areas of elevated risk only when a time lag of 20 years was considered. No significant results were found when using the residential locations at the time of diagnosis for cases and study enrollment for controls. The address at time of diagnosis is typically used in spatial analyses of disease risk. For diseases with long latencies, such as most cancers, using the residential location at time of diagnosis can miss important spatial signals. The cumulative spatial risk model detected a spatial signal in risk, but it was slightly attenuated compared with the risk surface estimated in the model with only the spatial component for a time lag of 20 years (Figure 8.3). This is not surprising considering that the spatial components for the time of diagnosis and early lags were not significant. For a disease with a long latency and environmental risk factors, adding spatial components for times close to diagnosis to the spatial component for a long latency could be effectively adding noise to the signal, particularly when there is relatively high population mobility.

The first simulation study showed that the GAM models with individual spatial risk components performed well when true risk was highly elevated inside either a circular area centered in the study area or a triangular area in a corner of the study area. The sensitivity was generally higher for the thin plate regression spline smoothers than for LOESS when the true risk area was circular, however, sensitivity was similar when the true risk area was triangular. The false positive rate was generally higher for the TPRS smoothers with the circular risk area, but was generally slightly higher for LOESS with the triangular risk area. The power to detect overall spatial variation in risk was higher with LOESS in both true risk scenarios when risk was elevated. The power to detect the true risk area was relatively high when the true risk area had an odds ratio of 2.0 or greater and was similar among the LOESS and TPRS smoothers. The first simulation study results tend to overall favor LOESS over TPRS. The simulation study results for cumulative spatial risk showed no meaningful difference in performance between the TPRS smoothers with and without a shrinkage penalty. Both smoothers had good sensitivity overall and similar power to detect overall spatial variation in risk. The smoothers behaved similarly with varying levels of population mobility. The first and second simulation study results demonstrate that in these scenarios the methods are more successful at detecting a local area of elevated risk than identifying overall spatial variation in risk.

Though the simulation study results are encouraging, there are limitations to this research. The ability to detect an area of true elevated risk depends on the assumption that study participation among eligible cases and controls is nondifferential over space. If there was differential participation in the study among cases and controls by space and time, then this differential participation could lead to the false detection of an area of elevated risk where there was different participation. It is possible to check for nondifferential study participation at the time of study enrollment if the residential addresses at time of enrollment for all eligible study participants are known (Shen et al., 2008; Wheeler et al., 2011). However, it is typically not possible to evaluate differential spatial participation among cases and controls at times years prior to study enrollment, as these addresses are usually unknown. In addition, we have not considered time-varying covariates in our models. A benefit of the flexibility of the generalized additive modeling framework is that time-dependent exposures, if known, could be included in a GAM that contains a spatial risk component. The generalized additive model approach is a convenient method to adjust for covariates while exploring unexplained spatial variation in disease risk.

References

Aamodt, G., Samuelsen, S.O., and Skrondal, A., 2006. A simulation study of three methods for detecting disease clusters. *International Journal of Health Geographics*, 5: 15.

Akaike, H., 1973. Information theory and an extension of the maximum likelihood principle, in *International Symposium on Information Theory*, edited by B. Petran and F. Csaaki. Budapest: 267–81.

Alexander, F., 1993. Viruses, clusters, and clustering of childhood leukemia: A new perspective? *European Journal of Cancer*, 29A: 1424–43.

Beale, L., Hodgson, S., Abellan, J., LeFevre, S., and Jarup, L., 2010. Evaluation of spatial relationships between health and the environment: The rapid inquiry facility. *Environmental Health Perspectives*, 118: 1306–12.

Bentham, G., 1988. Migration and morbidity: Implications for geographical studies of disease. *Social Science & Medicine*, 26: 49–54.

Bithell, J.F. and Vincent, T.J., 2000. Geographical variations in childhood leukemia incidence, in *Spatial Epidemiology: Methods and Applications*, edited by P. Elliot, J.C. Wakefield, N.G. Best and D.J. Briggs. New York: Oxford University Press. 317–32.

Blot, W.J., Fraumeni, J.F. Jr., Mason, T.J., and Hoover, R.N., 1979. Developing clues to environmental cancer: A stepwise approach with the use of cancer mortality data. *Environmental Health Perspectives*, 32: 53–8.

Chatterjee, N., Hartge, P., Cerhan, J.R., Cozen, W., Davis, S., Ishibe, N., Colt, J., Goldin, L., and Severson, R.K., 2004. Risk of non-Hodgkin's lymphoma and family history of lymphatic, hematologic, and other cancers. *Cancer Epidemiology, Biomarkers & Prevention*, 13(9): 1415–21.

Cleveland, W.S. and Devlin, S., 1988. Locally-weighted regression: An approach to regression analysis by local fitting. *Journal of the American Statistical Association*, 83: 596–610.

Craven, P. and Wahba, G., 1979. Smoothing noisy data with spline functions. *Numerische Mathematik*, 31: 377–403.

De Roos A.J., Davis, S., Colt J.S., Blair, A., Airola, M., Severson, R.K., Cozen, W., Cerhan, J.R., Hartge, P., Nuckols, J.R., and Ward, M.H., 2010. Residential proximity to industrial facilities and risk of non-Hodgkin lymphoma. *Environmental Research*, 110: 70–8.

Devesa, S., Blot, W., and Fraumeni, J. Jr., 1998. Changing patterns in the incidence of esophageal and gastric carcinoma in the United States. *Cancer Epidemiology, Biomarkers & Prevention*, 83: 2049–53.

Devesa, S., Grauman, D., Blot, W., and Fraumeni, J. Jr., 1999. Cancer surveillance series: Changing geographic patterns of lung cancer mortality in the United States, 1950 through 1994. *Journal of the National Cancer Institute*, 91: 1040–50.

Diggle, P.J., 2003. *Statistical Analysis of Spatial Point Patterns*. 2nd ed. London: Edward Arnold.

Elliott, P. and Savitz, D., 2008. Design issues in small-area studies of environment and health. *Environmental Health Perspectives*, 116: 1098–104.

Fraumeni, J. Jr. and Blot, W., 1977. Geographic variation in esophageal cancer mortality in the United States. *Journal of Chronic Diseases*, 30: 759–67.

Han, D., Rogerson, P.A., Nie, J., Bonner, M.R., Vena, J.E., Vito, D., Muti, P., Trevisan, M., Edge, S.B., and Freudenheim, J.L., 2004. Geographic clustering of residence in early life and subsequent risk of breast cancer (United States). *Cancer Causes and Control*, 15: 921–9.

Hastie, T., 2013. gam: Generalized additive models (1.09). [computer program]. Available at: http://cran.r-project.org/web/packages/gam/index.html [Accessed March 15, 2014].

Hastie, T. and Tibshirani, R., 1990. *Generalized Additive Models*. London: Chapman & Hall.

Jacquez, G., Kaufmann, A., Meliker, J., Goovaerts, P., AvRuskin, G., and Nriagu, J., 2005. Global, local and focused geographic clustering for case-control data with residential histories. *Environmental Health*, 4.

Jacquez, G.M., Meliker, J.R., AvRuskin, G.A., Goovaerts, P., Kaufmann, A., Wilson, M.L., and Nriagu, J., 2006. Case-control geographic clustering for residential histories accounting for risk factors and covariates. *International Journal of Health Geographics*. 5: 32.

Kelsall, J. and Diggle, P., 1995. Non-parametric estimation of spatial variation in relative risk. *Statistics in Medicine*, 14: 2335–42.

Kelsall, J. and Diggle, P., 1998. Spatial variation in risk of disease: A nonparametric binary regression approach. *Applied Statistics*, 47: 559–73.

Kulldorff, M., 1997. A spatial scan statistic. *Communications in Statistics: Theory and Methods*, 26: 1487–96.

Kulldorff, M., 2006. SaTScan: Software for the spatial and space-time scan statistics. [computer program] Information Management Services, Inc. Available at: http://www.satscan.org/ [Accessed March 15, 2014].

Linet, M.S., Devesa, S.S., and Morgan, G.J., 2006. The leukemias, in *Cancer Epidemiology and Prevention*, edited by D. Schottenfeld and J.F. Fraumeni Jr. 3rd ed. New York: Oxford University Press, 841–71.

Morton, L.M., Wang, S.S., Cozen, W., Linet, M.S., Chatterjee, N., Davis, S., Severson, R.K., Colt, J.S., Vasef, M.A., Rothman, N., Blair, A., Bernstein, L.,Cross, A.J., De Roos, A.J., Engels, E.A., Hein, D.W., Hill, D.A., Kelemen, L.E., Lim, U., Lynch, C.F., Schenk, M., Wacholder, S., Ward, M.H., Zahm, S.H., Chanock, S.J., Cerhan, J.R., and Hartge, P., 2008. Etiologic heterogeneity among non-Hodgkin lymphoma subtypes. *Blood*, 112(13): 5150–60.

Murray, E.T., Mishra, G.D., Kuh, D., Guralnik, J., Black, S., and Hardy, R., 2011. Life course models of socioeconomic position and cardiovascular risk factors: 1946 birth cohort. *Annals of Epidemiology*, 21(8): 589–97.

Murray, E.T., Southall, H., Aucott, P., Tilling, K., Kuh, D., Hardy, R., and Ben-Shlomo, Y., 2012. Challenges in examining area effects across the life course on physical capability in mid-life: Findings from the 1946 British Birth Cohort. *Health & Place*, 18: 366–74.

Ozonoff, A., Jeffery, C., Manjourides, J., White, L., and Pagano, M., 2007. Effect of spatial resolution on cluster detection: A simulation study. *International Journal of Health Geographics*, 6:52.

Paulu, C., Aschengrau, A., and Ozonoff, D., 2002. Exploring associations between residential location and breast cancer incidence in a case-control study. *Environmental Health Perspectives*, 110: 471–8.

R Development Core Team, 2010. R: A language and environment for statistical computing. [computer program] R Foundation for Statistical Computing. Available at: http:// www.r-project.org/ [Accessed March 15, 2014].

Rappaport, S.M. and Smith, M.T., 2010. Environment and disease risks. *Science*, 330: 460–61.

Richardson, D.B., Volkow, N.D., Kwan, M., Kaplan, R.M., Goodchild, M.F., and Croyle, R.T., 2013. Spatial turn in health research. *Science*, 339: 1390–92.

Sabel, C.E., Boyle, P.J., Raab, G., Loytonen, M., and Maasilta, P., 2009. Modelling individual space-time exposure opportunities: A novel approach to unravelling the genetic or environmental disease causation debate. *Spatial and Spatio-temporal Epidemiology*, 1: 85–94.

Shen, M., Cozen, W., Huang, L., Colt, J., De Roos A.J., Severson, R.K., Cerhan, J.R., Bernstein, L., Morton, L.M., Pickle, L., and Ward, M.H., 2008 Census and geographic differences between respondents and nonrespondents in a case-control study of non-Hodgkin lymphoma. *American Journal of Epidemiology*, 167: 350–61.

Silverman, D.T., Devesa, S.S., Moore, L.E., and Rothman, N., 2006. Bladder cancer, in *Cancer Epidemiology and Prevention*, 3rd ed, edited by D. Schottenfeld and J.F. Fraumeni Jr. New York: Oxford University Press, 1101–27.

Spitz, M.R., Wu, X., Wilkinson, A., and Wei, Q., 2006. Cancer of the lung, in D. Schottenfeld and J.F. Fraumeni Jr, eds. 2006. *Cancer Epidemiology and Prevention*, 3rd ed. New York: Oxford University Press, 638–58.

Strand, B.H., Murray, E.T., Guralnik, J., Hardy, R., and Kuh, D., 2010. Childhood social class and adult adiposity and blood-pressure trajectories 36–53 years: Gender-specific results from a British birth cohort. *Journal of Epidemiology & Community Health*, 66(6): 512–18.

Tong, S., 2000. Migration bias in ecologic studies. *European Journal of Epidemiology*, 16: 365–9.

Vieira, V., Webster, T., Weinberg, J., and Aschengrau, A., 2008. Spatial-temporal analysis of breast cancer in upper Cape Cod, Massachusetts. *International Journal of Health Geographics*, 7:46.

Vieira, V., Webster, T., Weinberg, J., Aschengrau, A., and Ozonoff, D., 2005. Spatial analysis of lung, colorectal, and breast cancer on Cape Cod: An application of generalized additive models to case-control data. *Environmental Health*, 4.

Wakefield, J. and Elliott, P., 1999. Issues in the statistical analysis of small area health data. *Statistics in Medicine*, 18: 2377–99.

Waller, L.A. and Gotway, C.A., 2004. *Applied Spatial Statistics for Public Health Data.* New York: John Wiley.

Waller, L.A., Hill, E.G., and Rudd, R.A., 2006. The geography of power: Statistical performance of tests of clusters and clustering in heterogeneous populations. *Statistics in Medicine*, 25: 853–65.

Webster, T., Vieira, V., Weinberg, J., and Aschengrau, A., 2006. Method for mapping population-based case-controls studies: An application using generalized additive models. *International Journal of Health Geographics*, 5.

Wheeler, D.C., 2007. A comparison of spatial clustering and cluster detection techniques for childhood leukemia incidence in Ohio, 1996–2003. *International Journal of Health Geographics*. 6:13.

Wheeler, D.C., De Roos, A.J., Cerhan, J.R., Morton, L.M., Severson, R., Cozen, W., and Ward, M.H., 2011. Spatial-temporal cluster analysis of non-Hodgkin lymphoma in the NCI-SEER NHL Study. *Environmental Health*, 10: 63.

Wheeler, D.C., Waller, L.A., Cozen, W., and Ward, M.H., 2012. Spatial-temporal analysis of non-Hodgkin lymphoma risk using multiple residential locations. *Spatial and Spatio-temporal Epidemiology*, 3(2): 163–71.

Wheeler, D.C., Ward, M.H., and Waller, L.A., 2012 Spatial-temporal analysis of cancer risk in epidemiologic studies with residential histories. *Annals of the Association of American Geographers*, 102(5): 1049–57.

Wild, C.P., 2005. Complementing the genome with an "exposome": The outstanding challenge of environmental exposure measurement in molecular epidemiology. *Cancer Epidemiology, Biomarkers & Prevention*, 14(8): 1847–50.

Winn, D.M., Blot, W.J., Shy, C.M., Pickle, L.W., Toledo, A., and Fraumeni, J.F. Jr., 1981. Snuff dipping and oral cancer among women in the southern United States. *New England Journal of Medicine*, 304: 745–9.

Wood, S.N., 2006. *Generalized Additive Models: An Introduction with R.* Boca Raton, FL: Chapman & Hall/CRC.

Wood, S.N., 2014. mgcv: Mixed GAM computation vehicle with GCV/AIC/REML smoothness estimation (1.7–28). Available at: http://cran.rproject.org/web/packages/mgcv/index.html [Accessed March 15, 2014].

Young, R.L., Weinberg, J., Vieira, V., Ozonoff, A., and Webster, T.F., 2010. A power comparison of generalized additive models and the spatial scan statistic in a case-control setting. *International Journal of Health Geographics*, 9: 37.

Young, R.L., Weinberg, J., Vieira, V., Ozonoff, A., and Webster, T.F., 2011. Generalized additive models and inflated type I error rates of smoother significance tests. *Computational Statistics and Data Analysis*, 55: 366–74.

Chapter 9
The Spatial Epidemiology of Mental Well-being in Dhaka's Slums

Oliver Gruebner, Mobarak Hossain Khan, Sven Lautenbach,
Daniel Müller, Alexander Krämer, Tobia Lakes, Patrick Hostert,
and Sandro Galea

The majority of the world's population now lives in urban areas (United Nations, 2012). Mental health problems are known to be prevalent in the rapidly urbanizing megacities of low-income countries and range at the third place in the list of the ten leading factors of the burden of disease in low-income countries in a projection for 2030 (Mathers and Loncar, 2006). Urbanization is most pronounced in Asia having the largest number of megacities, that is, cities with more than 10 million inhabitants worldwide (United Nations, 2012). There are several characteristics of rapidly urbanizing megacities of low-income countries that can contribute to poor mental health, including, social segregation, lack of infrastructure, and exposure to ongoing adversity and life stressors (Douglas, 2012). However, there is a paucity of data about the relation between characteristics of urban environments and mental health (United Nations, 2012; Riley et al., 2007).

Using a spatial epidemiological approach, this chapter assessed factors linked to mental well-being in the slums of Dhaka, the second fastest growing megacity in the world, which currently accommodates an estimated population of more than 15 million (United Nations, 2012), including at least 3.4 million slum dwellers (Angeles et al., 2009). Furthermore, we investigated the spatial variability of mental well-being for different population groups in several slums of Dhaka.

Specifically, we hypothesized (i) that socio-ecological environmental characteristics of informal settlements are associated with the mental well-being of slum dwellers after adjusting for personal factors such as age, gender and other diseases. We further investigated the hypotheses (ii) that mental well-being shows a significant spatial pattern (that is, spatial clustering) for different population groups, and (iii) that spatially auto-correlated socio-ecological factors relate to the spatial patterns of mental well-being.

Methodology

Sample

In 2005, the Centre for Urban Studies (CUS) together with MEASURE Evaluation identified approximately 4,900 slum clusters in Dhaka (Angeles et al., 2009). Building on this database in 2009, we first established minimum threshold values of 500 households and six acres per slum to select comparable slum settlements from the CUS survey.

Second, in order to achieve an adequate geographical distribution of the slum settlements, we selected administrative units that were non-adjacent to each other. In units with more than one slum, we randomly selected one of these settlements. We adapted our selection to account for slums in which the residents were evicted, or those converted into affluent residential areas or open spaces since the CUS survey in 2005. This way we identified nine urban slums, which can be assessed in Gruebner et al. (2011a).

Third, we collected data from 1,938 adults in these nine slums (male = 48% and female = 52%) aged 15–99 years after excluding some respondents < 15 years of age suitable for our analyses. Trained Bangladeshi university graduates performed face-to-face interviews and applied global positioning system (GPS) devices to record the location of each household interviewed. Please refer to Gruebner et al. (2012) for details on the sampling strategy of the respondents within the slums.

Fourth, we assessed additional ecological information for the surveyed households through geo-processing in GIS, detailed elsewhere (Gruebner et al., 2012; Gruebner, 2011). Briefly, we calculated vegetation and water coverage in 100 m buffers around GPS-located households from the cross-sectional health survey. In addition, distances from the survey-household coordinates to the nearest river, street and park gathered from GoogleMaps™ were calculated. Euclidean distances to the nearest major river, street and designated park areas were measured and categorized according to whether they were reachable within a walking distance of 1 km. Distance and area calculations were performed in the geographic information system ArcGIS version 9.3.1 (ESRI 2010).

Fifth, having produced the ecological variables in GIS, we merged this data to the health survey dataset and used principal component analysis (PCA) in SPSS (Version 17) to reduce our merged variable set to a smaller number of uncorrelated linear combinations of variables that contain most of the variance (Jolliffe, 2002). The principal components were based on the correlation matrix. Please again refer to Gruebner et al. for details (2012).

Variables

Explanatory variables
We used 14 principal components (PCs) from 41 variables found in Gruebner et al. (2012), which we subsequently used as covariates in the multivariable generalized linear regression analysis (cf. Table 9.1, Figure 9.1). The PCs represent the socio-ecological environment and individual health knowledge and behavior; they explained 59.5% of the variance in the data, ranging from 6.3% (housing quality) to 3.4% (personal health knowledge).

Health outcomes
We used the WHO-5 Well-being Index as a measure to assess self-rated mental well-being. The WHO-5 is a quick, reliable, and valid measure for assessing psychological well-being (Barua and Kar, 2010; Bonsignore et al., 2001; Delaney et al., 2009; Newnham et al., 2010; Bech et al., 2003). The WHO-5 assesses the indicators of depression by five questions rated on a 6-point forced-choice Likert scale (Likert, 1932), from 0 to 5. The rates were summed to a range from 0 to 25. Within that range, a raw score of < 13 suggested poor well-being. More details on the WHO-5 questionnaire can be found in (WHO Collaborating Centre for Mental Health Frederiksborg General Hospital, 2010). The WHO-5 has been successfully applied in both high-icnome (Henkel et al., 2003; Awata et al., 2007; Kessing et al., 2006; Liwowsky et al., 2009) and low-income countries (Barua and Kar, 2010; Momtaza et al., 2011; Saipanish et al., 2009). Although the WHO-5 was not yet validated in Bangladesh, it was found reliable and effective among elderly Indian communities (Barua and Kar, 2010), which are socio-economically similar to Bangladeshi communities.

Table 9.1   Explanatory variables used for this study

Level	Socio-ecological factor (explained variance)	Original variable (Pearson correlation coefficients)
Neighborhood **Physical environment**	Natural environment (4.3% percent)	○ Vegetation ratio 100 m around household (0.8) ○ Surface water ratio 100 m around household (-0.6) ○ Distance to nearest street (0.7)
	Flood non-affectedness (4.1%)	○ Distance to nearest river (0.5) ○ Is your area flood-affected? (0.7) ○ Does your area have a proper drainage system? (0.7)
Household	*Physical environment*	
	Housing quality (6.3%)	○ Monthly rent for the house (0.6) ○ Family has household item: gas burner (0.8) ○ Cooking material (0.9) ○ Housing construction material (0.5)
	Basic services (4.7%)	○ Distance to nearest park (-0.8) ○ Distance to nearest river (0.5) ○ Family has household item: electric fan (0.6) ○ Type of water supply (0.5)
	Sanitation (3.6%)	○ Type of toilet facility (0.7) ○ Type of garbage disposal (0.6)
	Housing sufficiency (3.6%)	○ Light sufficiency in the house (0.6) ○ Room is used also for purposes other than living (0.7) ○ Room is sufficient for family (0.5)
	Housing durability (3.5%)	○ Family has household item: refrigerator (0.7) ○ Is your house provisional or permanent? (0.8)
	Socio-economic environment	
	Household wealth (4.3%)	○ Family has household item: radio (0.6), TV (0.6), tape/CD/VCD (0.7) ○ How many rooms do you have? (0.5)
	Job satisfaction (4%)	○ Working hours per day (-0.4) ○ Do you think your job is harmful for your health? (0.8) ○ Do you like your job? (0.7)

(continued ...)

Table 9.1 *(concluded)*

Level		Socio-ecological factor (explained variance)	Original variable (Pearson correlation coefficients)
		Income generation (3.7%)	○ Do you have a job contract? (0.4) ○ Family members earning income (0.7) ○ Monthly family income (0.7) ○ Working hours per day (0.2)
	Social environment	Population density (5.2%)	○ Family size (0.8) ○ Persons sharing same meals (0.7) ○ Persons living in the same room (0.7)
Individual		Smoking behavior (4.8%)	○ Do you smoke cigarettes? (0.8) ○ Do you smoke inside your room? (0.8) ○ How many family members smoke? (-0.7)
		Environmental health knowledge (3.9%)	○ Do you think that: – polluted/logged water/garbage near the house spread disease and increase the risk of poor health? (0.8) – air pollution is bad for your health? (0.6)
		Personal health knowledge (3.4%)	○ Do you think that: – smoking tobacco is bad for your health? (0.7) – physical exercise is good for your health? (0.7)
		Community member Using bed net Education Married Migrant Age Gender Disease	Original variables were used

Note: These variables were obtained through the prior factor analysis. Details are available in Gruebner et al. (2012). In the table, we report the name of each socio-ecological factor and, in brackets, the explained variance. The original variables from health survey and geo-processing that were found to be correlated with these socio-ecological factors are also displayed, with Pearson correlation coefficients in brackets.

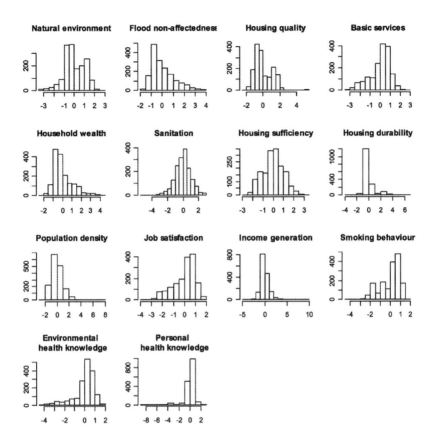

Figure 9.1   Descriptive statistics for the 14 socio-ecological variables used in this study

Note: On the x-axis the factor's value is depicted and on the y-axis the frequency.

Analytic Methods

Multivariable generalized linear regression analysis
Associations between a set of 22 independent variables (that is, socio-ecological variables, Table 9.1) and the health outcome mental well-being (WHO-5) were studied using generalized linear regression models. Since the response can be modeled as overdispersed count data, the assumption of a negative binomial distribution was applied.

Multivariable regression analysis was applied using the function "glm.nb" available within the MASS packet (Venables and Ripley, 2002) in the statistical programming environment R (R Development Core Team, 2013). We used the "stepAIC" algorithm, a stepwise regression with both backward and forward selection according to the Akaike Information Criterion (Akaike, 1974) in order to identify the best model for the relationship between mental well-being and the explanatory variables.

Spatial statistics for epidemiological studies
Spatial autocorrelation analysis was applied to summarize the degree to which persons with a similar health status tend to occur proximate to each other (that is, form spatial clusters) (Waller and Gotway, 2004). Multivariable spatial correlation analysis was used to gain information on the extent to which values for the well-being of one person observed at a given location show a systematic (more than likely under spatial randomness) association with another variable observed at the "neighboring" locations.

Neighborhood relationships
Spatial autocorrelation statistics depend on the definition of neighborhood relationships through which the spatial configuration of the sampled subpopulation was defined prior to analysis. Since the choice of the neighborhood relationship can influence the results (Schabenberger and Gotway, 2005), we explored various neighborhood definitions. Subsequently, we selected the neighborhood relationship based on Moran's I (see below). First, we used 30-, 60- and 90-meter fixed-band definitions, which treat every observation point inside that search radius as a neighbor. Second, we used the k-neighbor approach, which treats the nearest k observations as neighbors. We used k values of 3, 5, and 10. For both approaches, we used a binary weight matrix to assign weights to the neighbors. This binary weight matrix assigns a weight of unity for neighbors and zero for non-neighbors. All spatial analyses were performed in GeoDa (Anselin, 2003).

Global univariate spatial autocorrelation
We applied Moran's I (Moran, 1950; Moran, 1948) to account for the global spatial autocorrelation of similar and dissimilar WHO-5 scores of the nine slums of Dhaka. For the Moran's I statistic, the sum of covariations between the sites for the distance $d(i,j)$ was divided by the overall number of sites $W(i,j)$ within the distance class $d(i,j)$ (Fortin and Dale, 2005). Thus, the spatial autocorrelation coefficient for a distance class $d(i,j)$ was the average value of spatial autocorrelation at that distance. The actual value for Moran's I was then compared with the expected value under the assumption of complete randomization. Moran's I values may range from -1 (dispersed) to $+1$ (clustered). Under specific circumstances however, values outside of this range are possible, but these values are interpreted as if they are at the boundary of the range $[-1,1]$. A Moran's I value of 0 suggests complete spatial randomness. To verify that the value of Moran's I was significantly different from the expected value, we applied a Monte Carlo randomization test with 9,999 permutations. Data values were thereby reassigned among the N locations, providing a randomized distribution against which the observed value was judged. If the observed value of I was within the tails of this distribution, the assumption of independence among the observations could be rejected which can be seen as an indicator for the presence of significant spatial autocorrelation in the data (Cliff and Ord, 1981). For each combination of population group and slum, the neighborhood relationship with the highest Moran's I value was used for further analysis. We then selected for the mental health outcomes for each population group and chose from the nine slums the neighborhood relationship with the highest Moran's I for further bivariate analysis.

Global bivariate spatial correlation
Multivariable spatial correlation focuses on the extent to which values for one variable yk observed at a given location show an association with another variable yo observed at the neighboring locations (Anselin et al., 2002). This yields to the multivariable counterpart of a Moran-like spatial autocorrelation statistic, the bivariate coefficient of spatial correlation

(ibid.). We tested the 21 socio-ecological factors for spatial correlations with mental health. The significance was assessed, as in the univariate case, by means of a randomization approach (Anselin, 1995).

Local univariate spatial autocorrelation
To identify local patterns of spatial autocorrelation, we calculated the local univariate Moran's *I* for WHO-5 scores and for socio-ecological factors using the identified best neighborhood (as described above) and respective population group. We concentrated on the Anselin Local Moran's *I* statistic (Anselin, 1995; Schabenberger and Gotway, 2005). Unlike the global Moran's *I*, which has the same expected value for the entire study area, the expected value of local Moran's *I* varies for each sampling location because it is calculated in relation to its particular set of neighbors (Anselin, 1995). We calculated the significance of the local Moran's *I* using a randomization test on the Z-score with 9,999 permutations (Fortin and Dale, 2005). Positive spatial autocorrelation occurs when, for example, WHO-5 scores of residents living in one location are surrounded by similar WHO-5 scores of other residents in neighboring locations (low-low–LL, high-high–HH), thus forming a spatial cluster. Negative spatial autocorrelation appears when high WHO-5 scores are surrounded by low WHO-5 scores (HL) and vice versa (LH), that is, when spatial outliers occur (Anselin, 1995).

Local bivariate spatial correlation
In a next step, we calculated the bivariate local Moran's *I* for the WHO-5 scores for the best neighborhood and population group. This statistic provides an indication of the degree of linear association (positive or negative) between the values for one variable at a given location, and the average of another variable at neighboring locations (Anselin et al., 2002). A greater than indicated similarity under spatial randomness suggests a spatially similar cluster in the two variables. A dissimilarity greater than spatial randomness would imply a strong, local, negative relationship between the two variables (ibid.). The significance of the statistic was assessed by means of the permutation approach.

Results

Health Outcomes

Good mental well-being (WHO-5 scores ≥13) was found in 21% of women and in 25% of men as well as in 25% of the poorest (lower household wealth quintile) and 26% of the least poor (upper household wealth quintile) population groups. Slum dwellers rated their health mainly as "so-so" (56%), whereas 27% of the poorest and 30% of the least poor population groups reported good or excellent health status.

Socio-ecological Factors Determining Mental Well-being

We identified several factors from the socio-ecological environment and found that they had significant associations with mental well-being (Table 9.2). Furthermore, personal determinants like gender, age, and disease were significantly associated with mental well-being. We found a positive association between environmental health knowledge and mental well-being while controlling for gender. Furthermore, lower flood risk (flood non-

Table 9.2   Determinants of good mental well-being (WHO-5)

Level	Socio-ecological factor	Multivariable generalized linear regression	
		Coefficient	**95% CI LL / UL**
Neighborhood Physical environment	Natural environment	−0.06***	−0.08 / −0.03
	Flood non-affectedness	0.06***	0.04 / 0.09
Household · *Physical environment*	Housing quality	0.03*	0.01 / 0.06
	Basic services	---	---
	Sanitation	0.08***	0.06 / 0.11
	Housing sufficiency	0.07***	0.04 / 0.09
	Housing durability	0.07***	0.05 / 0.09
Socio-economic environment	Household wealth	---	---
	Job satisfaction	0.09***	0.06 / 0.11
	Income generation	0.08***	0.06 / 0.11
Social environment	Population density	−0.05***	−0.07 / −0.02
Individual	Smoking behavior	---	---
	Environmental health knowledge	0.11***	0.08 / 0.13
	Personal health knowledge	−0.03*	−0.05 / −0.004
	Community member	0.07	−0.02 / 0.15
	Using bed net	---	---
	Education	---	---
	Married	---	---
	Migrant	0.06	−0.02 / 0.15
	Age	−0.01***	−0.01 / −0.004
	Gender:		
	Female	Reference	
	Male	0.11***	0.06 / 0.16
	Having had a disease:		
	No	Reference	
	Yes	−0.22***	−0.28 / −0.16

Significance codes: <0.001 '***,' < 0.01 '**,' < 0.05 '*,' >0.05 '.' not in model '---'
CI: Confidence interval with LL= Lower limit and UL= Upper limit

Note: A multivariable generalized linear regression model was used assuming a negative-binomial distribution of the response variable, mental well-being (WHO-5 score). The explanatory variables (socio-ecological factor) are displayed in the table. A forward/backward model selection approach was used based on AIC. Note that, despite conceptualizing the socio-ecological factor at multiple levels, all analyses were performed at the individual level.

affectedness), better sanitation, better income generation ability, job satisfaction, and higher quality, sufficiency, and durability of housing were positively associated with mental well-being. A strong negative association was found for respondents who had suffered from any disease in the three months prior to the survey. Older age, better personal health knowledge, higher population density, and "natural environment," a regression factor including the amount of vegetation and water around households, as well as the distances to streets (cf. Table 9.1) were further found to be negatively associated with well-being. The model could explain 20% of the deviance.

Looking at the marginal effects, we found that the variables "natural environment," personal health knowledge, population density, age, and disease had a negative association with mental well-being, while the others had a positive marginal effect. It could be seen that income generation had the strongest marginal effect on mental well-being, when all other predictors are kept at their average value, followed by housing durability (cf. Figure 9.2, 9.3 and 9.4 below).

Spatial Patterns of Mental Well-being

We found the strongest global spatial clustering when the three nearest (sampled) neighbors were considered in the analysis (mean distances ranging from 9 to 11.4 meters) (Table 9.3). Beguntila and Bishil/Sarag were among the settlements with the highest values (cf. Table 9.4). In Beguntila, spatial clustering of well-being was most significant (p<0.001) among the young adult age group. In Abdullapur East and Kunipara, we found significant spatial clustering of WHO-5 scores only when using a large neighborhood of the 10 nearest neighbors of respondents (mean distances ranging from 35 to 57 meters). Using lower neighborhood definitions, that is, the 3 and 5 nearest neighbors revealed no significant spatial clusters. In addition, we found that within slums and within population groups, the strength and significance of spatial autocorrelation differed with the type of neighborhood relation (Gruebner and Khan et al., 2011a). For example, spatial autocorrelation among males in Bishil/Sarag decreased when more neighbors or longer distances were considered in the analysis; the same was true among young adults in Beguntila (Table 9.4 below). Therefore, the global univariate Moran's *I* of the response variable (WHO-5 scores) reflect the spatial variation at the scale of the settlements. Focusing on socio-ecological environmental factors with the global bivariate Moran's *I*, revealed a similar spatial pattern at the scale of the settlements.

Spatial Patterns of Socio-ecologic factors and Mental Well-being

Within the young adult age group in Beguntila, spatial clustering of housing quality and male gender was positively associated with spatial clustering of good well-being (cf. Table 9.4). In Bishil/Sarag, the strongest and most significant (p<0.001) global spatial clustering was detected among males. These clusters were positively associated with housing quality, sanitation, and environmental health knowledge; however, they (and those in Beguntila among young adults) were also negatively associated with selected features of the natural environment (as found in the multivariable regression model). To some extent, we also found that spatial clustering of flood non-affectedness, housing quality, and education were positively associated with spatial clustering of mental well-being among females in Bishil/Sarag and negatively with age. Within Bishil/Sarag, we also found spatial clustering of income generation to be correlated with spatial patterns of mental well-being among young adults.

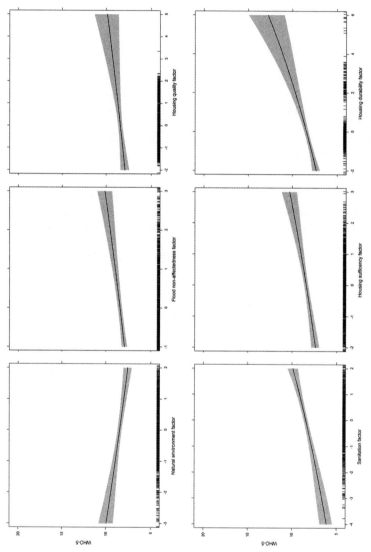

Figure 9.2   Marginal effects for one predictor in the multivariable generalized linear regression model in Table 9.1, while keeping all others at a constant value

Note: For the continuous variables, the mean has been taken. The effects are shown at the response scale, that is, the link function has already been inversed. In addition, the 95% confidence bands are shown, which provide an overview about the uncertainty behind the estimated marginal effect.

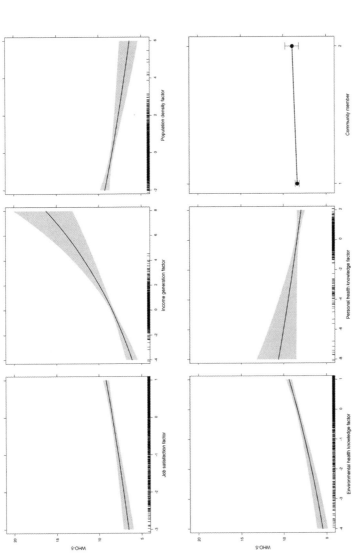

Figure 9.3 Marginal effects for one predictor in the multivariable generalized linear regression model in Table 9.1, while keeping all others at a constant value

Note: For continuous variables, the mean has been taken while for categorical variables the percentages of variables, which belong to a factor level have been taken. The effects are shown at the response scale, that is, the link function has already been inversed. In addition, the 95% confidence bands are shown, which provide an overview about the uncertainty behind the estimated marginal effect.

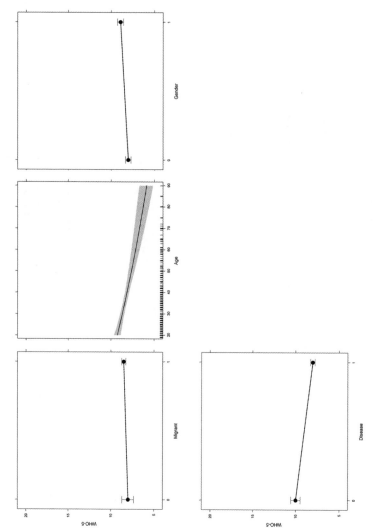

Figure 9.4   Marginal effects for one predictor in the multivariable generalized linear regression model in Table 9.1, while keeping all others at a constant value

Note: For continuous variables, the mean has been taken while for categorical variables the percentages of variables, which belong to a factor level have been taken. The effects are shown at the response scale, that is, the link function has already been inversed. In addition, the 95%confidence bands are shown, which provide an overview about the uncertainty behind the estimated marginal effect.

Table 9.3   Global univariate Moran's *I* values for different neighborhood relationships

Neighborhood	Beguntila	Bishil/Sarag				Abdullapur East	Kunipara	Adabar	Buhiapara
Relationship	Young adults	Males	Females	Young adults	Middle aged adults	Females	Middle aged adults	Young adults	Total sample
Nearest neighbors									
3 nn	0.16* *(8.5)*	0.19** *(11.4)*	0.12* *(10.8)*	0.12* *(9)*	.	.	.	.	.
5 nn	0.16** *(10.4)*	0.17** *(14.7)*	.	0.1* *(11.6)*	.	.	.	.	.
10 nn	0.13** *(14.2)*	0.01*** *(20.5)*	.	.	.	0.06* *(35.2)*	0.09* *(57)*	.	.
Fixed distance									
30 m	0.1***	.	.	.	.	.	.	.	.
60 m	0.05**	0.13***	.	0.05**	.	.	.	.	.
90 m	0.12**	0.12**	.	0.03*	0.09**	.	.	0.02*	0.02*

Significance levels: <0.001 '***', <0.01 '***', <0.05 '*', >0.05 '.'.

Note: Global Moran's *I* values for those slums and population groups, which were significant under a Monte Carlo test with 9,999 permutations (p<0.05). We only report positive Moran values, that is, those revealing global spatial clustering. For nearest neighbor-based distances, we report in parentheses the average distance-per-slum in meters. Note that the strongest values occur with three nearest neighbors. We thus used this neighborhood relationship in the subsequent bivariate Moran's *I* analysis.

Table 9.4   Global bivariate Moran's *I* values for the three nearest neighbors

Level	Socio-ecological factor	Beguntila Young adults n=115	Bishil/Sarag Young adults n=170	Females n=104	Males n=122
Global univariate Moran's *I* for WHO-5 scores		0.16*	0.12*	0.12*	0.19**
Neighborhood *Physical environment*	"Natural environment"	−0.19***	−0.16***	.	−0.21**
	Flood non-affectedness	.	0.13**	0.13*	.
Household *Physical environment*	Housing quality	0.13*	0.19***	0.14**	0.3***
	Basic services	.	.	.	.
	Household wealth	.	.	.	.
	Sanitation	.	.	.	0.18***
	Housing sufficiency	.	.	.	.
	Housing durability	.	.	.	.
Social and socio-economic environment	Population density	.	.	.	.
	Job satisfaction	.	.	.	.
	Income generation	.	0.1*	.	.
Individual	Smoking behavior	.	.	.	.
	Environmental HK	.	0.09*	.	0.13*
	Personal HK	.	.	.	.
	Community member	.	.	.	.
	Using bed net	.	.	.	.
	Education	.	.	0.11*	.
	Married	.	.	.	.
	Migrant	.	.	.	.
	Age	---	---	−0.12*	.
	Gender	0.12*	.	---	---

Significance levels: <0.001 '***,' < 0.01 '**,' < 0.05 '*,' >0.05 '.' not applicable '---,'
HK: Health knowledge

Note: The table displays socio-ecological factors that are significantly ($p<0.05$) spatially correlated with mental well-being (WHO-5 scores) of those population groups and slums in which strongest global spatial clustering of WHO-5 scores were found (cf. Table 9.4). Note that the WHO-5 scores among men in Bishil/Sarag are clustered most strongly, and that there is a strong negative spatial correlation with "natural environment" and a positive one with housing quality in this population group, which will be questioned in the discussion section

Since global spatial correlation was strongest and most significant in Bishil/Sarag, we concentrate on this slum in the following analysis. We found that for this settlement, well-being among men was spatially structured in a western and an eastern part, with poor well-being localized predominantly in the western area and good well-being in the eastern part of the settlement (cf. Figure 9.5). Furthermore, the local bivariate Moran's *I* statistic revealed that low WHO-5 scores were associated with poor housing quality in the western part of Bishil/Sarag, whereas high WHO-5 scores and better housing quality were clustered in the east. "Natural environment" was found to be negatively correlated with well-being: in the western area, a higher amount of "natural environment" could be found, together with poor well-being clusters. In contrast to the spatial clustering of mental well-being among men, both patterns of good and poor well-being appeared among women in the western part of Bishil/Sarag.

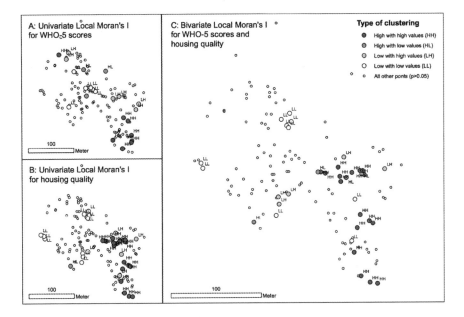

Figure 9.5   Mental well-being (WHO-5 scores) and housing quality of men in the slum settlement Bishil/Sarag

Note: Each dot on the map indicates a slum household (GPS point). The maps indicate significant (p<0.05) spatial clusters of high (HH) or low (LL) WHO-5 scores (A), housing quality (B), or similar values of both WHO-5 scores and housing quality (C), respectively. High values surrounded by low values (HL) and vice versa (LH) indicate outliers. The three nearest neighbors of a household were used in the statistics.

Discussion

There were three main findings from our study. First, several factors from the socio-ecological environmental were significantly associated with mental well-being in the slums of Dhaka. Second, we identified spatial patterns of mental well-being for different population groups. Third, we provided evidence that spatial patterns of the socio-ecological environment are associated with the spatial patterns of mental well-being.

Socio-ecological Environmental Factors and Mental Well-being

Individual level characteristics
We found that mental well-being was unequally distributed among the population, and younger, male, and more affluent dwellers enjoyed better mental well-being. This is in line with prior work (Rahman and Barsky, 2003; Izutsu et al., 2006; Ompad et al., 2007). We found that environmental health knowledge was positively associated with mental well-being, which may reflect a person's awareness of environmental threats. According to our understanding, such knowledge may trigger protective measures and eventual adaptation strategies of the local residents but one could also argue that people with more robust mental health are more likely to have better health knowledge. We further observed a negative

relationship between personal health knowledge and mental well-being, which may reflect a person's awareness of the effects of personal sedentary lifestyles and other activities that can cause poor health, such as smoking or less physical exercise. One explanation for this relationship could be that a higher awareness of health issues might have caused a tendency to be dissatisfied with the overall poor living conditions.

The built environment
We assessed several socio-ecological factors at the household level, relating in particular to the built environment. Unfavorable housing quality is thereby assumed to cause poor health by provoking asthma and other respiratory conditions, injuries, or psychological distress or by hindering child development (Vlahov et al., 2007). Good sanitation (that is, garbage disposal and the quality of the toilet facility) can decrease the risk of infectious diseases and other ailments, such as gastrointestinal diseases or respiratory diseases (ibid.). In accordance with these relationships, good sanitation was positively associated with mental well-being in the slums of Dhaka. Furthermore, the quality, sufficiency, and durability of housing were found to be positively associated with mental well-being. Each of these predictors, which represent the built environment could capture the socio-economic status (SES) of an individual or household that is well known to be associated with mental health (Galea et al., 2007; Ompad et al., 2007; Aneshensel, 2009). These SES predictors can define the frame of action within which a household can respond to health threats (Villagrán De León, 2006; Gruebner, Staffeld et al., 2011b). Hence, these SES factors could also be conceptualized as belonging to the socio-economic environment. In any case, these SES factors may shape the intrinsic ability of an individual or household to resist or cope with the impact of a possible physical or social event (Villagrán De León, 2006) and were therefore crucial determinants of mental well-being in our study.

The natural environment
The rapid urban expansion of Dhaka has facilitated a huge loss of prime agricultural areas and wetlands (Griffiths et al., 2010; Byomkesh et al., 2011), which are generally known to provide important provisioning and regulating Ecosystem Goods and Services (ESS) (Millennium Ecosystem Assessment 2005b), which support health in a variety of ways (Millennium Ecosystem Assessment 2005a). In Dhaka, for instance, water retention areas have been increasingly lost due to the widespread practice of earth in filling during ground construction. The loss of ESS regulation, combined with poor infrastructural planning, has thus led to deteriorating living conditions and increased environmental risks, particularly the risk of flooding (Caldwell, 2004). As slum dwellers in Dhaka are highly vulnerable to flood events (Braun and Aßheuer, 2011; Aßheuer et al., 2013), it is consequently quite understandable that not being affected by flooding was found to be positively associated with mental well-being in our study.

Having large areas of vegetation in the nearby neighborhood often increases the health-related quality of life, for example, by reducing heat stress induced through a local urban heat island effect (Bowler et al., 2010; Burkart and Endlicher, 2011; Uejio et al., 2011). Furthermore, urban green and park areas are typically considered to be recreational facilities for urban residents (Galea et al., 2005; Alberti, 2009). In Dhaka's slums, vegetation cover is scarce, and we therefore assumed a strongly positive association between nearby green areas and mental well-being. However, many of those areas that we had expected to improve living conditions and thus improve mental health turned out to be low-lying and regularly flooded areas. Combined with poor sanitation, open waste water drainage, and garbage disposal, such vegetation patches increase the risk for infectious disease (for example, diarrhea or worm infections) (Sclar et al., 2005). Our analysis thus identified ecosystem disservices (Lyytimaki and Sipila, 2009), which were negatively associated with mental health.

The socio-economic environment

Our study revealed a positive association of income generation and job satisfaction with mental well-being, describing the ability to generate income as well as satisfaction and safety at work. More than 80% of Dhaka's adult slum dwellers are engaged in the informal economy, which provides a means of survival for a substantial section of the workforce (Staffeld and Kulke, 2011; Kulke and Staffeld, 2009). The informal economy is often associated with unfavorable environments with regard to working and living conditions, pollutants, discrimination, exploitation, income, occupational safety, and legal and social security (Barten et al., 2008; Gruebner and Staffeld et al., 2011b). Against this background, it becomes clear why good income generation and job satisfaction showed up as important predictors for good mental well-being among Dhaka's slum dwellers.

The social environment

Population density was also an important factor for mental well-being in our study. We assume that in the slums of Dhaka, crowding put enormous stress on residents with consequent implications for mental health, possibly due to a lack of privacy. Other studies showed that social norms in densely populated urban areas may further support individual or group behaviors that affect health outcomes (for example, smoking, diet, exercise, or sexual behavior) (Galea et al., 2005).

Spatial Patterns of Mental Well-being

For the neighborhood defined by the three nearest (sampled) neighbors, we provided evidence that spatially clustered households/respondents contrast strongly with other households/respondents in different clusters with regard to their mental well-being. Hence, it can be stated that there are substantial health inequalities in slums. The investigated patterns showed a spatial dependence of mental well-being at one location on the mental well-being of a neighboring location (that is, they were spatially auto-correlated). A possible explanation is that good mental health in a neighborhood may support salutary effects in the social fabric, and vice versa (Gee and Payne-Sturges, 2004). In our study, the strength of spatial clustering decreased with increasing neighborhood size—regardless if the neighborhood was defined based on nearest neighbors or with a distance based approach. We demonstrated this by the mental well-being among males in Bishil/Sarag, but the pattern could also be revealed by the socio-ecological factors for this population group. This spatial pattern could also be verified among other population groups within the same slum and in Beguntila. Overall, it can be stated that such spatial patterns point to small-scale effects within the slums, indicating that autocorrelation effects and the spatial effects of socio-ecological factors take place at short distances. Moreover, our results provide evidence that model outcomes are sensitive to different definitions of neighborhood relation. Considering that spatial patterns of health status uncover health disparities and provide the basis for further analysis, our study helps to determine the feasibility of using a particular statistical method to avoid violating the assumption of data independence that underlies most non-spatial statistical approaches (Waller and Gotway, 2004; Schabenberger and Gotway, 2005; Bivand et al., 2008).

Spatial Patterns of Socio-ecological Factors and Mental Well-being

We were able to show that the association between individual characteristics such as age or gender and mental well-being in slums was often indirect and could be heavily influenced by other factors such as neighborhood socio-ecological characteristics. These factors are believed to shape the distinct vulnerabilities and the resilience of residents towards ill or good health. Many of those socio-ecological factors, which in the multivariable regression analysis were found to be associated with mental well-being, remained significant in the spatial bivariate analysis taking into consideration each slum separately. The example of Bishil/Sarag showed that the spatial distribution of for example higher values of "environmental disservices," such as a higher share of (as we presume polluted) green spaces and flood affectedness, as well as poor housing quality and poor sanitation were correlated with the spatial distribution of poor mental well-being and vice versa. Thus, we did not only provide evidence that certain socio-ecological factors in the same households were associated with the mental well-being of slum residents, but that socio-ecological factors prevalent in the immediate neighborhood of a respondent's home was also associated with mental well-being. In other words, neighborhood clusters, which constitute significantly better or poorer well-being in comparison to the other neighborhoods of a slum, also comprise similar socio-ecological factors that are related to their respective health status.

Conclusions

Knowledge of the spatial distribution and structure of one's health status may help us to understand a community's social fabric and its related mental health-determining factors, but most importantly, it allows for a more efficient and effective spatial allocation of scarce resources to target the alleviation of poverty and the improvement of living standards. Because our methodology provides evidence for spatial dependencies in epidemiological data, it might lead to more sophisticated spatial -epidemiological models that create a deeper understanding of functional relationships between the socio-ecological environment and mental health. Spatial epidemiological models could thus lead to improved rationales for public health interventions and might strengthen policy significance.

Given that mental health is related to physical health, our chapter provides insights for developing better health care and disease prevention programs in urban slum settlements of low-income countries.

References

Akaike, H., 1974. A new look at the statistical model identification. *IEEE Transactions on Automatic Control*, 19(6): 716–23.

Alberti, M., 2009. *Advances in Urban Ecology: Integrating Humans and Ecological Processes in Urban Ecosystems*, New York: Springer.

Aneshensel, C.S., 2009. Toward Explaining Mental Health Disparities. *Journal of Health and Social Behavior*, 50(4): 377–94.

Angeles, G., Lance, P., Barden-O'Fallon, J., Islam, N., Mahbub, AQM., and Nazem, N.I., 2009. The 2005 census and mapping of slums in Bangladesh: Design, select results and application. *International Journal of Health Geographics*, 8: 32.

Anselin, L., 2003. *GeoDa 0.9 User's Guide* 0 ed., Urbana-Champaign, Illinois, USA: Spatial Analysis Laboratory, University of Illinois.

Anselin, L., 1995. Local Indicators of Spatial Association—LISA. *Geographical Analysis*, 27(2): 93–115.

Anselin, L., Syabri, I., and Smirnov, O., 2002. Visualizing multivariate spatial correlation with dynamically linked windows L. Anselin and S. Rey, eds. *New Tools for Spatial Data Analysis: Proceedings of the Specialist Meeting*.

Aßheuer, T., Thiele-Eich, I., and Braun, B., 2013. Coping with the impacts of severe flood events in Dhaka's slums—the role of social capital. *Erdkunde*, 67(1): 21–35.

Awata, S. Bech, P. Yoshida, S. Suzuki, S. Yamashita, M. Ohara, A., Hinokio, Y., Matsuoka, and H., Oka, Y., 2007. Reliability and validity of the Japanese version of the World Health Organization-Five Well-Being Index in the context of detecting depression in diabetic patients. *Psychiatry and Clinical Neurosciences*, 61: 112–19.

Barten, F. et al., 2008. Contextualising workers health and safety in urban settings: The need for a global perspective and an integrated approach. *Habitat International*, 32(2): 223.

Barua, A. and Kar, N., 2010. Screening for depression in elderly Indian population. *Indian J Psychiatry*, 52(2): 150–53.

Bech, P., Olsen, L.R., Kjoller, M., and Rasmussen, N.K., 2003. Measuring well-being rather than the absence of distress symptoms: A comparison of the SF-36 Mental Health subscale and the WHO-Five well-being scale. *International Journal of Methods in Psychiatric Research*, 12(2): 85–91.

Bivand, R.S., Pebesma, E.J., and Gomez-Rubio, V., 2008. *Applied Spatial Data Analysis with R*, New York: Springer.

Bonsignore, M., Barkow, K., Jessen, F., and Heun, R., 2001. Validity of the five-item WHO Well-Being Index (WHO-5) in an elderly population. *European Archives of Psychiatry and Clinical Neuroscience.*, 251(Suppl 2): II27–31.

Bowler, D.E., Buyung-Ali, L., Knight, T.M., and Pullin, A.S., 2010. Urban greening to cool towns and cities: A systematic review of the empirical evidence. *Landscape and Urban Planning*, 97(3): 147–55.

Braun, B. and Aßheuer, T., 2011. Floods in megacity environments: Vulnerability and coping strategies of slum dwellers in Dhaka/Bangladesh. *Natural Hazards*: 1–17.

Burkart, K. and Endlicher, W., 2011. Human Bioclimate and Thermal Stress in the Megacity of Dhaka, Bangladesh: Application and Evaluation of Thermophysiological Indices Health in Megacities and Urban Areas, in A. Krämer, M.H. Khan, and F. Kraas, eds. *Health in Megacities and Urban Areas*. New York: Springer,: 153–70.

Byomkesh, T., Nakagoshi, N., and Dewan, A., 2011. Urbanization and green space dynamics in Greater Dhaka, Bangladesh. *Landscape and Ecological Engineering*: 1–14.

Caldwell, B., 2004. Global Environmental Change, Urbanization and Health. The case of rapidly growing Dhaka. *IHDP update (Magazine of the International Human Dimensions Programme on Global Environmental Change)*, 1:8–9.

Centre for Urban Studies, National Institute of Population Research and TrainingMEASURE Evaluation, 2006. *Slums of Urban Bangladesh: Mapping and Census, 2005*. Dhaka, Bangladesh and Chapel Hill, USA.

Cliff, A.D. and Ord, J.K., 1981. *Spatial Processes: Models and Applications*. London: Pion Limited.

Delaney, L., Doyle, O., McKenzie, K., and Wall, P. (2009). The distribution of wellbeing in Ireland. *Irish Journal of Psychological Medicine*, 26: 119–26 doi:10.1017/S0790966700000409

Douglas, I., 2012. Urban ecology and urban ecosystems: Understanding the links to human health and well-being. *Current Opinion in Environmental Sustainability*, 4(4): 385–92.

Elliott, P., Wakefield, J., Best, N., and Briggs, D., 2006. *Spatial Epidemiology: Methods and Applications*. Oxford: Oxford University Press.

ESRI (Environmental Systems Research Institute), 2010. ArcGIS 9.3.1.

Fortin, M-J. and Dale, M., 2005. *Spatial Analysis. A Guide for Ecologists*. Cambridge: Cambridge University Press.

Galea, S., Ahern, J., Nandi, A., Tracy, M., Beard, J., and Vlahov, D., 2007. Urban neighborhood poverty and the incidence of depression in a population-based cohort study. *Annals of Epidemiology*, 17(3): 171–9.

Galea, S., Freudenberg, N., and Vlahov, D., 2005. Cities and population health. *Social Science & Medicine*, 60(5): 1017–33.

Gee, G.C. and Payne-Sturges, D.C., 2004. Environmental Health Disparities: A Framework Integrating Psychosocial and Environmental Concepts. *Environ Health Perspect*, 112(17).

Griffiths, P., Hostert, P., Gruebner, O., and van der Linden, S., 2010. Mapping megacity growth with multi-sensor data. *Remote Sensing of Environment*, 114(2): 426–39.

Gruebner, O., 2011. *A Spatial Epidemiological Approach on Well-being in Urban Slums—Evidence from Dhaka, Bangladesh*. Berlin: Humboldt-Universität zu Berlin.

Gruebner, O., Khan, MMH., Lautenbach, S., Müller, D. Krämer, A., Lakes, T., and Hostert, P., 2012. Mental health in the slums of Dhaka—a geoepidemiological study. *BMC Public Health*, 12(1): 177. Available at: http://eutils.ncbi.nlm.nih.gov/entrez/eutils/elink.fcgi?dbfrom=pubmed&id= 22404959&retmode=ref&cmd=prlinks.

Gruebner, O., Khan, MMH., Lautenbach, S., Müller, D. Krämer, A., Lakes, T., and Hostert, P., 2011a. A spatial epidemiological analysis of self-rated mental health in the slums of Dhaka. *International Journal of Health Geographics*, 10(1): 36. Available at: http://eutils.ncbi.nlm.nih.gov/entrez/eutils/elink.fcgi?dbfrom=pubmed&id= 21599932&retmode=ref&cmd=prlinks.

Gruebner, O., Staffeld, R., Khan, M.M.H., Burkart, K., Krämer, A., and Hostert, P., 2011b. Urban health in megacities: Extending the framework for developing countries. *IHDP update (Magazine of the International Human Dimensions Programme on Global Environmental Change)*: 42–9.

Henkel, V., Mergl, R., Kohnen, R., Maier, W., Möller, H., and Hegerl, U., 2003. Identifying depression in primary care: A comparison of different methods in a prospective cohort study. *BMJ*, 326: 200–201.

Izutsu, T., Tsutsumi, A., Islam, A.M., Kato, S., Wakai, S., and Kurita, H. , 2006. Mental health, quality of life, and nutritional status of adolescents in Dhaka, Bangladesh: Comparison between an urban slum and a non-slum area. *Social Science & Medicine*, 63(6): 1477–88. Available at: http://eutils.ncbi.nlm.nih.gov/entrez/eutils/elink.fcgi?dbfrom=pubmed&id= 16765497&retmode=ref&cmd=prlinks.

Jolliffe, I.T., 2002. *Principal Component Analysis*, New York: Springer.

Kessing, L.V., Hansen, H.V., and Bech, P., 2006. General health and well-being in outpatients with depressive and bipolar disorders. *Nord Journal of Psychiatry*, 60(2): 105–56.

Kulke, E. and Staffeld, R., 2009. Informal Production Systems—the Role of the Informal Economy in the Plastic Recycling and Processing Industry in Dhaka. *Die Erde*, 140(1): 25–43.

Likert, R., 1932. A Technique for the Measurement of Attitudes. *Archives of Psychology*, 140.

Liwowsky, I., Krämer, D., Mergl, R., Bramesfeld, A., Allgaier, A.K., Pöppel, E., and Hegerl, U., 2009. Screening for depression in the older long-term unemployed. *Social Psychiatry and Psychiatric Epidemiology*, 44(8): 622–7.

Lyytimaki, J. and Sipila, M., 2009. Hopping on one leg—The challenge of ecosystem disservices for urban green management. *Urban Forestry & Urban Greening*, 8(4): 309–15.

Mathers, C.D. and Loncar, D., 2006. Projections of Global Mortality and Burden of Disease from 2002 to 2030. *PLoS Med*, 3(11), p.e442.

Millennium Ecosystem Assessment, 2005a. *Ecosystems and Human Well-being: Health Synthesis*. Geneva: WHO Press.

Millennium Ecosystem Assessment, 2005b. *Ecosystems and Human Well-being: Synthesis*, Washington D.C.: Island Press.

Momtaza, Y.A., Ibrahima,R., Hamida, T., and Yahayaa, N., 2011. Sociodemographic predictors of elderly's psychological well-being in Malaysia. *Aging & Mental Health*, 15(4): 437–45.

Moran,P.A.P.,1950.NotesonContinuousStochasticPhenomena.*Biometrika*,37(1–2):17–23.

Moran, P.A.P., 1948. The interpretation of statistical maps. *Journal of the Royal Statistical Society Series B*, 10: 243–51.

Newnham, E.A., Hooke, G.R. and Page, A.C., 2010. Monitoring treatment response and outcomes using the World Health Organization's Wellbeing Index in psychiatric care. *Journal of Affective Disorders*, 122(1–2): 133–8.

Ompad, D., Galea, S., Caiaffa, W., and Vlahov, D., 2007. Social Determinants of the Health of Urban Populations: Methodologic Considerations. *J Urban Health*, 84(0): 42–53.

Prince, M. Patel, V., Saxena, S., Maj, M., Maselko, J., Phillips, M.R., and Rahman, A., 2007. No health without mental health. *The Lancet*, 370(9590): 859–77.

R Development Core Team, 2013. *R: A Language and Environment for Statistical Computing*, Vienna, Austria: R Foundation for Statistical Computing.

Rahman, M.O. and Barsky, A.J., 2003. Self-reported health among older Bangladeshis: How good a health indicator is it? *Gerontologist*, 43(6): 856–63. Available at: http://eutils.ncbi.nlm.nih.gov/entrez/eutils/elink.fcgi?dbfrom=pubmed&id= 14704385&retmode=ref&cmd=prlinks.

Riley, L.W., Ko, A.I., Unger, A., and Mitermayer, R., 2007. Slum health: Diseases of neglected populations. *BMC International Health and Human Rights*, 7(1): 2.

Saipanish, R., Lotrakul, M., and Sumrithe, S., 2009. Reliability and validity of the Thai version of the WHO-Five Well-Being Index in primary care patients. *Psychiatry and Clinical Neurosciences*, 63: 141–6.

Schabenberger, O. and Gotway, C.A., 2005. Statistical Methods for Spatial Data Analyses. *Texts in Statistical Science*. Boca Raton: Chapman & Hall/CRC.

Sclar, E.D., Garau, P., and Carolini, G., 2005. The 21st century health challenge of slums and cities. *The Lancet*, 365(9462): 901–3.

Staffeld, R. and Kulke, E., 2011. Informal Employment and Health Conditions in Dhaka's Plastic Recycling and Processing Industry, in A. Krämer, M.H. Khan, and F. Kraas, eds. *Health in Megacities and Urban Areas*. New York: Springer: 209–19.

Uejio, C.K., Wilhelmi, O.V., Golden, J.S., Mills, D.M., Gulino, S.P., and Samenow, J.P., 2011. Intra-urban societal vulnerability to extreme heat: The role of heat exposure and the built environment, socioeconomics, and neighborhood stability. *Health & Place*, 17(2): 498–507.

United Nations, U.N., 2012. World Urbanization Prospects The 2011 Revision.

Venables, W.N. and Ripley, B.D., 2002. *Modern Applied Statistics with S*. New York: Springer.

Villagrán De León, J.C., 2006. Vulnerability: A Conceptual and Methodological Review. *(SOURCE) Studies Of the University: Research, Counsel, Education—Publication Series of UNU-EHS*, 4, p.64.

Vlahov, D., Freudenberg, N., Proietti, F., Ompad, D., Quinn, A., Nandi, V., and Galea, S., 2007. Urban as a Determinant of Health. *Journal of Urban Health*, 84(0): 16–26.

Waller, L. and Gotway, C., 2004. Applied Spatial Statistics for Public Health Data. *Wiley Series in Probability and Statistics*. New Jersey.

Weich, S., Brugha, T., King, M., McManus, S., Bebbington, P., Jenkins, R., Cooper, C., McBride, O., and Steward-Brown, S., 2011. Mental well-being and mental illness: Findings from the Adult Psychiatric Morbidity Survey for England 2007. *The British Journal of Psychiatry*, 199(1): 23–8.

WHO (World Health Organization) Collaborating Centre for Mental Health Frederiksborg General Hospital, 2010. WHO-Five Well-being Index (WHO-5). (27.04.2010).

Chapter 10
Space-Time Analysis of Late-Stage Breast Cancer Incidence in Michigan

Pierre Goovaerts and Maxime Goovaerts

Breast cancer is the most common cancer among women and is second only to lung cancer as a cause of cancer-related deaths in women (Jemal et al., 2004). Interpretation of cancer incidence and mortality rates in a defined population requires an understanding of multiple complex and interacting factors. For example, many studies in the literature report association between late-stage breast cancer diagnosis and covariates, such as socio-economic status, access to health care, marital status, ethnicity and neighborhood of residence (Farley and Flannery, 1989; Barry and Breen, 2005; MacKinnon et al., 2007).

Analyzing temporal trends in cancer incidence and mortality rates can provide a more comprehensive picture of the burden of the disease and generate new insights about the impact of various interventions (Potosky et al., 2001). It is widely accepted in the medical community that adherence to screening guidelines leads to the early detection of breast cancer and that, in turn, leads to decreased mortality rates. In a recent study, Summers et al. (2010) showed that in California early breast cancer detection rates reached a plateau at 70% in the mid-1990s in spite of ongoing efforts to promote adherence to recommended breast cancer screening guidelines. The analysis of temporal trends outside a spatial framework is however unsatisfactory. For example, Figure 10.1 reveals that the decrease in breast cancer late-stage diagnosis observed for white women in Michigan since 1985 encompasses significant geographical disparities. Indeed, it has long been recognized that there is significant variation among U.S. counties and states with regard to the incidence of cancer (Cooper et al., 2001; Wang et al., 2008). Visualizing, analyzing and interpreting these geographical disparities in temporal trends should bring important information and knowledge that will benefit substantially cancer epidemiology, control and surveillance and help reducing these disparities.

Figure 10.1 (below) allows only a visual assessment of the temporal trends in percentage of late-stage diagnosis for breast cancer. A quantitative analysis, including the detection of the timing and extent of significant changes in time series of health outcomes, can be conducted using joinpoint regression, also known as piecewise linear regression (Kim et al., 2000). The basic idea is to model the time series using a few continuous linear segments (see an example for Wayne County in Figure 10.3). Line segments are joined at points called joinpoints which represent the timing for a statistically significant change in rate trend (for example, 1990 and 1997 for breast cancer late-stage diagnosis in Wayne County, Figure 10.3). The number of joinpoints, as well as the parameters of the piecewise linear regression, are estimated through an iterative procedure that tests whether models of increasing complexity (that is, including more joinpoints) provide a significantly better goodness-of-fit than simpler models. The tests of significance use a Monte Carlo permutation method.

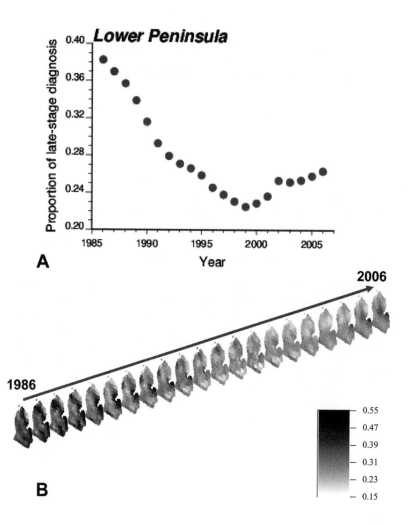

Figure 10.1   Two different representations of the Michigan space-time breast cancer dataset

Note: (A) time series of annual proportions of late-stage diagnosis recorded over the Lower Peninsula, and (B) three-dimensional display of annual county-level maps of noise-filtered rates of late-stage diagnosis.

There have been a few applications of joinpoint regression in cancer research, and it is now increasingly used to characterize long-terms trends in cancer mortality in the U.S. (Jemal et al., 2008) and foreign countries (Yang et al., 2003; Qiu et al., 2009; La Vecchia et al., 2010). Joinpoint regression is also used to project future cancer burden through the prediction of the future number of new cancer cases or deaths (Pickle et al., 2007). Most studies were however aspatial and conducted at the National level or for a single cancer registry. One main challenge with the spatialization of joinpoint regression is the need to conduct modeling over smaller geographical units where health outcomes can become less reliable because of small population sizes. To examine how socio-economic disparities in mammography use and associated changes in five breast cancer indicators varied over time, Schootman et al. (2010) used smoothed rates that accounted for spatial relationships among counties, leading to more stable measures of disparities. Goovaerts (2013) also proposed to apply joinpoint regression to smoothed rates in his county-level analysis of temporal trends in prostate cancer late-stage diagnosis across Florida. A sensitivity analysis showed that kriging-based noise-filtering improved the fit by the joinpoint regression models (that is, lower residual variability and smaller bias) compared to the modeling of raw rates.

Using techniques that were recently introduced in the field of health geostatistics and medical geography (Goovaerts, 2009, 2013), this chapter explores how the county-level proportions of breast cancer diagnosed late in Michigan Lower Peninsula changed yearly over the period 1985–2007. Despite the limitations associated with a county-level analysis (that is, results can be difficult to interpret because of potentially wide heterogeneity within a county), the present study represents a substantial improvement over most analyses of temporal trends which are usually aspatial. In addition, aggregation of individual level data at the county-level allowed the use of a fine temporal resolution (that is, year) which would not be possible for finer spatial resolutions (for example, ZIP codes or census tracts) because of rate instability caused by the small number problem.

Material and Methods

The Data

The space-time methodology will be illustrated using invasive breast cancer cases, diagnosed during the calendar years 1985 through 2007 in Michigan. Approximately 92% of these records, which were compiled by the Michigan Cancer Surveillance Program (MCSP), were successfully geocoded at residence at time of diagnosis. The present study focused on cases diagnosed for white women 84 year old and younger residing in the Lower Peninsula of Michigan. Out of the 109,694 women diagnosed with breast cancer during that time period, 27.3% of cases were defined as late-stage (that is, regional and distant metastatic cancer) according to the SEER General Summary Stage classification (Young et al., 2001).

Individual level cases were aggregated at the county-level to allow the analysis of temporal fluctuations across space. Following other studies (Goovaerts and Xiao, 2012; Goovaerts, 2013), proportions (rates) of late-stage diagnosis were computed over a 3-year moving window to reduce random fluctuations, yielding for each county a times series spanning 1986 through 2006. Rates were directly age-adjusted using the 109,694 women as reference population distribution. In sparsely populated counties, all four age categories ([0,50], [51,64], [65,74], [75,84]) might not be observed every single year. If only a single age category was missing, the statewide rate for that age category and year was used for the standardization. If more than one age category was empty, the rate was considered as missing. The time-averaged map of Figure 10.2(A) shows significant geographical disparities across the Lower Peninsula. On average over the period 1985–2007, the proportion of late-stage diagnosis was higher in the mid-eastern region identified as The Thumb of Michigan. Larger rates were also observed in the Southern counties at the border with Ohio and Indiana.

Two covariates that according to the literature (for example, Barry and Breen, 2005; MacKinnon et al., 2007; Wang et al., 2008) could potentially explain differences in the

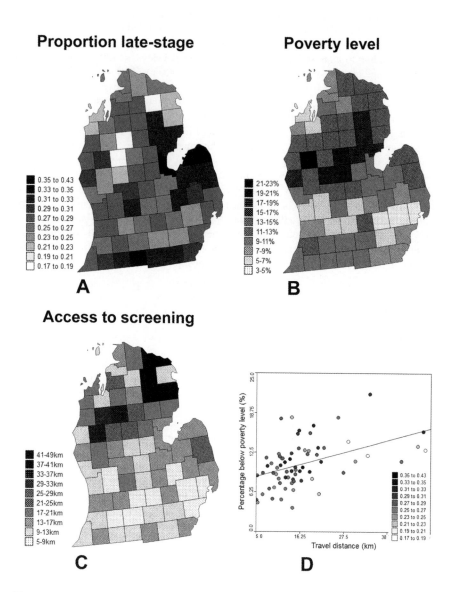

**Figure 10.2   Maps of three county-level quantities averaged over the entire
             time period**

Note: (A) proportion of breast cancer late-stage diagnosis, (B) percentage of habitants living below
poverty level, and (C) average travel distance between any census tract and the closest mammography
clinic. The scatterplot (D) illustrates the weak relationship between the two covariates and the
proportion of late-stage diagnosis.

prevalence of breast cancer late-stage diagnosis were considered: annual percentage of habitants living below the federally defined poverty line, and access to screening. Census tract level poverty data were downloaded from the U.S. Census Bureau for three decades; 1980, 1990, and 2000. Each cancer case was assigned a poverty level based on the census tract of residence and the year of diagnosis (that is, 1980 poverty level for years ≤ 1985, 1990 poverty level for the period [1986, 1995], and 2000 poverty level for years > 1995). Information about mammography facilities in Michigan (that is, address, date of creation, date of last action on the facility, and status: active or inactive) was obtained from the Michigan Department of Community Health. Network distance and travel time from census tract centroids to the closest mammography facilities were estimated for each year using the Network Analyst Extension in ArcInfo 9.3 to quantify spatial access to health care (Wang et al., 2008). This computation was based on the 2010 road network and despite ignoring potential changes in the network over the period 1985–2007 these travel distances were deemed more realistic than time-invariant Euclidian distances. In absence of individual level information on access to screening and residential history, only in-state mammography facilities were considered and the computation of travel distances was based on the residence at time of diagnosis and the closest mammography clinic.

Figures 10.2b and 10.2c show the values of the two census tract level covariates after averaging at the county-level and over the entire time period [1985, 2007]. The Northern part of the Peninsula is less densely populated and these more rural counties tend to have a higher proportion of habitants living below the poverty line and also a longer travel distance to the closest mammography facility. The moderate relationship between these two covariates is illustrated by their scatterplot (Figure 10.2D). Each dot in the scatterplot represents a county and is colored according to its proportion of late-stage diagnosis. No obvious relationship stands out except that larger proportions of late-stage diagnosis seem to be better correlated with greater poverty levels than with longer travel distances.

Spatial Analysis: Noise-filtering using Binomial Kriging

Let $\{z(v_a;t),\ \alpha=1,\ \ldots,N,\ t=1,\ \ldots,T\}$ be the observed proportions or rates of late-stage diagnosis recorded at T different time periods (for example, years) for a given number N of geographical units or areas v_a (that is, counties here). Each observation $z(v_a;t)$ is computed as the ratio $d(v_a;t)/n(v_a;t)$, where $n(v_a;t)$ is the total number of cases at time t, $d(v_a;t)$ of which were diagnosed late. Analyzing directly the rates $z(v_a;t)$ might lead to misleading conclusions since in sparsely populated counties the number of breast cancer cases recorded in a single year is likely too small to compute reliable estimates of late-stage diagnosis rates. It is thus recommended to first filter the noise due to the so-called small number problem.

The noise-filtered rate for a given area v_a and time t, denoted $\hat{r}(v_a;t)$, was estimated as a linear combination of the kernel rate $z(v_a;t)$ and the rates observed in $(n-1)$ neighboring entities v_i at that time t:

$$\hat{r}\left(v_{\alpha};t\right) = \sum_{i=1}^{n}\lambda_{it}z\left(v_{i};t\right) \quad (1)$$

The associated prediction error variance, commonly known as the kriging variance, is computed as:

$$\sigma_K^2\left(v_\alpha;t\right) = \bar{C}_t\left(v_\alpha,v_\alpha\right) - \sum_{i=1}^{n}\lambda_{it}\bar{C}_t\left(v_i,v_\alpha\right) - \mu_t\left(v_\alpha\right) \quad (2)$$

The kriging weights λ_{it} used in both equations (1) and (2) are computed by solving for each time t the following system of linear equations, known as "binomial kriging" system (Webster et al., 1994; Goovaerts, 2009):

$$\sum_{j=1}^{n}\lambda_{jt}\left[\bar{C}_t\left(v_i,v_j\right) + \delta_{ij}\frac{a_t}{n\left(v_i;t\right)}\right] + \mu_t\left(v_\alpha\right) = \bar{C}_t\left(v_i,v_\alpha\right) \quad i=1,\ldots,n$$

$$\sum_{j=1}^{n}\lambda_{jt} = 1 \tag{3}$$

where $\delta_{ij}=1$ if i=j and 0 otherwise, $a_t=m_t^*(1-m_t^*)-\bar{C}_t(v_i,v_i)$, and m_t^* is the population-weighted average of the N rates $z(v_\alpha;t)$. The quantity $a_t/n(v_i;t)$ is an error variance term that increases the variance $\bar{C}_t(v_i,v_i)$ of the units with small population size $n(v_i;t)$ the most. Thus, smaller weights are assigned to less reliable late-stage rates based on fewer cases.

The area-to-area covariance terms $\bar{C}_t(v_i,v_j)$ and $\bar{C}_t(v_i,v_\alpha)$ are numerically approximated by averaging the point-support covariance $C_t(\mathbf{h})$ computed between any two locations discretizing the areas v_i, v_j or v_α. The point-support covariance $C_t(\mathbf{h})$, or equivalently the point-support semivariogram $\mu_t(\mathbf{h})$, cannot be estimated directly from the observed rates, since only areal data are available. Thus, only the regularized semivariogram can be estimated using the following population-weighted estimator (Goovaerts, 2005):

$$\hat{\gamma}_t\left(\mathbf{h}\right) = \frac{1}{2\displaystyle\sum_{\alpha,\beta}^{N(\mathbf{h})} n\left(v_\alpha;t\right)n\left(v_\beta;t\right)} \sum_{\alpha,\beta}^{N(\mathbf{h})}\left\{n\left(v_\alpha;t\right)n\left(v_\beta;t\right)\left[z\left(v_\alpha;t\right)-z\left(v_\beta;t\right)\right]^2\right\} \quad (4)$$

where $N(\mathbf{h})$ is the number of pairs of areas (v_α,v_β) whose population-weighted centroids are separated by the vector $\mathbf{h}$. The different spatial increments $[z(v_\alpha;t)-z(v_\beta;t)]^2$ are weighted by the product of their respective population sizes to assign more importance to the more reliable data pairs. Derivation of a point-support semivariogram from the experimental semivariogram $\hat{\gamma}_t(\mathbf{h})$ computed from areal data is called "deconvolution," an operation that is conducted using an iterative procedure (Goovaerts, 2008).

Spatial analysis of temporal trends

Following Goovaerts (2013), a spatial analysis of temporal trends can be conducted by fitting a joinpoint regression model to each time-series $\{\hat{r}(v_\alpha;t), t=1, \ldots,T\}$. Joinpoint regression models each time series as a sequence of linear segments. In its log-linear version, the segmented regression model for any unit v_α is written as:

Wayne County

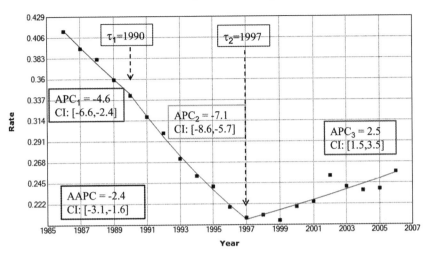

Figure 10.3   **Annual proportions of breast cancer late-stage cases (white females 84 years and younger) that were diagnosed over the period 1985–2007 within Wayne County**

Note: The segmented regression model (solid lines) includes two joinpoints (τ) that correspond to years of statistically significant changes in rate trend: 1990 and 1997. The estimate and 95% confidence intervals of the annual percent change (APC) are computed for each segment, whereas the average annual percent change (AAPC) refers to the entire time period.

$$\mathrm{Log}(\hat{r}(v_a;t)) = m(v_a;t) + \varepsilon(v_a;t) \qquad t=1,\dots,T \quad (5)$$

where $\varepsilon(v_a;t)$ is the residual for the t-th time, and the regression mean $m(v_a;t)$ is defined as a succession of (K_a+1) linear segments over the time interval $[a,b]$: $[a,\tau_{1a}] \dots (\tau_{ka},\tau_{k+1a}] \dots (\tau_{Ka}, b]$. The parameter τ_{ka} is the timing (joinpoint) for a statistically significant change in the slopes of two successive segments.

For example, Figure 10.3 shows how the proportion of breast cancer cases diagnosed late in Wayne County (Detroit) changed yearly between 1986 and 2006 ($T=21$). The observed time series was fitted with a regression model that includes two joinpoints: $\tau_1=1990$ and $\tau_2=1997$. The rate has been decreasing since the first year 1986. This decline accelerated in 1990 (that is, steeper slope) before the rate started increasing in 1997; see parameters listed in Figure 10.3.

The unknowns in Equation (5) include the number and values of the joinpoints, as well as the regression parameters (for example, slopes of linear segments). They are estimated using a two-step procedure: 1) a grid search method (Lerman, 1980) is conducted over the set of possible joinpoints, and 2) at each step of the search the regression parameters and their standard errors are estimated by weighted least-square regression using the following criterion:

$$Q = \sum_{t=1}^{T} w(v_\alpha;t)\Big(log(\hat{r}(v_\alpha;t)) - m(v_\alpha;t)\Big)^2 \quad (6)$$

The weights account for the fact that the variance of the residuals $\varepsilon(v_\alpha;t)$ may vary with time (heteroscedasticity) and they were set to the inverse of the Binomial kriging variance: $1/\sigma_K^2(v_\alpha;t)$. The correlation among residuals was accounted for using an autorrelation error model automatically fitted to the data (Kim et al., 2000).

The number K of joinpoints is estimated through an iterative procedure that tests whether models of increasing complexity (that is, including more joinpoints) provide a significantly better goodness-of-fit than simpler models (Kim et al., 2009). The tests of significance use a Monte Carlo permutation procedure described in Kim et al. (2000). To reduce the number of solutions and the computational time, a maximum number of joinpoints is typically specified (that is, $K_{max}=3$ here). To keep joinpoints from getting too close together or too close to either end of the time series, a minimum number of observations between joinpoints is also required and was set to 4 in the present study. This minimum number allowed the computation of the standard error of the slope parameters and the associated p-values.

Trends in health outcomes over a specified time interval $[\tau_k, \tau_{k+1}]$ are usually described by the annual percent change (APC) that can be derived from the slope of the regression model over that time interval. For the example of Wayne County, the APC is particularly large for the period [1990, 1997]: the proportion of late-stage diagnosis declined by 7.1% every year (Figure 10.3). Like other regression parameters, confidence intervals can be computed for each APC and one can test whether an APC is significantly different from zero (Kim et al., 2000). Figure 10.3 indicates that changes were significant for all three time periods: none of the 95% confidence intervals $[L_{APC}, U_{APC}]$ includes zero.

The trend over the entire time series $[a,b]$ can be summarized using the average annual percent change (AAPC) which is computed as the weighted average of the APC's from the joinpoint model. This measure is valid even if the joinpoint model indicates that there were changes in trends during those years (Clegg et al., 2009). Like for the APC, a $(1-\alpha)$ confidence interval can be computed and if it contains zero, then there is no evidence to reject the null hypothesis that the true AAPC is zero at the significance level of α.

An alternative approach to the post-processing of joinpoint regression results is to analyze directly the time series of kriged estimates, for example through clustering to aggregate counties with similar behaviors and facilitate the visualization of the main temporal patterns present in the study area. In this chapter a cluster analysis was conducted to aggregate Michigan counties based on the similarity in their temporal trends of the proportion of late-stage diagnosis. A common approach is to apply a clustering algorithm (for example, complete linkage, kth-Nearest-Neighbor) to a matrix of dissimilarities $d_{\alpha\beta}$ that quantifies the difference between any two pair of geographical units $v_{\alpha'}$ and v_β. The dissimilarity was here identified with the Euclidean distance between any two time series. The Ward's minimum variance hierarchical method was used as clustering algorithm since it has been shown to give the best recovery of cluster structure (Milligan, 1981).

Temporal Disparities

Instead of looking at geographical disparities in temporal trends, one could also explore temporal trends in the extent of these geographical disparities. For example, Goovaerts

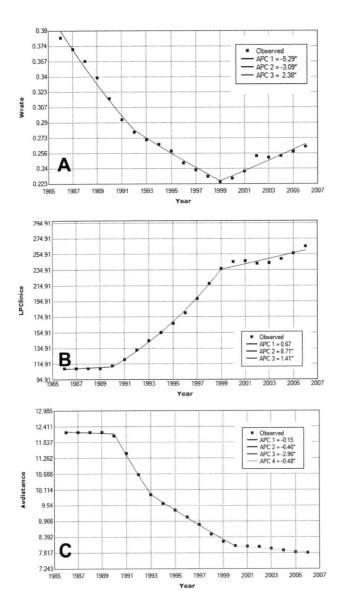

Figure 10.4   Time series, with the joinpoint regression model fitted, for three quantities

Note: (A) annual proportions of late-stage diagnosis recorded over the Lower Peninsula, (B) number of mammography clinics in operation in the Lower Peninsula, and (C) average travel distance between a census tract centroid and the closest clinic. Values were computed over a 3-year moving window to reduce random fluctuations.

(2013) proposed to compare for every time period t (that is, year) the 95% confidence intervals of the APC for any two adjacent counties $(v_a, v_{a'})$ and to count the number of times these two intervals do not overlap. The number of pairs of counties with non-overlapping confidence intervals for any given time t can be used to measure the extent of geographical disparities that existed at that time:

$$D(t)= \sum_{\alpha=1}^{N} \sum_{\alpha'=\alpha+1}^{N} I\left(CI\left(v_\alpha;t\right);CI\left(v_{\alpha'};t\right)\right) \quad (7)$$

where the indicator function $I=1$ if the following condition on the upper bounds (U) and lower bounds (L) of the two confidence intervals CI are met: $U(v_a;t) < L(v_{a'};t)$ or $L(v_a;t) > U(v_{a'};t)$.

Software

Joinpoint regression was conducted using the public-domain Joinpoint Regression Program 4.0.1 December 2013 (Kim et al., 2000) developed at the U.S. National Cancer Institute, NCI (http://surveillance.cancer.gov/joinpoint). Binomial kriging was performed using the commercial software SpaceStat 3.6 (BioMedware, Inc, 2013). The three-dimensional display of county-level time series in Figure 4.4 was created using the Stanford Geostatistical Modelling Software (Remy, Boucher and Wu, 2008) 3D visualization panel and FORTRAN programs developed to format the data. All other computations, including the calculation of disparity statistics, were accomplished using FORTRAN programs developed by the first author.

Results and discussion

State-wide temporal trends

Figure 10.4a shows the joinpoint regression model that was fitted to the time series of proportion of breast cancer late-stage diagnosis displayed in Figure 10.1a. Experimental proportions were averaged over the Lower Peninsula and 3-year time moving windows to increase stability. The general trend is similar to what was observed for the densely populated Wayne County (Figure 10.3): significant decrease until the late nineties followed by a significant increase. The regression model includes two joinpoints: $\tau_1=1992$ and $\tau_2=1999$, which represents a two year shift relatively to Wayne County: $\tau_1=1990$ and $\tau_2=1997$. Another difference is that the rate of decline in the Lower Peninsula slowed down in the nineties while it accelerated for Wayne County during that same period.

Interestingly, the sudden increase in proportion of late-stage diagnosis happened at the same time ($t_2=1999$) as the opening of new mammography facilities (Figure 10.4b) started slowing down, leading to a much slower decline in travel time to the closest mammography clinic (Figure 10.4c). This similarity in timing of events might however be fortuitous since the proportion of late-stage diagnosis started declining in 1985, well before the number of clinics in operation started increasing steadily (a two-fold increase between 1990 and 1999).

One advantage associated with the analysis of rates of late-stage diagnosis versus incidence or mortality rates is that the underlying population (that is, total number of breast cancer cases) is known with great accuracy. On the other hand, the computation of annual

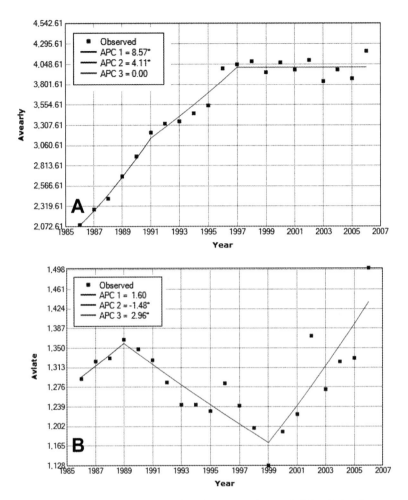

Figure 10.5 **Time series, with the joinpoint regression model fitted, for the annual number of cases that were diagnosed early and late over the Lower Peninsula**

Note: (A) early, (B) late. Values were computed over a 3-year moving window to reduce random fluctuations.

incidence or mortality rates is based on the number of women within each age category that lived within each county for that particular year, which is more difficult to estimate from decennial census data. The interpretation of temporal trends in late-stage diagnosis might however be more challenging because they reflect trends in several quantities, such as number of diagnosed cases and number of early-stage diagnosis. For the particular case of breast cancer, the introduction of screening mammography in the United States has been associated with a doubling in the number of cases of early-stage breast cancer that were detected annually from 1976 through 2008 (Bleyer and Welch, 2012). The same authors showed that this substantial increase was largely caused by the detection of in

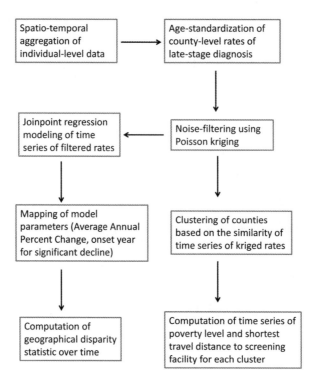

Figure 10.6   Flowchart describing the different steps of the space-time analysis of county-level rates of breast cancer late-stage diagnosis

situ cancers that would have never led to clinical symptoms (over diagnosis). The amount of time diagnosis is advanced due to screening (lead time) decreases as the screening procedure becomes more accessible. Another important factor to explain temporal trends in breast cancer is the excess diagnoses associated with hormone-replacement therapy in the late 1990s and early 2000s (Bleyer and Welch, 2012).

To interpret the trend in late-stage diagnosis displayed in Figure 10.4a, time-series of numbers of early-stage and late-stage cases were created for the Lower Peninsula and fitted using joinpoint regression. As expected, the number of early-stage diagnosis increased sharply in the eighties (APC = 8.57%) following the introduction of screening (Figure 10.5a). This increase decelerated in the nineties (APC = 4.11%) before a plateau was reached in 1997. On the other hand, the number of late-stage diagnosis also increased in the eighties, likely a result of the introduction of screening (Figure 10.5b). The positive impact of early detection of breast cancer translated into a decrease in number of late-stage cases during the nineties. The rate of decrease (APC = −1.48%) is however much smaller than the rate of increase of early-stage diagnosis (APC = 4.11–8.57%), indicating that the decline in percentage of late-stage diagnosis in Figure 10.4a was mainly driven by an increase in the total number of breast cancer cases. The combination of an increase in the number of late-stage diagnosis with a plateau in early-stage detection explains the significant increase in percentage of late-stage diagnosis experienced after 1999.

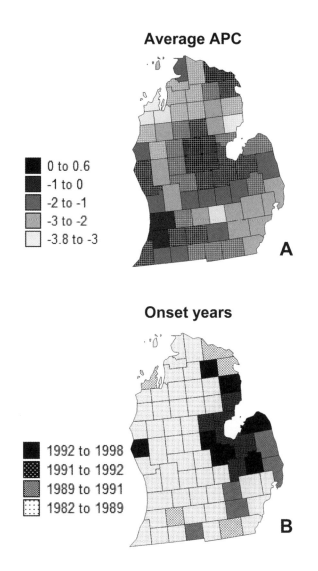

Average APC

0 to 0.6
-1 to 0
-2 to -1
-3 to -2
-3.8 to -3

A

Onset years

1992 to 1998
1991 to 1992
1989 to 1991
1982 to 1989

B

Figure 10.7   Maps of two parameters of the joinpoint regression models fitted to Michigan county-level time series of proportions of breast cancer late-stage diagnosis

Note: (A) average annual percent change (APPC) over the period 1985–2007, cross-hatched counties denote AAPC not significantly different from zero, and (B) onset year for a significant decline in proportion of late-stage diagnosis.

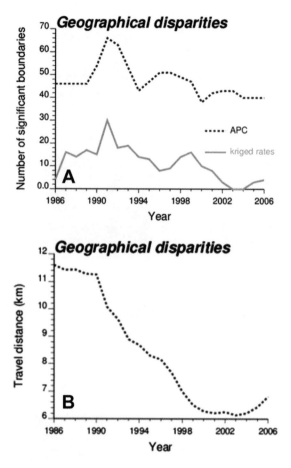

Figure 10.8   **(A) Number of boundaries (that is, pairs of adjacent counties) with significant differences in kriged estimates or APC values as a function of time. (B) Average difference in travel distances to the closest screening facilities recorded for adjacent counties over time**

Space-time joinpoint regression

The flowchart in Figure 10.6 describes the different steps of the analysis. Yearly county-level rates of late-stage diagnosis were noise-filtered using binomial kriging and the semivariogram computed for that year. A joinpoint regression model was then fitted to each county-level time series. The geographical variability of temporal trends is summarized using two statistics that are mapped in Figure 10.7: the average annual percent change (AAPC) and the joinpoint corresponding to the first significant decline in proportion of late-stage diagnosis (that is, negative APC significantly different from zero). The annual rate of decrease in breast cancer late-stage diagnosis and the onset years vary greatly across Michigan. Most counties with non-significant AAPC are located in northern and middle Michigan (Figure 10.7a), which is explained to some extent by smaller population sizes that result in wider confidence intervals for the AAPC. Spatial trends are more pronounced for the onset years: the first significant decline in proportion of late-stage diagnosis started much later on the Eastern border of the State along Lake Huron, in particular in the Thumb area where late-stage diagnosis has been more prevalent over the years (Figure 10.2a).

The extent of geographical disparities was analyzed through time by comparing for each year the 95% confidence intervals of the APC estimate (or the rate kriging estimate) for any two adjacent counties and flagging as significant all edges and years where the two intervals do not overlap. For both parameters, the number of significant county boundaries (Equation 7, Figure 10.8a) peaked in the early 1990s when the number of mammography clinics in operation started increasing steadily (Figure 10.4c). Such a temporal trend suggests the existence of geographical disparities in access to screening, in particular as it began more widely available. In order to quantify these geographical disparities the minimum travel distances to screening recorded within any pair of adjacent counties were compared for each year. The absolute difference between travel distances was averaged over all 153 county boundaries and its change as a function of time is plotted in Figure 10.8b. The sharpest drop from one year to the next (1.2km) was recorded in 1991, which confirms the above interpretation. Even more interesting is the increase in geographical disparities recorded at the end of the time period. As fewer new clinics were being opened (Figure 10.4c), between-county disparities in access to screening started increasing and incidentally during that period the proportion of late-stage diagnosis rose again.

Figure 10.9 shows the results of a cluster analysis to aggregate Michigan counties based on the similarity in their temporal trends of the proportion of late-stage diagnosis. The analysis of the dendogram led to the identification of five clusters displayed in Figure 10.9a. The cross-hatched cluster corresponds to the group of counties with the highest prevalence of late-stage diagnosis over the 21 years (dashed line in Figure 10.9b), whereas the dark cluster includes counties with the lowest proportion of late-stage diagnosis. Individual time series are also displayed according to their cluster allocation in Figure 10.9c. Each column corresponds to a particular county ordered according to their FIPS codes and each pixel to a particular year; the gray scale indicates the proportion of late-stage diagnosis.

The cluster map in Figure 10.9a bears some similarity with the map of average annual percent change (Figure 10.7a) and the map of time-averaged rates (Figure 10.2a). In particular, Cluster #5 with the largest proportion of late-stage diagnosis corresponds to the Thumb area. Figures 10.9b–c showcases much contrasted temporal trends among clusters. The largest spread among the five time series occurred in the early nineties (peak for Cluster #5 and low for Cluster #1), which confirms the findings in Figure 10.8a regarding the timing for the maximum extent of geographical disparities. A steady decline in late-stage diagnosis started in the mid-eighties for Central Michigan and the Northern tip of the peninsula (Cluster #3) whereas a steep decline followed an increase for the Thumb area (Cluster #5) and counties hosting the cities of Ann Arbor and Traverse City (Cluster #4). It is noteworthy that the decline was so sharp for Cluster #4 that despite the significant initial increase in proportion of late-stage diagnosis the average annual percent of change over the period 1986–2006 is still significantly negative for several of these counties (Figure 10.7a). In addition, the time series for this cluster does not display the recent increase in late-stage diagnosis observed for the other clusters and the State of Michigan in general. Finally, the proportion of late-stage diagnosis started declining early for counties in Clusters #2, yet at a moderate pace, which agrees with the non-significant AAPC detected in Figure 10.7a for half of the cluster (11 counties out of 21).

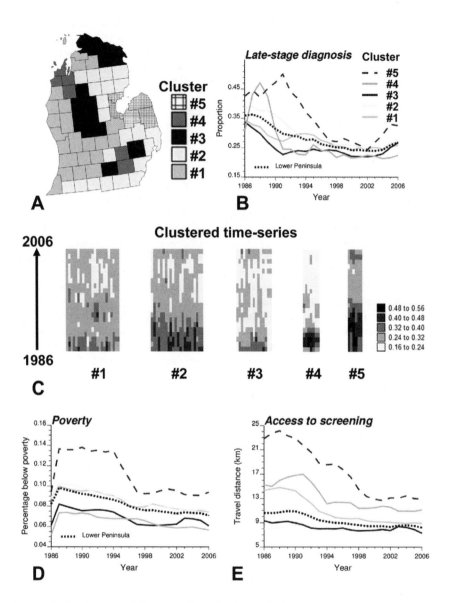

Figure 10.9   Results of the space-time cluster analysis

Note: (A) Grouping of counties based on the similarity of their temporal trends in proportion of late-stage diagnosis, (B) Time series of proportion of late-stage diagnosis for each of the five clusters and Michigan Lower Peninsula (dotted line), (C) individual time series of noise-filtered rates of late-stage diagnosis displayed as vertical strings and ordered according to their FIPS code and allocation to one of the five clusters. Bottom plots show the time series of two covariates (driving distance, poverty level) for each of the five clusters and Michigan Lower Peninsula (dotted line). The same line and shading codes are used for the time series in plots B, D, and E.

Potential relationships between clustering results and the two covariates were explored by computing for each cluster the poverty level and shortest travel distance experienced by cases diagnosed in these clusters. Figures 10.9d–e clearly illustrate the lower socio-economic status (that is, highest proportion of habitants living below poverty level) and longer travel distance for counties that are part of Cluster #5 characterized by the largest percentage of late-stage diagnosis. In particular, the average travel distance steadily decreased during the nineties along with the percentage of late-stage diagnosis. Low poverty levels and short travel distances were recorded for Cluster #3 which has the smallest proportion of late-stage diagnosis on average over the entire time period. The ranking of the other three clusters is in better agreement with the poverty level than the travel distance. This confirms the results of a previous study conducted in three Michigan counties (Goovaerts, 2010) where proximity to clinics was shown to have almost no impact on incidence rates of late-stage diagnosis for affluent neighborhoods [0–5% poverty] and poor neighborhoods [10–15% poverty]. Another factor that needs to be factored in is that fact that in the mid-1980s few insurance plans covered mammography screening, which likely reinforced the impact of socio-economy statistics on access to screening (McElroy et al., 2006).

Conclusions

The case study emphasized how the proportion of late-stage diagnosis for a common disease, such a breast cancer, can change dramatically over time (that is, 40% decline over 20 years) and display striking geographical disparities within a single State (county-level percentages of change ranging from -3.8% to 0.8% per year). Thus, a comprehensive picture of the burden of cancer and the impact of various interventions can only be achieved through the simultaneous incorporation of the spatial and temporal dimensions in the visualization and analysis of health outcomes and putative covariates.

This chapter described a methodology that starts with the estimation of reliable disease rates through a geostatistical noise-filtering followed by the modeling of temporal trends using joinpoint regression. Trend model parameters, such as annual percentages of change or onset year for significant change in disease rates, can be mapped to visualize geographical disparities in temporal trends. An alternative approach is to analyze directly the time series of kriged estimates, for example through clustering to aggregate counties with similar behaviors and facilitate the visualization of the main temporal patterns present in the study area.

Both approaches were used to explore spatio-temporal disparities in the incidence of late-stage diagnosis for breast cancer in Michigan Lower Peninsula over the period 1985–2007. At the State-level the proportion of late-stage diagnosis declined significantly at a rate of 3 to 5.3% per year until 1999 when it started rising again at a significant rate of 2.38% per year. This decline was to a large extent caused by the significant increase in newly diagnosed breast cancer cases following the introduction of screening mammography. Temporal trends greatly vary among counties and geographical disparities peaked in the early 1990s when the number of mammography clinics in operation started increasing steadily. The first significant decline in proportion of late-stage diagnosis started much later on the Eastern border of the State along Lake Huron, in particular in the Thumb area where late-stage diagnosis has been more prevalent over the years and both access to screening and socio-economic status are less favorable.

The present study was mainly methodological and these results warrant further exploration of how the substantial increase in the number of mammography clinics impacted the prevalence of late-stage diagnosis, in particular in areas of lower socio-economic status. This will require the computation and analysis of county-level time series of incidence rates for breast cancer and its early-stage diagnosis. Also, most preliminary screenings and referrals for breast cancer detection are conducted at primary care doctors' offices; hence primary care access might be a much more important predictor of late-stage risk than is travel time to mammography. For example, Wang et al. (2008) reported that in Illinois poor spatial access to primary health care is more strongly associated with late-stage diagnosis than is spatial access to mammography. This alternative measure of access to screening should be investigated in the future along with the impact of level of educational attainment since women who are more educated are more likely to ask their physician for a referral or to self-refer (Wells and Horm, 1992).

As longer time series and individual level data become widely available, there will be a growing need for algorithms and software to visualize, process and summarize large space-time databases of health outcomes and putative covariates.

References

Barry, J. and Breen, N., 2005. The importance of place of residence in predicting late-stage diagnosis of breast or cervical cancer. *Health & Place*, 11: 15–29.

Bleyer, A. and Welch, H.G., 2012. Effect of three decades of screening mammography on breast-cancer incidence. *The New England Journal of Medicine*, 367: 1998–2005.

BioMedware, Inc., 2013. *SpaceStat User Manual* version 3.6.

Clegg, L.X., Hankey, B.F., Tiwari, R., Feuer, E.J. and Edwards, B.K., 2009. Estimating average annual percent change in trend analysis. *Statistics in Medicine*, 28: 3670–82.

Cooper, G.S., Yuan, Z., Jethva, R.N. and Rimm, A.A., 2001. Determination of county-level prostate carcinoma incidence and detection rates with Medicare claims data. *Cancer*, 92: 102–9.

Farley, T.A. and Flannery, J.T., 1989. Late-stage diagnosis of breast cancer in women of lower socioeconomic status: Public health implications. *American Journal of Public Health*, 79: 1508–12.

Goovaerts, P., 2005. Simulation-based assessment of a geostatistical approach for estimation and mapping of the risk of cancer, in Leuangthong, O. and Deutsch, C.V. eds. *Geostatistics Banff 2004*. Dordrecht: Kluwer Academic Publishers: 787–96.

Goovaerts, P., 2008. Kriging and semivariogram deconvolution in presence of irregular geographical units. *Mathematical Geosciences*, 40(1): 101–28.

Goovaerts, P., 2009. Combining area-based and individual level data in the geostatistical mapping of late-stage cancer incidence. *Spatial and Spatio-temporal Epidemiology*, 1: 61–71.

Goovaerts, P., 2010. Visualizing and testing the impact of place on late-stage breast cancer incidence: A non-parametric geostatistical approach. *Health and Place*, 16: 321–30.

Goovaerts, P., 2013. Analysis of geographical disparities in temporal trends of health outcomes using space-time joinpoint regression. *Journal of Applied Earth Observation and Geoinformation*, 22: 75–85.

Goovaerts, P. and Xiao, H., 2012. The impact of place and time on the proportion of late-stage diagnosis: The case of prostate cancer in Florida, 1981–2007. *Spatial and Spatio-temporal Epidemiology*, 3: 243–53.

Jemal, A., Clegg, L.X., Ward, E., Ries, L.A., Wu, X., Jamison, P.M., Wingo, P.A., Howe, H.L., Anderson, R.N. and Edwards, B.K., 2004. Annual report to the nation on the status of cancer, 1975–2001, with a special feature regarding survival. *Cancer*, 101: 3–27.

Jemal, A., Thun, M.J., Ries, L.A., Howe, H.L., Weir, H.K., Center, M.M., Ward, E., Wu, X.C., Eheman, C., Anderson, R., Ajani, U.A., Kohler, B., and Edwards, B.K., 2008. Annual report to the nation on the status of cancer, 1975–2005, featuring trends in lung cancer, tobacco use, and tobacco control. *Journal of the National Cancer Institute*, 100: 1672–94.

Kim, H.J., Fay, M.P., Feuer, E.J., and Midthune, D.N., 2000. Permutation tests for joinpoint regression with applications to cancer rates. *Statistics in Medicine*, 19: 335–51.

Kim, H.J., Yu, B., and Feuer, E.J., 2009. Selecting the number of change-points in segmented line regression. *Statistica Sinica*, 19(2): 597–609.

La Vecchia, C., Bosetti, C., Lucchini, F., Bertuccio, P., Negri, E., Boyle, P., and Levi, F., 2010. Cancer mortality in Europe, 2000–2004, and an overview of trends since 1975. *Annals of Oncology*, 21(6): 1323–60.

Lerman, P.M., 1980. Fitting segmented regression models by grid search. *Applied Statistics*, 29: 77–84.

MacKinnon, J.A., Duncan, R.C, Huang, Y., Lee, D.J., Fleming, L.E., Voti, L., Rudolph, M., and Wilkinson, J.D., 2007. Detecting an association between socioeconomic status and late stage breast cancer using spatial analysis and area-based measures. *Cancer Epidemiology Biomarkers & Prevention*, 16: 756–62.

McElroy, J.A., Remington, P.L., Gangnon, R.E., Hariharan, L., and Andersen, L.D., 2006. Identifying geographic disparities in the early detection of breast cancer using a geographic information system. *Preventive Chronic Disease* [e-journal], 3(1), Available through CDC: http://www.cdc.gov/pcd/issues/2006/jan/05_0065.htm [Accessed January 24 2014].

Milligan, G.W., 1981. A review of Monte Carlo tests of cluster analysis. *Multivariate Behavorial Research*, 16(3): 379–407.

NCI, 2012. *Joinpoint Regression Program*, Version 4.0.1. December 2012; Statistical Research and Applications Branch, National Cancer Institute.

Pickle, L.W., Hao, Y., Jemal, A., Zou, Z., Tiwari, R.C., Ward, E., Hachey, M., Howe, H.L., and Feuer, E.J., 2007. A New Method of Estimating United States and State-level Cancer Incidence Counts for the Current Calendar Year. CA: *A Cancer Journal for Clinicians*, 57: 30–42.

Potosky, A.L., Feuer, E.J., and Levin, D.L., 2001. Impact of screening on incidence and mortality of prostate cancer in the United States. *Epidemiologic Review*, 23(1): 181–6.

Qiu, D., Katanoda, K., Marugame, T., and Sobue, T., 2009. A Joinpoint regression analysis of long-term trends in cancer mortality in Japan (1958–2004). *International Journal of Cancer*, 124: 443–8.

Remy, N., Boucher, A., and Wu, J., 2008. *Applied Geostatistics with SGeMS: A User's Guide*. New York: Cambridge University Press.

Schootman, M., Lian, M., Deshpande, A.D., Baker, E.A., Pruitt, S.L., Aft, R., Jeffe, D.B., 2010. Temporal trends in geographic disparities in small-area breast cancer incidence and mortality, 1988 to 2005. *Cancer Epidemiology Biomarkers & Prevention*, 19(4): 1122–31.

Summers, C., Saltzstein, S.L., Blair, S.L., Tsukamoto, T.T., and Sadler, G.R., 2010. Racial/ethnic differences in early detection of breast cancer: A study of 250,985 cases from the California Cancer Registry. *Journal of Women's Health*, 19(2): 203–7.

Wang, F., McLafferty, S., Escamilla, V., and Luo, L., 2008. Late-stage breast cancer diagnosis and health care access in Illinois. *Professional Geographer*, 60: 54–69.

Webster, R., Oliver, M.A., Muir, K.R., and Mann, J.R., 1994. Kriging the local risk of a rare disease from a register of diagnoses. *Geographical Analysis*, 26: 168–85.

Wells, B.L. and Horm, J.W., 1992. Stage at diagnosis in breast cancer: Race and socioeconomic factors. *American Journal of Public Health*, 82: 1383–5.

Yang, L., Parkin, D.M., Li, L., and Chen, Y., 2003. Time trends in cancer mortality in China: 1987–1999. *International Journal of Cancer*, 106: 771–83.

Young, J.L. Jr., Roffers, S.D., Ries, L.A.G., Fritz, A.G., and Hurlbut, A.A. eds., 2001. *SEER Summary Staging Manual—2000: Codes and Coding Instructions*. National Cancer Institute. National Institutes of Health Pub. # 01–4969, Bethesda, MD.

SECTION 4
Exposure

Chapter 11
A Method for Reducing Classical Error in Long-Term Average Air Pollution Concentrations from Incomplete Time-Series Data

Matthew D. Adams and Pavlos S. Kanaroglou

Air pollution exposure negatively affects human health including reduced cognitive function (Hutter et al., 2013), reduced lung function (Wallner et al., 2012), early childhood cancer (Ghosh et al., 2013), increased low and underweight births (Padula et al., 2012), and mortality and morbidity due to cardiovascular and respiratory diseases (Hoek et al., 2013). Observational epidemiologic studies that use ambient air pollution concentrations assigned to research subjects are the primary method of determining the association between long-term air pollution exposure and health effects. Hoek et al. (2013) reviewed long-term air pollution exposure studies of health effects and cardio-respiratory mortality, including such research studies as the Harvard Six Cities, American Cancer, German Cohort, California Teachers and the Nurses' Health Study. These studies were fundamental in identifying possible associations and health effect outcomes to varying levels of air pollution.

Current air pollution studies assess exposure over space, because air pollution in cities is long-known to vary spatially and that a single value or monitor is not representative (Goldstein and Landovitz, 1977a, 1977b); however, the number of air pollution monitoring stations is typically limited, often due to their high-cost, and concentration data must be spatially-interpolated to provide values for each individuals location. Kriging or land use regression modeling are commonly employed to spatially-interpolate air pollution concentrations at unmonitored locations (Jerrett et al., 2005; Kanaroglou et al., 2005; Kumar, 2009). The effectiveness of these spatially refined estimates is dependent on a well-designed spatial sampling approach, which, in general, should maximize the probability of capturing the spatial variability (Delmelle, 2014). For a primer on spatial sampling see Delmelle (2014), who introduces two-dimensional sampling, geostatistical sampling, and second-phase sampling. More in-depth information on spatial sampling for a spatially correlated phenomena can be found in (Griffith, 2005).

Two types of error, Berkson and classical, affect observational epidemiologic studies (Heid et al., 2004). Classical error occurs when multiple measurements, commonly in time-series data, do not represent the true value of interest because of the type of monitoring strategy. In other cases, this occurs because of missing values in the time series. These effects may combine to bias the estimated effect measure upwards or downwards (Armstrong, 1998). Berkson error reduces the study's power, which results in increased confidence intervals on the coefficients. This error occurs when subjects are assigned a group-average exposure (Armstrong, 1998), for example by assigning all residents within 2 km of a pollution monitor the same value.

Attempts to reduce classical error (in estimating a long-term mean) caused by incomplete datasets have included adjusting data values based on a fixed-location continuous monitor

(Larson et al., 2009; Adams et al., 2012; Kanaroglou et al., 2013; Hoek et al., 2002). This practice entails determining if air pollution conditions in the study area are above or below average by comparing the current concentration at a fixed-location continuous monitor to its long-term mean. The incomplete data are then adjusted to account for the above or below average conditions. After the adjustment, all observations in the incomplete dataset should be closer to the true long-term mean, which when averaged together should reduce classical error in estimating the long-term mean concentration. The adjustment method requires that regional phenomena affect air pollution concentrations uniformly in the study area. If this occurs, when one monitor is observing higher than average concentrations, all monitors should be observing higher than average concentrations. Thus, if monitoring times were biased, in that, data were collected mainly during above or below average conditions, this bias would be reduced.

Researchers increasingly use mobile monitoring to collect data on air pollution concentrations (Adams et al., 2012; Kanaroglou et al., 2013; Larson et al., 2009; Reggente et al., 2010); this concentration data is often collected to supplement exisitng monitoring networks that have few air pollution monitors. Mobile monitoring is different from traditional fixed-location continuous air pollution monitoring because the monitors are designed for rapid relocation, which researchers take advantage of in an attempt to reduce Berkson error by monitoring at various locations. This strategy, produces incomplete time-series datasets that are prone to give rise to classical error.

Our study evaluates an adjustment formula, which is designed to reduce classical error when one uses incomplete time-series datasets to estimate a long-term mean concentration. Evaluation is based on a set of incomplete datasets derived through sampling from a database of observations obtained with continuous fixed-location monitors. The error is determined by calculating the actual long-term mean, which is estimated from the entire time-series, with both the adjusted and unadjusted incomplete datasets. It is critical that researchers understand the effect that mobile monitoring may have on the assignment of ambient air pollution concentrations to subjects in their research and identify any approaches to reduce Berkson and classical error. Understanding the effect of the adjustment method will help ensure appropriate study designs.

Methods

Observed Air Pollution Data

The Paris, France, air pollution monitoring network was selected for study, which consists of 67 air pollution monitors. We focus on observations from 2012, which was a year that Paris' air pollution concentrations often exceeded guidelines for particulate matter, nitrogen dioxide, ozone, and benzene; along with highly variable weather conditions. In January to March, meteorological conditions were conducive to escalated air pollution episodes. Pollution concentrations reduced in the next months because of cool and wet weather, which continued through the fall (Airparif, 2013). The temporal variability of Paris' air pollution concentrations in 2012 provides us with a suitable dataset for testing the proposed adjustment method, because, without temporal variability, even incomplete datasets would adequately represent the long-term mean, and no adjustment would be necessary.

This study examined the effect of the adjustment method on particulate matter 10 microns or smaller in aerodynamic diameter (PM_{10}) concentrations, which were observed

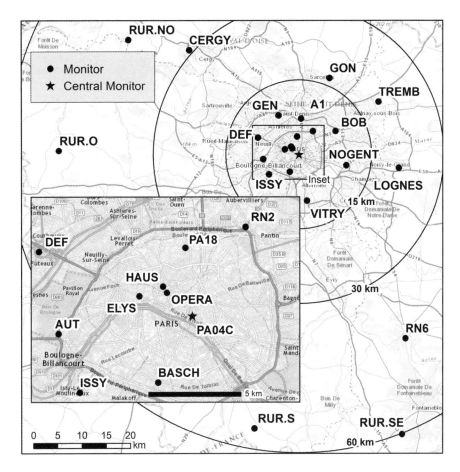

**Figure 11.1   Study area map of Paris, France with monitors identified by their
ID. Circles with radii of 15, 30 and 60 km are included and center on
the Paris Centre monitor. An inset is included of the downtown region**

at 24 locations with fixed-location continuous monitors and reported as hourly averages. We
present a map of the locations in Figure 11.1. Monitors were located in the following four
land use types: urban, peri-urban, rural, and transportation focused locations. All monitors
in the network are within 60 km of the central monitor (PA04C), located in the city center,
which will be used to determine any adjustments to the data.

During 2012, each fixed monitor recorded 8,784 hourly-observations. Missing,
erroneous, or incomplete data in these time-series ranged between 1.6% and 7.4% (mean
= 3.3%). The network's central monitor's missing data were filled by down-filling with
the previous hours' values. Down-filling was chosen over spatial interpolation for this
monitor to not increase its correlation with the other monitors; increased correlation would
increase the datasets similarity and artificially increase the effectiveness of the adjustment
method. When a datum in the time-series was missing for the central monitor, the down-
filling would replace the missing value with the previous hour's record. The other stations'

missing data were filled by spatially interpolating a value with the other monitoring stations' data, excluding the central location. The spatial interpolation method was inverse distance squared interpolation.

Time-Series Correlation

The adjustment method requires that monitors in the study area be temporally correlated. Without correlation between monitors the application of this or a similar adjustment method would be in vain. We identify the level of correlation with Pearson's r correlation coefficient, which is calculated between each monitor and the adjustment monitor (Paris Centre monitor). Log-values of the data were used because the data were distributed log-normal. To investigate if the correlation between monitors were associated to the distance between monitors or the station type, we regressed the correlation coefficients against distance to the Paris Centre monitor while controlling for the different monitoring land use types. If distance in the model was significant, it would indicate the adjustment approach is biased based on distance to the central monitor.

Adjustment method

In our adjustment method, we term the fixed-location continuous monitor used to adjust the incomplete datasets as the adjustment monitor. Our adjustment method is a linear adjustment defined with the following formula:

$$O_A = O_R \div \frac{log_e(S_h + e)}{log_e(S_L + e)}$$

where, O_R is the air pollution observation to be adjusted, O_A is the air pollution observation following adjustment, S_H is the concurrent hourly observation at the adjustment monitor, and S_L is the long-term arithmetic mean at the adjustment monitor. We add the base of the natural logarithms (e) as a constant to all values in the adjustment monitor's dataset to ensure the lowest value after the logarithm is taken is not less than one. Zero values in S_h would be indivisible, and the adjustment formula would fail. Paris Centre is used as the adjustment monitor because of its central location in the city. We limit the influence of extreme values when adjusting data by using log-values of the adjustment monitor's data.

Computer Simulation to Generate Incomplete Observations

We test our adjustment approach with a computer simulation to generate incomplete observations of air pollution time-series data. These incomplete data were obtained by sampling the Paris, France, time-series data. We analyzed the adjustment method for three periods of interest, which included one-week, one-season, and one year; one-season and one-year are general periods of interest in epidemiological studies, and one-week was chosen to explore the method on a shorter period. The geometric-mean was chosen to represent the long-term mean for a period of interest, because, the monitoring stations' data were distributed log-normal, and it is a better expected value of the data than the arithmetic mean.

Mobile monitoring that collects incomplete time-series data consists of two parameters that detail how sampling is conducted at a single location, which include the number of repeated observations (the number of times a monitor is set-up at a location) and the length of each observation. Regular relocation is common in mobile monitoring (Adams et al., 2012; Kanaroglou et al., 2013). Our computer simulation generated the incomplete observations varying those two parameters. We first stipulated that the total number of sampling hours (total sample hours) be less than one-third of the total hours in the period of interest, which ensures that the monitor would be able to observe three locations during the period of interest. For the one-week simulation, samples lengths included 1, 2, 4, 8, 16, and 32 hours. For the one-season (2,184 hours) simulation, sample lengths included additional sample lengths of 64, 128, 256 and 512 hours. The one-year (8,784 hours) simulation included all the sample lengths used for one-week and one-season with the addition of 1,024 and 2,048 hours. Sample counts began at one, and were doubled until the total number of hours sampled would be greater than one-third of the period of interest.

The process of the computer simulation for the selection of incomplete datasets follows:

1. Define the simulation parameters:
   a. Sample Length (*SL*)
   b. Sample Count (*SC*)
   c. Period of Interest (*POI*)
2. Randomly choose a time-period equal to the length of the *POI*. For the one-year simulation, this step is skipped as one-year cannot be varied. The one-week and one-season periods of interest are selected by a time-period with the correct number of continuous hours, in that, they could begin at any hour within the year with sufficient hours remaining.
3. Choose one monitor at random and select data for the time-period, excluding the Paris Centre monitor.
4. Calculate the long-term mean from the entire dataset obtained in 3.
5. Select *SC* samples of *SL* length without repetition from the dataset obtained in step 3 to generate the incomplete sample.
6. Apply the adjustment method to the incomplete sample.
7. Determine the percent error in estimating the long-term mean for both the adjusted and unadjusted incomplete sample.

For each combination of sample length, sample count, and period of interest, we repeated the simulation 50,000 times.

Statistical Analysis

We determined if the adjustment method reduced classical error using statistical analysis. The statistical significance comparisons were conducted with the student's *t-test* comparing the adjusted error and unadjusted error calculated for each of the 50,000 simulations. This was conducted for each combination of the period of interest, sample length, and sample count; $\alpha = 0.05$. Air pollution monitoring data and the error data were distributed log-normal; appropriate transformations were used to satisfy the assumptions of the statistical tests. Throughout the results, we refer to the percent error in calculating the long-term mean concentration from the unadjusted incomplete dataset or the adjusted incomplete dataset as the unadjusted error and adjusted error, respectively. All statistical analysis and simulations were conducted in R (R Core Team, 2013).

Results

Monitor Correlation

The minimum correlation between any monitor with the Paris Centre monitor was $r = 0.7$, with a maximum of $r = 0.91$, and a mean $r = 0.83$ (s.d. = 0.05). Table 11.1 presents the pairwise-correlations between all monitors with the adjustment monitor. It also includes the Euclidean distance between each monitor and the adjustment monitor, and each monitor's minimum, maximum, and geometric-mean air pollution concentrations.

Table 11.1   Paris air pollution monitors' correlation and distance to the Paris Central Monitor, the monitor's land use type, and the minimum, maximum and geometric-mean values for their time-series of data

Monitor ID	Pearson's *r*	Distance to Central Paris Monitor (m)	Type	Concentration		
				Min	Max	Geo- Mean
PA18	0.91**	3,611	Urban	1	166	21.7
VITRY	0.88**	9,437	Urban	1	172	21.5
ISSY	0.88**	7,326	Urban	1	128	20.4
OPERA	0.87**	1,851	Transportation	1	282	29.7
DEF	0.87**	8,984	Urban	0	172	21.9
BASCH	0.87**	3,972	Transportation	4	154	35.8
NOGENT	0.86**	9,968	Urban	0	108	17.9
GON	0.86**	16,609	Urban	0	132	21
GEN	0.86**	8,880	Peri-Urban	0	180	21.5
BOB	0.86**	8,850	Urban	1	144	20.7
TREMB	0.85**	19,528	Peri-Urban	2	122	21.1
HAUS	0.84**	2,229	Transportation	0	165	28.9
LOGNES	0.83**	20,782	Urban	1	134	19.3
ELYS	0.83**	3,072	Transportation	3	174	35.4
CERGY	0.83**	30,790	Urban	0	126	20.1
RN2	0.82**	5,496	Transportation	3	220	35.6
RN6	0.80**	43,046	Transportation	0	245	26.6
AUT	0.80**	7,316	Transportation	6	423	43.2
RUR.O	0.79**	49,491	Rural	1	101	18.5
RUR.SE	0.76**	60,163	Rural	0	105	16.8
RUR.NO	0.74**	42,115	Rural	0	95	13.9
RUR.S	0.73**	55,823	Rural	1	107	17.3
A1	0.70**	7,236	Transportation	7	282	49.4
PA04C	N/A	0	Urban	2	154	23.7

p < 0.01 **

A linear regression model with the dependent variable of correlation between each monitor and the adjustment monitor, using the predictor variables of (1) distance between the monitors and (2) dummy variables for each of the land use types, identified only one significant variable, which was rural land use; rural monitors' correlation with the Paris Centre monitor were significantly lower ($p < 0.05$) than the other monitors. The distance between any monitor to the Paris Centre monitor was not a significant factor in the linear regression model. Further analysis excluded the rural monitors because of their significantly lower correlations to the Paris Centre monitor, and research suggests that rural and urban ambient air pollution should be examined separately because of different causal factors and resulting air pollution conditions (Pedersen et al., 2013).

The one-week period of interest simulations demonstrated, for all combinations of sample lengths and counts, significantly reduced the classical error for calculating the long-term mean from incomplete datasets. Table 11.2 presents the unadjusted classical error and the amount of reduction in the classical error by applying the adjustment method.

Table 11.2   Error results from the computer simulation for the one-week period of interest

		Sample Length					
		1	2	4	8	16	32
Sample Count	1	24.76 (7.77)	24.22 (8.21)	23.29 (8.93)	21.14 (8.54)	18.64 (8.02)	15.25 (6.07)
	2	18.76 (5.36)	18.29 (5.7)	17.28 (5.92)	15.3 (5.15)	12.55 (3.84)	
	4	15.9 (4.72)	15.19 (4.72)	13.83 (4.28)	11.68 (3.07)		
	8	16.63 (5.95)	15.14 (5.53)	12.65 (3.92)			
	16	17.45 (7.24)	14.61 (5.78)				
	32	15.69 (6.54)					

Note: The classical error when estimating the long-term mean from the unadjusted incomplete dataset, and the reduction in classical error from the adjustment method is included in parenthesis. All reductions were statistically significant (p <.05).

The one-season and one-year periods of interests' statistical evaluations identified that the adjustment method did not always significantly reduce the mean, in some cases additional error was introduced. Results for the one-season period of interest are found in Table 11.3, and the one-year results are presented in Table 11.4.

Table 11.3   Error results from the computer simulation for the one-season period of interest

					Sample Length					
		1	2	4	8	16	32	64	128	256	512
	1	66 (17)	64 (18)	63 (20)	60 (20)	55 (19)	51 (17)	45 (12)	37 (6)	31 (2)	21 (−4)
	2	50 (10)	49 (11)	47 (11)	44 (10)	41 (9)	38 (6)	32 (2)	26 (−2)	19 (−6)	
	4	39 (5)	39 (5)	37 (5)	35 (5)	32 (2)	28* (0)	23 (−4)	17 (−6)		
	8	38 (6)	37 (6)	36 (5)	32 (3)	27* (0)	23 (−3)	16 (−6)			
Sample Count	16	42 (11)	40 (9)	36 (7)	31 (2)	25 (−1)	17 (−5)				
	32	44 (13)	39 (9)	33 (4)	27* (0)	17 (−5)					
	64	41 (11)	34 (5)	27* (1)	18 (−5)						
	128	34 (5)	28 (1)	18 (−5)							
	256	28 (1)	18 (−5)								
	512	18 (−5)									

$p > 0.05*$

Note: The classical error for estimating the long-term mean from the unadjusted incomplete dataset is presented, and the reduction in classical error from the adjustment method is included in parenthesis. Cells with light gray backgrounds are not statistically significantly different, and cells with dark gray backgrounds are when the adjustment method significantly increased the error.

Discussion

A city's diverse urban structure and local meteorology create spatially varying air pollution concentrations, which is one causal factor for different levels of ambient air pollution exposure across the population. Another factor is a person's movement throughout the city in a day. This issue of spatial variability has been known for a while, and that research concluded that a single monitor cannot represent an entire city (Goldstein and Landovitz, 1977a; Goldstein and Landovitz, 1977b). Currently, many monitoring locations are established in a city to capture the variability of air pollution exposure. The air pollution data obtained from the Paris, France, monitoring network were spatially variable with yearly-mean concentrations ranging from 14 µg/m3 to 50 µg/m3, which allowed us an effective study of the adjustment method. We feel these results are generalizable because of the similarity in concentration to many of areas, such as Greece (Sfetsos and Vlachogiannis, 2010; Grivas and Chaloulakou, 2006), Italy (Badaloni et al., 2013), Germany (Liu et al., 2013), in general Western Europe (Vienneau et al., 2013), Canada (Brook et al., 1997) and the United States (Samet et al., 2000).

Table 11.4   Error results from the computer simulation for the one-year period of interest

Sample Count		Sample Length											
		1	2	4	8	16	32	64	128	256	512	1024	2048
	1	333 (61)	332 (67)	340 (71)	311 (70)	309 (74)	285 (63)	270 (50)	235 (31)	210 (24)	160* (−5)	124* (−5)	87 (−7)
	2	260 (32)	255 (33)	252 (38)	251 (40)	229 (32)	214 (26)	196 (15)	170* (−2)	136 (−15)	96 (−22)	69 (−17)	
	4	215 (31)	210 (33)	211 (36)	203 (35)	193 (29)	175 (21)	154 (10)	121* (−6)	92 (−23)	55 (−24)		
	8	243 (104)	235 (103)	228 (101)	230 (99)	216 (91)	179 (76)	150 (49)	111* (5)	62 (−18)			
	16	306 (191)	289 (175)	269 (162)	273 (154)	218 (127)	179 (89)	136 (26)	67 (−13)				
	32	303 (196)	296 (192)	286 (163)	225 (134)	202 (104)	151 (34)	67 (−12)					
	64	314 (206)	298 (173)	231 (138)	201 (103)	154 (38)	70 (−9)						
	128	306 (183)	235 (139)	196 (92)	159 (38)	71 (4)							
	256	230 (135)	192 (85)	161 (38)	73 (5)								
	512	194 (86)	160 (38)	70* (2)									
	1024	159 (36)	69* (−1)										
	2048	69* (−2)											

$p > 0.05$*

Note: The classical error for estimating the long-term mean from the unadjusted incomplete dataset is presented, and the reduction in classical error from the adjustment method is included in parenthesis. Cells with light gray backgrounds are not statistically significantly different, and cells with dark gray backgrounds are when the adjustment method significantly increased the error.

The Paris, France, monitors were for the most part, temporally correlated, satisfying the main requirement of the adjustment method. Four primary land use types exist in this monitoring network, which included rural, peri-urban, urban, and transportation; rural monitors were significantly lower in their correlation with the Paris Centre monitor. Transportation-related pollution does not affect these areas significantly compared to urban areas, because of the low population and traffic density in rural regions. Our removal of these locations aligns to the thought that in epidemiological studies rural and urban areas should be assessed independently because of differing air pollution concentrations and respiratory health and exposure factors (Pedersen et al., 2013).

Recently, mobile monitoring technologies have been used to study the variability of air pollution in a city (Kanaroglou et al., 2013; Larson et al., 2009). The incomplete datasets that are observed with mobile monitoring may introduce classical error; data adjustment methods have been applied with the purpose to reduce classical error (Larson et al., 2009;

Adams et al., 2012; Kanaroglou et al., 2013; Hoek et al., 2002). We identify in our research that many different sampling parameters affect the amount of error for incomplete datasets, which include the number of samples obtained, and the length of the samples. Every simulation indicated that incomplete time-series datasets exhibit classical error, which occurred when either the unadjusted or adjusted observations were used to estimate the long-term mean concentration. When we compare a particular combination of sample count and sample length across all three periods of interest, the amount of error is positively correlated with the length of the period of interest, for example, an incomplete observation of two samples of two continuous hours resulted in 18%, 49 %, and 255 % average error for the unadjusted data for one-week, one-season, and one-year periods of interest respectively. The increased variability in meteorology that occurs with a longer period of interest is the probable cause for the increase in classical error; our findings are in agreement with the rational of controlling for seasonality when modeling air pollution (Chen et al., 2010; Pandey et al., 2014).

The total number of sampling hours consists of the sample count multiplied by the sample length. When we examine all the combinations of sample count and sample length that are a multiple of a particular total number of sampling hours, within a particular period of interest, we find that the least error occurs when the sample count and length are in the middle of the range of values tested. If we examine all the combinations that consist of 32 total sampling hours for the one-week period of interest, the lowest error occurs with four-samples of eight hours each, the second lowest error occurs with eight-samples of four hours each, and the highest error occurs with 32 samples of one hour each. This result has an implication for the design of monitoring programs that collect incomplete time-series data samples, which is that monitoring should occur with a balance between the number of observations and the length of each observation. Neither the adjusted or unadjusted data deviate from this finding.

When considering the use of an adjustment method, the total sampling time is important. If the total number of hours is a small portion of the period of interest, we find the adjustment method beneficial; however, as the portion sampled increases towards one-third of the period of interest, the utility of the adjustment method diminishes and the adjustment method may increase the classical error. Based on our findings the research that has incorporated an adjustment method would have benefited (Larson et al., 2009; Adams et al., 2012; Kanaroglou et al., 2013; Hoek et al., 2002).

Conclusions

It is apparent from our results that the optimal method for minimized classical error is to obtain the entire time-series with a fixed-location monitor. We understand this is not always possible and incomplete datasets may be the only option, to reduce classical error with these circumstances, we suggest the following guidelines:

(A) The total monitoring time should be equally divided by the number of samples and the sample length.
(B) When less than one-quarter of all possible observations are obtained, it is likely useful to employ an adjustment method.
(C) When possible, an evaluation similar to this study should be conducted on more than two monitors' historical data to provide an estimate of the classical error.

(D) If no data from historic monitoring are available, locate adjustment stations in the different primary land uses, for example urban and rural.

Classical error will not be eliminated by following our guidelines; however, it should be reduced. Our findings indicate that researchers who are using incomplete datasets have challenging decisions for sampling design that extend beyond the choice of locations for mobile monitoring.

References

Adams, M.D., De Luca, P., Corr, D., and Kanaroglou, P.S. 2012. Mobile Air Monitoring: Measuring Change in Air Quality in the City of Hamilton, 2005–2010. *Social Indicators Research*, 108(2): 351–64.

Airparif 2013. Air quality assessment network in the Paris Region. Accessed April 2013 from: http://www.airparif.asso.fr/_pdf/publications/bilan-2012-anglais.pdf.

Armstrong, B.G. 1998. Effect of measurement error on epidemiological studies of environmental and occupational exposures. *Occupational and Environmental Medicine*, 55(10): 651–6.

Badaloni, C., Ranucci, A., Cesaroni, G., Zanini, G., Vienneau, D., Al-Aidrous, F., De Hoogh, K., Magnani, C., and Forastiere, F. 2013. Air pollution and childhood leukemia: A nationwide case-control study in Italy. *Occupational and Environmental Medicine*, 70(12): 876–83.

Brook, J., Dann, T., and Burnett, R. 1997. The relationship among TSP, PM10, PM2.5, and inorganic constituents of atmospheric particulate matter at multiple Canadian locations. *Journal of the Air & Waste Management*, 47: 2–19.

Chen, L., Bai, Z., Kong, S., Han, B., You, Y., Ding, X., Du, S., and Lie, A. 2010. A land use regression for predicting NO2 and PM10 concentrations in different seasons in Tianjin region, China. *Journal of Environmental Sciences*, 22(9): 1364–73.

Delmelle, E.M. 2014. Spatial Sampling, in M.M. Fischer and P. Nijkamp (eds), *Handbook of Regional Science*: *1385–1399*. Berlin and Heidelberg: Springer Berlin Heidelberg.

Ghosh, J.K.C., Heck, J.E., Cockburn, M., Su, J., Jerrett, M., and Ritz, B. 2013. Prenatal exposure to traffic-related air pollution and risk of early childhood cancers. *American Journal of Epidemiology*, 178(8): 1233–9.

Goldstein, I. and Landovitz, L. 1977a. Analysis of air pollution patterns in New York City—I. Can one station represent the large metropolitan area? *Atmospheric Environment (1967)*, 11(1): 47–52.

Goldstein, I. and Landovitz, L. 1977b. Analysis of air pollution patterns in New York City—II. Can one aerometric station represent the area surrounding it? *Atmospheric Environment (1967)*, 11(1): 53–7.

Griffith, D. 2005. Effective geographic sample size in the presence of spatial autocorrelation. *Annals of the Association of American Geographers*, 95(4): 740–60.

Grivas, G. and Chaloulakou, A. 2006. Artificial neural network models for prediction of PM10 hourly concentrations, in the Greater Area of Athens, Greece. *Atmospheric Environment*, 40(7): 1216–29.

Heid, I.M., Küchenhoff, H., Miles, J., Kreienbrock, L., and Wichmann, H.E. 2004. Two dimensions of measurement error: Classical and Berkson error in residential radon exposure assessment. *Journal of Exposure Analysis and Environmental Epidemiology*, 14(5): 365–77.

Hoek, G., Krishnan, R.M., Beelan, R., Peters, A., Ostro, B., Brunekreef, B., and Kaufman, J.D. 2013. Long-term air pollution exposure and cardio-respiratory mortality: A review. *Environmental Health: A Global Access Science Source*, 12(1): 43.

Hoek, G., Meliefste, K., Cyrys, J., Lewné, M., Bellander, T., Brauer, M., Fischer, P., Gehring, U., Heinrich, J., van Vliet, P., and Brunekreef, B. 2002. Spatial variability of fine particle concentrations in three European areas. *Atmospheric Environment*, 36(25): 4077–88.

Hutter, H.P., Haluza, D., Piegler, K., Hohenblum, P., Fröhlich, M., Scharf, S., Uhl, M., Damberger, B., Tappler, P., Kundi, M., Wallner, P., and Moshammer, H. 2013. Semivolatile compounds in schools and their influence on cognitive performance of children. *International Journal of Occupational Medicine and Environmental Health*, 26(4).

Jerrett, M., Arain, A., Kanaroglou, P., Beckerman, B., Potoglou, D., Sahsuvaroglu, T., Morrison, J., Giovis, C. 2005. A review and evaluation of intraurban air pollution exposure models. *Journal of Exposure Analysis and Environmental Epidemiology*, 15(2): 185–204.

Kanaroglou, P., Jerrett, M., Morrison, J., Beckerman, B., Arain, M.A., Gilbert, N.L., and Brook, J.R. 2005. Establishing an air pollution monitoring network for intra-urban population exposure assessment: A location-allocation approach. *Atmospheric Environment*, 39(13): 2399–409.

Kanaroglou, P.S., Adams, M.D., De Luca, P.F., Corr, D., and Sohel, N. 2013. Estimation of sulfur dioxide air pollution concentrations with a spatial autoregressive model. *Atmospheric Environment*, 79: 421–7.

Kumar, N. 2009. An Optimal Spatial Sampling Design for Intra-Urban Population Exposure Assessment. *Atmospheric Environment*, 43(5): 1153.

Larson, T., Henderson, S.B., and Brauer, M. 2009. Mobile monitoring of particle light absorption coefficient in an urban area as a basis for land use regression. *Environmental Science & Technology*, 43(13): 4672–8.

Liu, C., Flexeder, C., Fuertes, E., Cyrys, J., Bauer, C-P, Koletzko, S., Hoffmann, B., von Berg, A., and Heinrich, J. 2013. Effects of air pollution on exhaled nitric oxide in children: Results from the GINIplus and LISAplus studies. *International Journal of Hygiene and Environmental Health*. DOI: 10.1016/j.ijheh.2013.09.006

Padula, AM, Mortimer, K, Hubbard, A., Lurmann, F., Jerrett, M., and Tager, IB 2012. Exposure to traffic-related air pollution during pregnancy and term low birth weight: Estimation of causal associations in a semiparametric model. *American Journal of Epidemiology*, 176(9): 815–24.

Pandey, B., Agrawal, M., and Singh, S. 2014. Assessment of air pollution around coal mining area: Emphasizing on spatial distributions, seasonal variations and heavy metals, using cluster and principal component analysis. *Atmospheric Pollution Research*, 5.

Pedersen, M., Siroux, V., Pin, I., Charles, M.A., Forhan, A., Hulin, A., Galineau, J., Lepeule, J., Giorgis-Allemand, L, Sunyer, J, Annesi-Maesano, I., and Slama, R. 2013. Does consideration of larger study areas yield more accurate estimates of air pollution health effects? An illustration of the bias-variance trade-off in air pollution epidemiology. *Environment International*, 60C: 23–30.

R Core Team 2013. R: A language and environment for statistical computing. R Foundation for Statistical Computing, Vienna, Austria. http://www.R-project.org/.

Reggente, M., Mondini, A., Ferri, G., Mazzolai, B., Manzi, A., Gabelletti, M., Dario, P., and Lilienthal, A.J. 2010. The dustbot system: Using mobile robots to monitor pollution in pedestrian area. *Proc. of NOSE*, 23: 273–8.

Samet, J.M., Dominici, F., Curriero, F.C., Coursac, I., and Zeger, S.L. 2000. Fine particulate air pollution and mortality in 20 U.S. cities, 1987–1994. *New England Journal of Medicine*, 343(24): 1742–9.

Sfetsos, A., and Vlachogiannis, D. 2010. A new approach to discovering the causal relationship between meteorological patterns and PM10 exceedances. *Atmospheric Research*, 98(2–4): 500–11.

Vienneau, D., de Hoogh, K., Bechle, M.J., Beelen, R., van Donkelaar, A., Martin, R., Millet, D., Hoek, G., and Marshall, J.D. 2013. Western European land use regression incorporating satellite- and ground-based measurements of NO2 and PM10. *Environmental Science & Technology*, 47: 13555−64.

Wallner, P., Kundi, M., Moshammer, H., Piegler, K., Hohenblum, P., Scharf, S., Fröhlich, M., Damberger, B., Tappler, P., and Hutter, H-P 2012. Indoor air in schools and lung function of Austrian school children. *Journal of Environmental Monitoring: JEM*, 14(7): 1976–82.

Chapter 12
The Geographic Distribution of Metal Dust Exposure in Syracuse, NY

Daniel A. Griffith

Two series of papers furnish spatial analyses of elevated blood lead levels (BLLs) and of the more comprehensive sample of soil metals for the City of Syracuse, NY. A more recent outcome of the research project producing these studies is the establishment of field methods for mapping urban metal distributions in house dusts and surface soils (Johnson et al., 2005). Another is establishment of the track-in on footwear transfer mechanism for mass transport of soil indoors (Hunt, Johnson and Griffith, 2006). A third dust data study describes spatial patterns of non-carpeted floor dust loading in Syracuse homes (Johnson et al., 2009), whereas a fourth assesses risk remaining from fine particle contaminants after vacuum cleaning of hard floor surfaces (Hunt et al., 2008). This chapter extends these latter studies by integrating the dust, BLL, soils, and socio-economic/demographic data, and completing a comprehensive spatial analysis of these data with the objective of exposure analysis/visualization and assessment of environmental risk. The need for this extension is to further our understanding of soil as a toxic metals reservoir, and as an important pathway of human environmental contamination exposure (see Wuana and Okieimen, 2011).

Three years of geo-referenced data collection fieldwork[1] were undertaken in the City of Syracuse (see Johnson et al., 2005; Griffith, Johnson and Hunt, 2009; Hunt et al., 2012). This effort produced a collection of 1,585 unique location outdoor soil samples tagged to 426 unique house indoor dust samples.[2] A NITON XL-700-series X-ray fluorescence (XRF) instrument was used in a chemistry laboratory to measure arsenic (As), cobalt (Co), copper (Cu), iron (Fe), mercury (Hg), manganese (Mn), molybdenum (Mo), nickel (Ni), lead (Pb), rubidium (Rb), selenium (Se), strontium (Sr), zinc (Zn), and zirconium (Zr) in these soil samples, based on 120-s testing time and NIST 2711 standard reference materials (SRM); measurements are in milligrams of metal per kilogram of soil, or ppm. The collection of dust samples essentially followed EPA wipe test protocols (USEPA, 1995); loading measurements are milligrams of metal per square foot of floor space. The dust loadings are for cadmium (Ca), Co, Cu, Fe, Ni, Mn, Pb, and Zn. Figure 12.1 presents a flowchart summarizing the data compilation methodology for this study. Figure 12.2

---

1   Individual house and BLL addresses were linked to Syracuse cadastral map parcels. Selected houses had four soil samples taken, one in the backyard, one in the front yard, one by the street, and one by the drip line; each soil sample was geocoded with a global positioning system (GPS) unit.

2   The three original separate datasets were reduced in size by eliminating houses with incomplete soil and dust sample, and without BLL, measurements. Consequently, 2,042 of the 3,627 parcels with soil samples, 62 of the 488 residences with dust samples, and 10,339 of the 13,708 children with BLL measurements were removed during construction of the combined dataset.

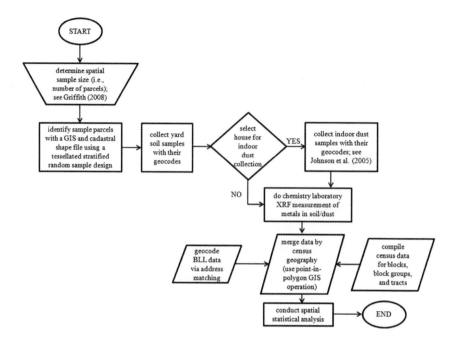

Figure 12.1   Flowchart articulating the data compilation methodology for this study

portrays the geographic distribution of house dust samples with their matched soil samples concurrently with the parcel boundaries for Syracuse, NY.[3]

The purpose of this chapter is to summarize features of these data, particularly as they relate to pediatric blood lead levels and census published socio-economic/demographic variables. Census data are used, and hence an ecological correlation study is undertaken, because individual level socio-economic and demographic data are not available; the data collection funded and approved by the Syracuse University Institutional Review Board (IRB) was for only soil and dust samples, not household characteristics. The goal of this exercise is, as a case study, to shed light on relationships between the geographic distribution of metals across Syracuse, NY, and especially their covariations with BLLs in order to better contextualize a public health concern.

Relationships Within and Between House Soils and Dust Samples: The Parcel Level

Matching houses with their nearby yard soil samples yields a dataset comprising 1,557 soil samples and 417 houses[4] in which dust was successfully collected (see Figure 12.2).

---

3   Eight houses were located on parcels with incomplete GIS shapefile boundaries. These data were retained by manual editing of the attribute table (see Figure 12.3c). One house was surveyed twice; its coordinates and measurements were averaged.

4   915 samples were taken from 433 houses, with one of these houses being surveyed twice.

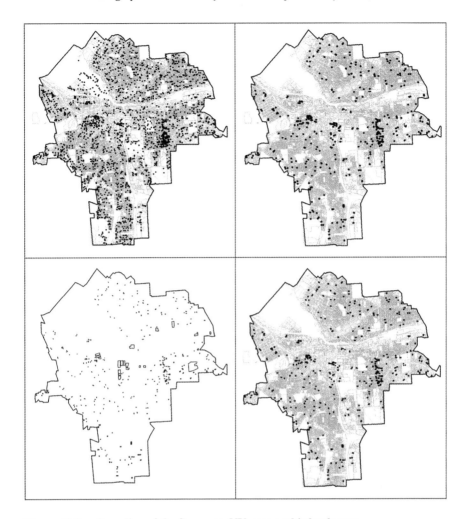

Figure 12.2   Sampling of the Syracuse, NY, geographic landscape

Note: Black dots denote house locations, and gray lines denote parcel boundaries. Top left (a): soil sample house locations superimposed on a parcel map. Top right (b): dust sample house locations superimposed on a parcel map. Bottom left (c): parcels in which sampled houses are located. Bottom right (d): matched soil and dust sample house locations superimposed on a parcel map.

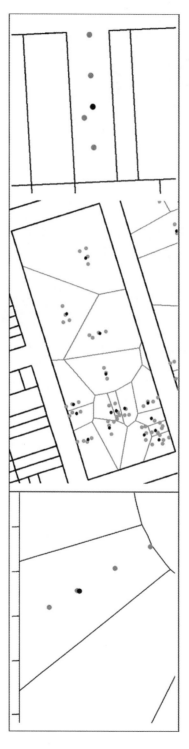

Figure 12.3   Zoom ins from Figure 12.2a: sample parcels in Syracuse, NY

Note: Black dots denote dust sampled house locations, and gray dots denote soil sample locations. Left (a): the typical geographic sampling of a parcel. Middle (b): partitioning of soil samples for allocation to houses on parcels with multiple houses by Thiessen polygons (gray lines denoted polygon boundaries). Right (c): a specimen parcel having a corrupt boundary

Figures 12.3a and 12.3b portray two typical situations. Each house for which dust samples were collected also tends to have four collected yard soil samples typically positioned near its front street and drip line, and in its back yard. In some cases, one or more of these soil samples yielded negligible or questionable assays results, resulting in fewer usable samples. In other cases, supplemental soil samples were collected primarily near streets, resulting in five, six or eight (parcels at the corner of two streets) samples. One house that was surveyed twice has seven soil samples. The frequency distribution of soil samples by house is as follows:

# samples	1	2	3	4	5	6	7	8
frequency	2	18	92	291	10	1	1	2

Metal measure geometric means were calculated to summarize sets of parcel soil samples. Not surprisingly, the distribution of a metal's concentration in a soil or dust sample is skewed, partly because it has a lower bound of 0, which is reinforced by a slightly positive lower bound determined by the detection level of a chemical analysis. Preliminary descriptive statistical analyses reveal that transforming these skewed distributions toward a bell-shaped curve is achievable with a logarithmic or inverse Box-Cox power transformation.[5] Table 12.1 summarizes the necessary transformations. Except for Zr, which is approximately normally distribution[6] in its raw measurement form, the frequency distributions for all of the other soil or dust measures significantly differ from a normal frequency distribution, with a reported Shapiro-Wilk diagnostic statistic probability of < 0.0001. All measures benefit from the application of a Box-Cox power transformation—even the probability for Zr increases—although some of the transformed distributions still deviate significantly from a bell-shape curve.

Table 12.1   **Estimated Box-Cox power transformation and normal distribution goodness-of-fit statistics for the metals: parcel geography**

Metal	Sample type	Box-Cox power transformation parameter		Shapiro-Wilk statistic probability
		exponenta	translation	
Pb	soil	0	−9.31	0.0062
	dust	0	0.02	0.0243
Zn	soil	0	15.41	0.1283
	dust	0	2.33	0.1668
Cu	soil	0	8.00	0.0005
	dust	0	0.07	<0.0001
Mn*	soil	0	124.05	0.0002
	dust	0	−0.10	0.0007
Ni	soil	−0.22	2.41	0.0017
	dust	−0.18	0.15	0.0003

(continued ...)

---

5   All inverse transformation values are multiplied by –1 to preserve the direct relationships between the raw and the transformed variables.

6   Shapiro-Wilk diagnostic statistic probability of 0.1824.

Table 12.1   (*concluded*)

Metal	Sample type	Box-Cox power transformation parameter		Shapiro-Wilk statistic probability
		exponent^a	translation	
Co	soil	0	1.43	0.0016
	dust	0	0.01	<0.0001
Fe	soil	−0.41	35383.13	<0.0001
	dust	−0.24	16.29	0.0228
Mo	soil	0	2.15	<0.0001
Zr	soil	−0.50	1916.27	0.5967
Sr	soil	−4.30	173.81	0.0129
Rb	soil	−1.13	178.86	0.0040
Se	soil	−4.06	20.70	0.1016
As	soil	−0.77	5.02	0.0003
Hg	soil	0	1.96	0.5282
Ca	dust	−0.43	224.83	

* denotes a marked difference exists between the optimal exponent for the soil and dust samples.
0 denotes a logarithmic transformation; transformed variables with a negative exponent are multiplied
by −1 to preserve the direct relationship between them and their raw variable counterparts.

One question about two sets of measurements for the same parcels asks whether or not
common dimensions span them. This is a classical canonical correlation analysis (CCA)[7]
problem because the research question of interest asks what the data dimensions are, if any,
that span the two datasets. One comparison here is between CCA results for the raw and
the transformed metal measures. CCA uncovers three significant dimensions in both cases
(Table 12.2). However, the third dimension in the transformed data fails to truly span both
sets of measures, based upon a substantively meaningful correlation of |0.4|.[8] The Box-
Cox power transformations also produce a simpler structure, although Pb is a dominant
dimension for both datasets. For the transformed dataset, the first canonical dimension may
be labeled a lead dimension, with Pb in both sets of measurements correlating strongly with
this dimension. The conspicuous covariation of Pb and Zn in soil, and Pb in house dust,
is consistent with lead-based paint usage. The correlation between this dimension and soil
As is consistent with the usage of arsenic-treated wood. The presence of soil As also is
consistent with road traffic and combustion sources. The Box-Cox power transformations
separate Fe from Pb. Both CCAs separate soil and dust Zn. Meanwhile, the second
canonical dimension may be labeled landscaping-sealed-house: likely sources are decaying
plant matter for Rb, fertilizer for Fe, the absence of air conditioning for Zn, and the absence
of carpet wear/rubbery underlay for Ca. If Ca is negatively correlated, then Rb and Fe
should be positively correlated, with this dimension. The local bedrock is the most likely
source for Zr. Mo may well result from some point source pollution, such as the combustion
of fossil fuels by power plants.

---

7   Canonical correlation analysis is a multivariate statistical technique that handles
multicollinearity in a set of response variables as well as in a set of covariates. Accordingly, it focuses
on correlations between linear combinations of variables in the two data sets. As such, it sometimes is
referred to as a double-barreled principal components analysis.
8   With a Bonferroni adjustment coupled with a two-tailed test using a 1% level of significance,
any correlation coefficient greater than 0.18 in absolute value is significant.

Table 12.2   Canonical structure: significant canonical dimensions at a 5% level of significance: parcel geography

Metal	Sample	Canonical variate #1	Canonical variate #2	Canonical variate #3
Raw data				
Mo	soil	0.0278	−0.1683	0.1281
Zr	soil	−0.3188	−0.3250	0.3663
Sr	soil	**0.9318**	−0.0208	0.2527
Rb	soil	−0.2330	0.2494	−0.2703
Pb	soil	0.1912	**0.7502**	0.2974
Se	soil	0.0008	0.1342	−0.2497
As	soil	−0.0104	0.2879	0.3406
Hg	soil	−0.0784	−0.1794	**−0.4435**
Zn	soil	0.2026	**0.5304**	0.3428
Cu	soil	0.1313	0.1070	0.1804
Ni	soil	0.3964	−0.0931	0.0291
Co	soil	0.0011	0.0297	−0.0669
Fe	soil	−0.1414	**0.4845**	0.1894
Mn	soil	−0.1781	−0.0131	0.2854
Zn	dust	−0.0254	−0.1790	**0.8448**
Cu	dust	**0.9513**	−0.1361	0.1303
Mn	dust	0.0572	0.0666	0.0530
Pb	dust	**0.4628**	**0.7379**	0.3222
Ni	dust	0.0846	0.1340	0.0730
Co	dust	0.0276	0.1992	0.3533
Ca	dust	0.1426	0.0574	0.2227
Fe	dust	0.0285	0.0643	0.0943
Canonical correlation		0.5897	0.3323	0.3107
Box-Cox power transformed data				
Mo	soil	−0.1942	**0.5250**	**−0.5479**
Zr	soil	−0.2710	**−0.4954**	−0.0012
Sr	soil	0.2534	−0.2878	0.0827
Rb	soil	0.0290	**0.4371**	**0.4413**
Pb	soil	**0.9670**	0.0961	−0.0282
Se	soil	0.1368	0.3117	0.1770
As	soil	0.4575	**0.1879**	−0.2606
Hg	soil	−0.1529	0.3088	0.0880
Zn	soil	**0.8001**	0.2397	0.0881
Cu	soil	0.2508	0.0697	−0.2928
Ni	soil	0.1296	−0.1456	−0.0100
Co	soil	0.0787	0.1640	0.1370
Fe	soil	0.3541	**0.4434**	0.3259
Mn	soil	0.0199	0.2346	−0.1409
Zn	dust	0.1591	**−0.4736**	−0.2773
Cu	dust	0.1669	−0.0292	−0.1175
Mn	dust	0.2784	0.0267	−0.1130

(continued ...)

Table 12.2   (*concluded*)

Metal	Sample	Canonical variate #1	Canonical variate #2	Canonical variate #3
Box-Cox power transformed data				
Pb	dust	**0.8459**	−0.0047	−0.0456
Ni	dust	0.2108	−0.0647	−0.0973
Co	dust	0.3205	0.0457	0.0043
Ca	dust	0.2925	−0.4797	0.1100
Fe	dust	0.3497	−0.0220	0.3024
Canonical correlation		0.6893	0.3525	0.2852

Note: bold denotes a prominent correlation coefficient.

One important finding here is that a linear relationship exists between individual house outside soil and inside dust Pb measurements (Figure 12.4). In other words, a potential exists for predicting indoor dust PB loadings from yard soil Pb concentrations. Given the dispersed nature of the geographic sample (Figure 12.2c), spatial autocorrelation should play a minor role here. This contention was evaluated by visually inspecting semivariograms, constructed both with SAS[9] and with the Geostatistical Wizard in ArcGIS, for the seven metals having both yard soil (Griffith, 2008) and house dust measures.

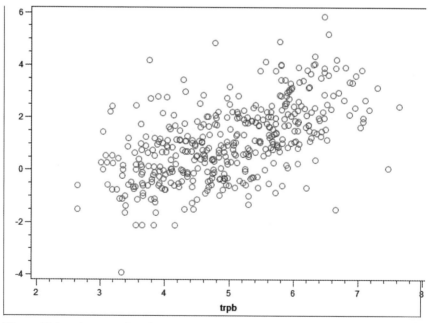

Figure 12.4   A scatterplot of the relationship between Box-Cox power transformed soil (horizontal axis) and dust (vertical axis) measurements

---

9   SAS is a widely used statistical software package.

Figure 12.5   Sampled census geography of the Syracuse, NY, geographic landscape
Note: Top left (a): black dots denote residential locations of children for whom pediatric BLLs were measured, superimposed on a parcel map (gray denotes parcel boundaries). Top right (b): census blocks in which dust sampled houses are located. Bottom left (c): census block groups with those highlighted in which dust sampled houses are located. Bottom right (d): census tracts with those highlighted in which dust sampled houses are located

Relationships Between House Dust Samples and Demographic and Pediatric BLL Variables: The Census Block Level

Because much of the BLL data are from the late 1990s, and the soil and dust data were collected in the early 2000s, socio-economic and demographic data were retrieved from United States Census 2000 sources. The 2000 census block data contain demographic variables such as counts for Hispanic (that is, ethnic group), racial groups (that is, Caucasian, Black, Asian), total population, and number of children. The area of blocks (in square kilometers) was calculated with ArcMap 10, the primary module of ESRI's

Table 12.3 **Estimated transformation parameters and normal distribution goodness-of-fit statistics for the demographic census block geography**

Manly transformation		
variable	Parameter estimate	Shapiro-Wilk statistic probability
% children	−0.01	0.0067
Logistic transformation		
% Hispanic	0.40	< 0.0001
% Caucasian	0.73	< 0.0001
% Black	0.84	< 0.0001
% Asian	0.18	< 0.0001

Box-Cox power transformation			
	Exponenta	Translation	
Density	−1.94	7862.47	< 0.0001
Average BLL Pb	−0.37	0.01	< 0.0001
Maximum BLL Pb	0	0.18	< 0.0001
Standard deviation of BLL Pb	−1.63	12.15	< 0.0001

Note: 0 denotes a logarithmic transformation; transformed variables with a negative exponent are multiplied by −1 to preserve the direct relationship between them and their raw variable counterparts.

(Environmental Systems Research Institute's) geographic information system software package, ArcGIS.[10] All counts were converted to percentages,[11] except for total population, which was converted to population density with the ArcMap area calculation. Each of these variables was transformed so that its frequency distribution better mimicked a bell-shaped curve. The Manly (1976) transformation of the following functional form best modifies the percentage of children: $e^{-\alpha\,children/population}$. The following transformation best modifies the other percentages: $\mathrm{LN}\left[\dfrac{p+\delta}{100+\delta-p}\right]$, where p denotes the percentage, LN denotes natural logarithm, and δ is a translation parameter shifting the percentages along the horizontal number line. Density was transformed with a standard Box-Cox power transformation. The normality diagnostic statistic improved in all cases, but remained significant (that is, the transformed frequency distributions significantly differ from a bell-shaped curve).

The pediatric BLL data (Griffith et al., 1998) have 3,369 cases located in the sampled census blocks (Figure 12.5a). Twenty-six of the dust sampled houses reside in census blocks without BLL data. The remaining sampled houses are located in census blocks with from 1 to 109 BLL measures. Ecological BLL measures for the blocks include average, maximum, and standard deviation of parts of Pb per micro-deciliter of blood. These measures also were subjected to a Box-Cox power transformation (see Table 12.3). As before, all measures benefit from the application of a suitable transformation, although the transformed distributions still deviate significantly from a bell-shape curve.

Table 12.4 summarizes CCA results for this level of census geography. For comparison purposes, the demographic dataset is analyzed separately and combined with pediatric BLL measures. Again, CCA uncovers three significant dimensions in both cases. However, the

---

10   A tool to view, edit, create, analyze, and map geospatial data.
11   Three houses were removed from the analysis because they are located in census blocks with zero population.

third dimension in the expanded nonmetals dataset fails to truly span both sets of measures, once more based upon a substantively meaningful correlation of |0.4|. The first canonical dimension relates most of the metals to the set of demographic variables. It appears to be a summary dimension. The second canonical dimension relates the BLL measures directly to Cu and Pb in house dust, and inversely to % Hispanic. This dimension may be labeled lead contamination, and is sensible because lead and copper plumbing fixtures are one potential source of house dust metals, and pediatric lead ingestion in urban areas tends to covary with poverty and minority racial status (although directly rather than inversely).

Table 12.4   Canonical structure: significant canonical dimensions at a 5% level of significance: block census geography

Variable	Canonical variate #1	Canonical variate #2	Canonical variate #3
Metals and demographic variables co-dimensions			
Zn	0.6601	**0.2598**	**0.4419**
Cu	−0.3325	**0.4245**	0.2744
Mn	**−0.4240**	**0.4590**	**0.4080**
Pb	−0.3787	**0.8124**	0.3472
Ni	**0.4788**	−0.1491	**−0.4967**
Co	−0.2663	**0.4430**	0.3702
Ca	**0.8107**	−0.3171	−0.0603
Fe	**0.5316**	−0.2817	**−0.5392**
% children	**−0.6268**	−0.0038	**0.5570**
% Hispanic	**−0.6604**	**−0.5803**	0.3300
% Caucasian	**0.9386**	0.0684	0.1648
% Black	**−0.8616**	−0.0642	−0.1873
% Asian	−0.0633	0.3068	−0.1400
Density	**0.7028**	−0.1154	**−0.4237**
Canonical correlation	0.5417	0.3245	0.2040
Metals and demographic plus BLL variables co-dimensions			
Zn	**−0.6923**	0.1618	0.1877
Cu	**−0.4184**	**0.4589**	0.2052
Mn	**−0.4653**	0.3135	−0.0446
Pb	**−0.4904**	**0.7813**	−0.0926
Ni	**0.5398**	−0.2357	−0.3006
Co	−0.2953	0.3664	−0.2936
Ca	**0.8289**	−0.1534	0.2176
Fe	**0.5834**	−0.2644	−0.1172
% children	**−0.6198**	−0.0507	0.3189
% Hispanic	**−0.5962**	**−0.5691**	0.2003
% Caucasian	**0.8938**	0.2500	0.1209
% Black	**−0.8249**	−0.2148	−0.0341
% Asian	−0.0616	0.1677	**−0.5391**
Density	**0.6861**	0.0165	−0.0427
Average BLL Pb	−0.3289	**0.5969**	0.2041
Maximum BLL Pb	−0.2043	**0.5105**	−0.0868
Standard deviation of BLL Pb	−0.2157	**0.5550**	0.0485
Canonical correlation	0.5591	0.3827	0.2526

Note: Bold denotes a prominent correlation coefficient.

Relationships Between House Dust Samples and Socio-Economic/Demographic and Pediatric BLL Variables: The Census Block Group Level

Employing census block group data in an ecological analysis allows socio-economic variables to be included with demographic variables, as well as strength to be borrowed from nearby BLL measures. Syracuse was partitioned into 147 census block groups for 2000 data tabulations. The dust sampled houses occupy 116 of these block groups (Figure 12.5c). The pediatric BLL data have 13,011 cases located in the sampled census block groups, whose numbers of BLL measures range from 3 to 500. Table 12.5 summarizes the transformations that result in frequency distributions that better mimic a bell-shaped curve; the standard deviation of BLLs is untransformed. Again, although the transformed data still have significant normality diagnostic statistics, the frequency distribution of each of the transformed variables is closer to that of a normal distribution.

Table 12.5   Estimated transformation parameters and normal distribution goodness-of-fit statistics for the demographic block group census geography

Manly transformation		
variable	Parameter estimate	Shapiro-Wilk statistic probability
% children	−0.01	< 0.0001
% Hispanic	−0.33	< 0.0001
Average family size	−0.77	< 0.0001
Female/male ratio	−0.79	< 0.0001
Logistic transformation		
% Caucasian	0.21	< 0.0001
% Black	−0.08	0.0004
% Asian	0.11	< 0.0001
% renter	7.00	< 0.0001
Box−Cox power transformation		
	Exponenta	Translation	
Density	0.50	−600.34	< 0.0001
Average BLL Pb	0	1.80	< 0.0001
Maximum BLL Pb	0	22.70	< 0.0001
No transformation		
Standard deviation of BLL Pb		< 0.0001

Note: 0 denotes a logarithmic transformation; transformed variables with a negative exponent are multiplied by −1 to preserve the direct relationship between them and their raw variable counterparts. Gray shading denotes no parameter estimates.

Table 12.6 summarizes CCA results for this level of geography. Once more, CCA uncovers three significant dimensions. And, as before, the third dimension fails to truly span both sets of measures, having a very small canonical correlation of 0.25. The first canonical dimension relates most of the metals to the expanded set of socio-economic/demographic variables.[12] As with the census block geography, it appears to be a summary dimension. The

---

12   More variables are publicly available for the coarser resolution census areal units.

second canonical dimension relates the BLL measures directly to Pb and Co in house dust, and inversely to a female/male ratio. This dimension may be labeled lead contamination, and exhibits a persistence of a Pb dimension with increasingly coarser geographic resolution. Figure 12.6 portrays relationships between the spatially adjusted transformed variables and the uncovered canonical dimensions.

Table 12.6   Canonical structure: significant canonical dimensions at a 5% level of significance for metals and demographic plus BLL variables co-dimensions: block group census geography

Variable	Canonical variate #1	Canonical variate #2	Canonical variate #3
Spatially unadjusted			
Zn	**0.6595**	−0.0300	0.2377
Cu	0.3466	0.2447	−0.0780
Mn	**0.4181**	0.3205	0.2237
Pb	**0.4374**	**0.8148**	0.1061
Ni	**−0.5366**	−0.0756	−0.2207
Co	0.3066	**0.4130**	0.3797
Ca	**−0.8019**	−0.2314	0.0830
Fe	**−0.5759**	−0.3672	−0.3994
% children	**0.6534**	0.0638	0.0185
% Hispanic	**0.6898**	−0.2575	0.2584
% Caucasian	**−0.8804**	0.0018	0.2979
% Black	**0.7991**	−0.0100	−0.3653
% Asian	−0.0131	0.2618	0.0737
Density	**0.4162**	0.3132	0.1128
Average family size	**0.4904**	0.2984	**0.4049**
% renters	**0.7464**	0.0735	0.0311
Female/male ratio	0.1028	**−0.4826**	−0.2959
Average BLL Pb	**0.7209**	**0.4760**	0.0207
Maximum BLL Pb	0.2496	**0.4796**	0.0696
Standard deviation of BLL Pb	0.3241	**0.4205**	0.0034
Canonical correlation	0.2500	0.5727	0.5517
Spatially adjusted			
Zn	**−0.6307**	0.2831	
Cu	−0.1986	**0.4259**	
Mn	−0.2566	**0.5904**	
Pb	−0.2499	**0.9219**	
Ni	**0.5314**	−0.3929	
Co	−0.2624	**0.7198**	
Ca	0.6601	−0.4153	
Fe	0.4062	−0.5738	
% children	−0.3776	0.2240	
% Hispanic	−0.4475	0.4253	
% Caucasian	0.5913	0.2499	
% Black	−0.4902	−0.1056	

(continued ...)

Table 12.6   (*concluded ...*)

Variable	Canonical variate #1	Canonical variate #2	Canonical variate #3
Spatially adjusted			
% Asian	−0.0321	0.1600	
Density	−0.6127	0.2255	
Average family size	−0.0566	0.3659	
% renters	−0.4769	0.0548	
Female/male ratio	−0.3703	−0.3997	
Average BLL Pb	−0.3954	0.4518	
Maximum BLL Pb	0.1140	0.5435	
Standard deviation of BLL Pb	0.0416	0.4618	
Canonical correlation	0.3585	0.2693	

Note: Bold denotes a prominent correlation coefficient; gray background denotes a non-significant dimension.

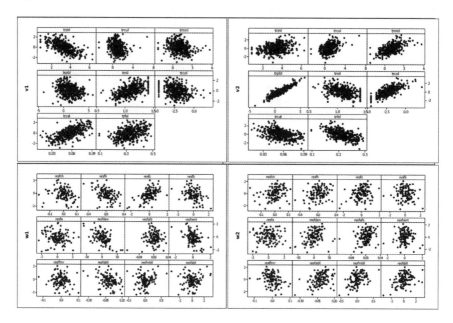

Figure 12.6   Scatterplots of transformed variables with canonical dimensions: census block groups resolution

Figure 12.5c suggests that spatial autocorrelation may play a role in this data analysis. Eigenvector spatial filter (Griffith, 2003) descriptions of the transformed socio-economic/demographic as well as BLL variables were constructed, and the aspatial components were used in a CCA for comparative purposes. The 166 census block groups have 454 rook adjacency links,[13]

---

13   For completeness, the single island census block group was connected to its closest neighboring block group.

and render 30 eigenvectors portraying positive spatial autocorrelation and having a relative Moran Coefficient of at least 0.25. Between 9 and 22 of these eigenvectors account for spatial autocorrelation latent in the individual variables. Table 12.7 summarizes the amount of redundant information contained in these variables that is attributable to spatial autocorrelation; most of the variables contain weak-to-moderate positive spatial autocorrelation.

Accounting for latent spatial autocorrelation dramatically alters the dimensions spanning the two datasets (see Table 12.6). Foremost, only one dimension is significant. It appears to be a rather weak (that is, the canonical correlation is only 0.36) poverty dimension; it is inversely correlated with percentage of minorities and renters, and with density, and positively correlated with percentage of whites. But the dust metals part of the dimension essentially is non-lead. What might be labeled a lead dimension (that is, the second canonical dimension) is not significant. This finding is quite interesting, given that only a modest level of detected positive spatial autocorrelation is accounted for here. Nevertheless, the comparative results show that it inflates some and deflates others of the pairwise correlations among the variables.[14] This inflation/deflation is dependent upon the map pattern components of the autocorrelation. The result here is an altering of multicollinear in the two datasets in such a way that some metals shift from the first to the second dimension. The BLL correlations with the second dimension essentially stay the same.

Table 12.7   **Selected statistics for the eigenvector spatial filtering descriptions of the block group level transformed socio-demographic and BLL variables**

Variable	Moran coefficient	Number of vectors	R^2
% children	0.38099	13	0.55381
% Hispanic	0.47214	15	0.59387
% Caucasian	0.38125	22	0.83442
% Black	0.38065	22	0.83482
% Asian	0.43236	9	0.54348
Density	0.72034	10	0.29835
Average family size	0.37792	12	0.46567
% renters	0.47056	14	0.52770
Female/male ratio	0.46942	11	0.53245
Average BLL Pb	0.38545	13	0.66902
Maximum BLL Pb	0.46774	10	0.36237
Standard deviation of BLL Pb	0.40899	9	0.40483

Relationships Between House Dust Samples and Socio-Economic/Demographic and Pediatric BLL Variables: The Census Tract Level

Employing census tract data in an ecological analysis allows additional socio-economic variables to be included (for example, poverty rate), as well as additional strength to be borrowed from nearby BLL measures. Syracuse was partitioned into 57 census tracts for 2000 data tabulations. The dust sampled houses occupy 52 of these tracts

---

14   Only the relative signs are relevant because each extracted dimension is except for a multiplicative factor of .

(Figure 12.5d). The pediatric BLL data have 13,011 cases located in the sampled census tracts, whose numbers of BLL measures range from 3 to 1,522. Table 12.8 summarizes the transformations that result in frequency distributions that better mimic a bell-shaped curve; six of the variables are untransformed. Now all retained untransformed and transformed variables have frequency distributions that do not deviate significant from a bell-shaped curve at least at the 1% level of significance.

Figure 12.5d suggests that spatial autocorrelation may play a role in this data analysis, too, because the entire set of geographically contiguous census tracts contains sampled data. As before, eigenvector spatial filter (Griffith, 2003) descriptions of the transformed socio-economic/demographic as well as the BLL variables were constructed, and the aspatial components were subjected to a CCA for comparative purposes. The 52 census tracts have 224 rook adjacency links, and render 14 eigenvectors portraying positive spatial autocorrelation and having a relative Moran Coefficient of at least 0.25. Between 3 and 8 of these eigenvectors account for spatial autocorrelation latent in the individual variables. Table 12.9 summarizes the amount of redundant information contained in these variables that is attributable to spatial autocorrelation; as with the block group census geography, most of the variables contain weak-to-moderate positive spatial autocorrelation.

Table 12.8   Estimated transformation parameters and normal distribution goodness-of-fit statistics for the demographic census tract geography

Manly transformation			
variable	Parameter estimate		Shapiro-Wilk statistic probability
% Hispanic	−0.24		0.5693
Female/male ratio	−1.05		0.0400
Logistic transformation			
% Caucasian	−4.71		0.7842
% Black	−0.81		0.4396
% Asian	0		0.1359
Box-Cox power transformation			
	Exponenta	Translation	
Density	0.69	−1787.37	0.9528
Poverty rate	0.55	−0.01	0.9902
No transformation			
% children			0.1638
% renter			0.7810
Average family size			0.1042
Average BLL Pb			0.2671
Maximum BLL Pb			0.1581
Standard deviation of BLL Pb			0.8855

Note: 0 denotes a logarithmic transformation; transformed variables with a negative exponent are multiplied by −1 to preserve the direct relationship between them and their raw variable counterparts. Gray shading denotes no parameter estimates.

Table 12.9   Selected statistics for the eigenvector spatial filtering descriptions of the tract level transformed socio-demographic and BLL variables

Variable	Number of vectors	Moran coefficient	R²
% children	6	0.57746	0.63036
% Hispanic	8	0.64705	0.56921
% Caucasian	8	0.48996	0.75901
% Black	7	0.49094	0.78161
% Asian	8	0.47992	0.61150
Density	4	0.34243	0.31249
Average family size	6	0.57989	0.59239
% renters	4	0.46962	0.38734
Female/male ratio	2	0.29756	0.20097
Poverty rate	7	0.45938	0.48898
Average BLL Pb	7	0.59032	0.62711
Maximum BLL Pb	3	0.63349	0.29155
Standard deviation of BLL Pb	7	0.58144	0.50459

Accounting for latent spatial autocorrelation again dramatically alters the dimensions spanning the two datasets (see Table 12.10). One of the three significant dimensions disappears. The first dimension appears to be more of a summary one, whereas no Pb dimension is uncovered. Figure 12.7 portrays relationships between the spatially adjusted transformed variables and the uncovered canonical dimensions. Census tract areal units may be too coarse for a meaningful analysis of house level data.

Table 12.10   Canonical structure: significant canonical dimensions at a 5% level of significance for demographic and BLL variables co-dimensions: tract census geography

Variable	Canonical variate #1	Canonical variate #2	Canonical variate #3
Spatially unadjusted			
Zn	**0.5948**	−0.3756	−0.0632
Cu	**0.4595**	−0.0034	**−0.6668**
Mn	**0.5584**	0.0323	−0.0612
Pb	**0.7443**	**0.5171**	−0.0271
Ni	**−0.5532**	0.2068	−0.0183
Co	**0.4715**	0.1853	0.1794
Ca	**−0.8475**	0.1526	0.1776
Fe	**−0.7348**	−0.0398	0.0812
% children	**0.6392**	−0.2635	0.0045
% Hispanic	**0.5647**	**−0.4974**	0.2549
% Caucasian	**−0.7502**	0.3772	0.1332
% Black	**0.6601**	−0.3410	−0.2207
% Asian	0.0776	0.2660	0.1884
Density	**0.5425**	0.2949	0.0215
Average family size	**0.7549**	0.0268	0.1321
% renters	**0.6954**	−0.2626	0.1762

(continued ...)

Table 12.10 (*concluded*)

Variable	Canonical variate #1	Canonical variate #2	Canonical variate #3
Spatially unadjusted			
Female/male ratio	−0.2149	**−0.6824**	−0.1536
Poverty rate	**0.9001**	−0.2794	0.0422
Average BLL Pb	**0.8405**	0.0743	0.1394
Maximum BLL Pb	0.1866	0.1099	**−0.5778**
Standard deviation of BLL Pb	**0.5215**	−0.0160	−0.0328
Canonical correlation	0.6130	0.5015	0.2634
Spatially adjusted			
Zn	**0.4659**	**−0.4942**	
Cu	**0.4918**	−0.1975	
Mn	**0.5682**	−0.1241	
Pb	**0.8562**	0.3011	
Ni	**−0.5522**	**0.4117**	
Co	**0.5501**	0.0088	
Ca	**−0.7932**	**0.4018**	
Fe	**−0.7410**	0.1224	
% children	0.2849	−0.0658	
% Hispanic	**0.7607**	−0.2819	
% Caucasian	**−0.5391**	**0.5132**	
% Black	**0.4356**	**−0.5143**	
% Asian	0.1317	0.2710	
Density	**0.4259**	0.3045	
Average family size	**0.4609**	0.3489	
% renters	**0.6739**	−0.2314	
Female/male ratio	**−0.4867**	**−0.6759**	
Poverty rate	**0.7986**	−0.3032	
Average BLL Pb	**0.5523**	0.2020	
Maximum BLL Pb	0.1196	0.3408	
Standard deviation of BLL Pb	0.1304	**0.4840**	
Canonical correlation	0.4837	0.4207	

Note: Bold denotes a prominent correlation coefficient; gray background denotes a non-significant dimension.

Summary, Conclusions, and Implications

In summary, this chapter presents an analysis of house dust samples collected across the City of Syracuse, NY, for four levels of geographic resolution. These data are noisy, messy, and dirty. Attempting to summarize yard soil and house dust metal content with relatively few single-visit samples results in considerable noise in the data (see Figure 12.4). Relating parcel level data to neighborhood data defined in terms of census geography (for example, block, block group, and tract) also introduces noise into the dataset, but through areal aggregation of attributes that may well suffer from the MAUP. The data are messy primarily

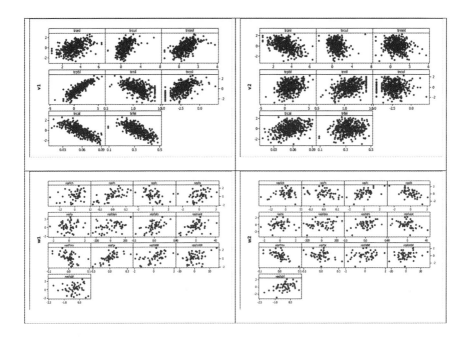

Figure 12.7   Scatterplots of transformed variables with canonical dimensions: census tracts resolution

because the sample size of houses varies across census areal units, and because assaying of some soil and dust samples failed to yield useful results. The fieldwork goal was to secure four yard soil samples per house; this goal was achieved in only 70% of the cases. Finally, the data are dirty because most of the variables are skewed, and require nonlinear analysis and outlier modification through the use of variable transformations (for example, the common Box-Cox power or Manly transformation). Data for the coarser geographic resolutions also contain moderate-to-positive spatial autocorrelation.

Conclusions stemming from the data analysis summarized in this chapter include the following:

1. finer geographic resolutions suggest the presence of a Pb dimension across the sets of metals measurements as well as the pediatric BLL measures;
2. the most suitable geographic resolution for the analysis of soil, dust, and blood Pb may well be approximately the census block group level;
3. the socio-economic/demographic, metals, and BLL measures analyzed here all display weak-to-moderate positive spatial autocorrelation across the coarser geographic resolutions;
4. spatial autocorrelation latent in coarser geographic resolutions may obscure true relationships; and,
5. findings reported here are not in complete agreement with Rasmussen (2007), who contends that indoor dust metal concentrations are not predicted accurately using outdoor soil metal concentrations.

More definitive conclusions require a more intensive spatial sampling design, one allowing better geographic coverage of parcels (see Figure 12.3) and with socio-economic/ demographic and BLL data collected for inhabitants of each sampled house. Nevertheless, the findings summarized here help better contextualize public health concerns associated with Pb contamination of urban environments.

Findings reported in this chapter imply that additional future work needs to be devoted to studying the geography of yard soil and indoor dust covariations, especially at the census block level. A more intense sampling of houses would support analysis of the spatial autocorrelation component at this geographic resolution. Sampling also needs to be done across the months of a year (see Johnson, McDade and Griffith, 1996). One unanswered question asks about the relationship between Table 12.4 results and spatial autocorrelation in the data. Establishing whether or not an indoor-outdoor Pb dimension exists requires completion of a variety of replicate studies. And to establish whether or not elevated BLLs result from such a dimension, if it exists, requires the execution of clinical trials at the household level.

Finally, although this chapter builds on tessellated stratified spatial sampling design work undertaken for the yard soils data collection (for example, Griffith, 2005), as for any sampling design that has been implemented, recognized possible improvements to this design warrant some discussion. Constraints placed on sampling in an urban area compromise a design like the one used for this study because yard soil and, especially, house dust are not continuous across an urban landscape: yards are defined by parcels, which make access for sampling purposes discrete, whereas by their very nature, houses are discrete units. This lumpiness of the geographic population under study limits the designing of an optimal spatial sampling scheme (see Rogerson et al., 2004). It introduces sparseness of geographic coverage in parts of a city because of varying parcel sizes, land owners/ renters declining to participate in a study (that is, parcels are inaccessible), and safety issues associated with some neighborhoods (for example, the southside of Syracuse; see Figure 12.2a). In a recent spatial sampling project conducted in Dallas, TX, this last factor was mitigated in part by partnering data collectors with local residences (which increased the cost of the sampling). Another limitation of the current design resulted from a resources constraint; 488 houses were insufficient for acquiring an excellent geographic coverage of household dust. This issue would be exacerbated if, as mentioned previously, the sampling also had been stratified by time. The remedy here is securing more funding from a granting agency, frequently a near impossible task. The sampling exercise also could be enhanced by including a survey of household attributes, including the collection of pediatric blood samples. Both would add substantial cost to such a study, as well as considerably more IRB oversight, while replicating at least some of the existing data available from a county health department and the United States Census Bureau. The design also could have been improved if it could have avoided collecting soil and dust samples whose metal contents were unassayable, an impossibility during the conducting of fieldwork. Nevertheless, although the sampling design could be improved in theory, because of limitations such as those noted here, most of these improvements are not possible in practice.

References

Griffith, D., 2003. *Spatial Autocorrelation and Spatial Filtering: Gaining Understanding through Theory and Scientific Visualization*. Berlin: Springer Verlag.

Griffith, D., 2005. Effective geographic sample size in the presence of spatial autocorrelation. *Annals, Association of American Geographers*, 95: 740–60.

Griffith, D., 2008. Geographic sampling of urban soils for contaminant mapping: How many samples and from where. *Environmental Geochemistry and Health*, 30: 495–509.

Griffith, D., Johnson, D., and Hunt, A., 2009. The geographic distribution of metals in urban soils: The case of Syracuse, NY. *GeoJournal*, 74: 275–91.

Griffith, D., Doyle, P., Wheeler, D., and Johnson, D., 1998. A tale of two swaths: Urban childhood blood lead levels across Syracuse, NY. *Annals, Association of American Geographers*, 88: 640–65.

Hunt, A., Johnson, D., and Griffith, D., 2006. Mass transfer of soil indoors by track-in on footwear. *The Science of the Total Environment*, 370: 360–71.

Hunt, A., Johnson, D., Griffith, D., and Zitoon, S., 2012. Citywide distribution of lead and other elements in soils and indoor dusts in Syracuse, NY. *Applied Geochemistry*, 27: 985–94.

Hunt, A., Johnson, D., Brooks, J., and Griffith, D., 2008. Risk remaining from fine particle contaminants after vacuum cleaning of hard floor surfaces. *Environmental Geochemistry and Health*, 30: 597–611.

Johnson, D., McDade, K., and Griffith, D., 1996. Seasonal variation in pediatric blood lead levels in Syracuse, NY. *Environmental Geochemistry and Health*, 18(2): 81–8.

Johnson, D., Hunt, A., Griffith, D., Hager, J., Brooks, J., Stella-Levinsohn, H., Lanciki, A., Lucci, R., Prokhorova, D., and Blount, S., 2009. Geographic patterns of non-carpeted floor dust loading in Syracuse, New York (USA) homes. *Environmental Geochemistry and Health*, 31: 353–63.

Johnson, D., Hager, J., Hunt, A., Blount, S., Griffith, D., Ellsworth, S., Hintz, J., Lucci, R., Mittiga, A., Prokhorova, D., Tidd, L., Millones, M., and Vincent, M., 2005. Field methods for mapping urban metal distributions in house dusts and surface soils of Syracuse, NY, USA. *Science in China* (Series C: Life Sciences), 48 (supplement): 192–9.

Manly, B., 1976. Exponential data transformations. *The Statistician*, 25: 37–42.

Rasmussen, P., 2007. Measurement of metal bioaccessibility in urban household dust and corresponding garden soils, in USEPA, ed. *Proceedings* of the ISEA Bioavailability Symposium: Use of In Vitro Bioaccessibility/Relative Bioavailability Estimates for Metals in Regulatory Settings: What Is Needed? Durham, NC: National Exposure Research Lab, no page numbers.

Rogerson, P., Delmelle, E., Batta, R., Akella, M., Blatt, A., and Wilson, G., 2004. Optimal sampling design for variables with varying spatial importance. *Geographical Analysis*, 36: 177–94.

USEPA, 1995. *Residential Sampling for Lead: Protocols for Dust and Soil Sampling*. Final Report EPA 747/R-95–002a. Washington, DC: U.S. Environmental Protection Agency, Office of Pollution Prevention and Toxics, Technical Programs Branch [online]. Available at http://1.usa.gov/1ECSv3o [Accessed 25 March 2013].

Wuana, R. and Okieimen, F., 2011. *Heavy Metals in Contaminated Soils: A Review of Sources, Chemistry, Risks and Best Available Strategies for Remediation*, International Scholarly Research Network Ecology [online]. Available at doi:10.5402/2011/402647 [Accessed 31 January 2014].

Chapter 13

Participatory and Ubiquitous Sensing for Exposure Assessment in Spatial Epidemiology

Michael Jerrett, Colleen E. Reid, Thomas E. McKone, and Petros Koutrakis

This chapter builds on a framework for exposure science adapted from the National Academy of Science (NAS) report "Exposure Science in the 21st Century" (NRC, 2012). This framework reflects a broad view of the role of exposure science in human and ecosystem health protection. The major elements of this expanded framework are identified, such as sources of stressors, environmental intensity (such as pollutant concentrations), time–activity and behavior, contact of stressors and receptors (which we term the "exposure moment"), and outcomes of contact (see Figure 13.1). This framework also highlights the role of upstream human and natural factors, and demonstrates the roles of both external and internal environments within exposure science while maintaining that exposure is measured at some boundary between the source and receptor. Here dose represents the amount of material that passes or otherwise has influence across the boundary to come into contact with the target system, organ, or cell to produce an outcome. This framework also recognizes the feedbacks inherent in exposure science such as how a diseased person changes behavior to influence their subsequent exposure. The normative goals of exposure assessment underlying this framework are to understand human health effects of stressors and to minimize contact with those stressors. A nascent goal is to understand the effects of beneficial or salutogenic stressors (such as contact with green space) for public health protection.

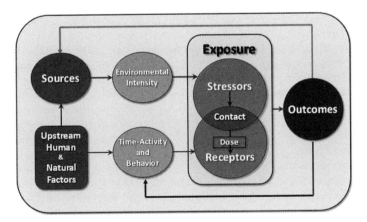

Figure 13.1 Core elements of exposure science
Source: Reprinted with permission from *Exposure Science in the 21st Century: A Vision and a Strategy*, 2012 by the National Academy of Sciences, Courtesy of the National Academies Press, Washington, D.C. (NRC, 2012).

Exposure science is integral to spatial epidemiology. The main focus of many epidemiology studies traditionally has been on a single chemical, biologic, or physical stressor. Efforts to characterize exposure in epidemiologic studies have typically tracked ambient conditions by assigning some measure of exposure from these ambient conditions to an individual through their home address. Although important health risks have been revealed, reliance on a proxy method such as this may impart large exposure-measurement error. Depending on the exposure-error type, health effect estimates may be attenuated and biased toward a null result, obscuring the true benefits of control measures (Zeger et al., 2000).

Innovations in science and technology provide opportunities to overcome limitations that have led to exposure measurement errors and will guide exposure science in the 21st century to deliver knowledge on current and emerging environmental health challenges. These innovations also provide the opportunity to understand multiple exposures from different media in time and space.

The major societal investments occurring in telemedicine, wireless communications, satellite technologies, and related logistics fields have resulted in technological advances that have and will have beneficial spin offs for exposure assessment in epidemiological studies. These innovations are now spurring fields known as "ubiquitous," "embedded" and "participatory" sensing that have substantial relevance to the future of exposure science.

We define ubiquitous sensing as a network of sensors, such as a dense array of air pollution or water contamination monitors, that have wide spatial coverage in urban areas. Similarly a ubiquitous system may rely on remote-sensing instruments that continuously supply information on particular phenomena, such as surface temperature or aerosol optical depth, which has virtual global coverage. Participatory sensing is defined as a means of obtaining detailed information on personal and population exposures via volunteers who supply this data often in exchange for useful information that might allow them to better understand and prevent the harmful exposures they face (Burke et al., 2006). Embedded sensors are those that are integrated into physical structures such as public transit systems or along roadways. These categories, however, should not be considered mutually exclusive as a participatory study with the use of cell phones combined with global positioning systems (GPS) and sensors, for instance, can provide information that could be considered ubiquitous, and ubiquitous sensing campaigns may involve embedded sensors.

This chapter provides a discussion of the various scientific and technological advances that will shape exposure assessment in the 21st century. How these new technologies relate to the NAS framework for exposure science is shown in Figure 13.2. New developments in geographic information technologies are leading to adoption of exposure information obtained from satellites via remote sensing (RS), and improved information on people's location and physical activity obtained with GPS and related geolocation technologies. Remote sensing and GPS data can be integrated into geographic information systems (GIS) that operate either through stand-alone computing platforms or through the World Wide Web. This chapter then discusses the increasing development of personal environmental monitors that are getting smaller and more easily transportable and also less expensive – allowing an increased number of sensor nodes in monitoring campaigns. We also document the increased use of sensors that are already in smartphones for use in ubiquitous and participatory sensing.

As ubiquitous sensing technologies become more widely available and more accurate, the need for models will remain, but the focus may shift from interpolation between sparse monitoring points to exploitation of massive datasets. In this context, models can provide both estimates of individual exposures and estimates of uncertainty in those exposure assignments—a process that can better address measurement error in health analyses (for

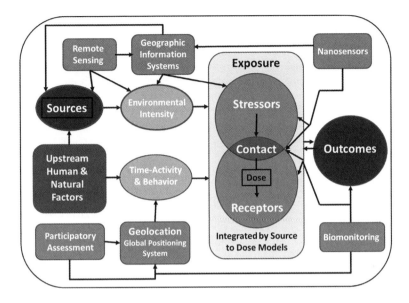

Figure 13.2   Selected scientific and technologic advances considered in relation to the NAS conceptual framework

Source: Reprinted with permission from *Exposure Science in the 21st Century: A Vision and a Strategy*, 2012 by the National Academy of Sciences, Courtesy of the National Academies Press, Washington, D.C. (NRC, 2012).

example, Molitor et al., 2007). The convergence of these scientific methods and technologies raises the possibility that in the near future embedded, ubiquitous, and participatory sensing systems will facilitate individual level exposure assessments on large populations.

We do not discuss here the use of biomonitoring or the associated fields of "-omics" technologies that can identify and quantify all elements of a particular cellular type such as proteomics (proteins) and metabolomics (metabolites). These topics are addressed, however, in the NRC report on exposure science (NRC, 2012). The potential exists for these technologies to influence spatial epidemiology through an understanding of internal dose across large populations that could be linked to information on outcomes that come from the whole population such as cancer registries or the National Death Index. Despite recent developments of comparing health outcomes and environmental exposure from biomonitoring data from large population surveys, such as the environment-wide association study by Patel, Bhattacharya, and Butte et al. (2010), there are currently no published studies that assess the spatial exposure of large populations with these technologies for use in epidemiological investigations.

The chapter concludes with discussion of future challenges and trends. Participatory, embedded, and ubiquitous sensing offer tremendous promise for improved exposure assessment, yet many challenges remain before these advances fundamentally change the way exposures are assessed in epidemiology. Protection of personal privacy, analysis of the voluminous "big" data generated by the sensors, and integration with other emerging methods from molecular epidemiology represent critical areas for research and development. These are active areas of research that will require extensive work to resolve over the next five to 10 years.

Geographic Information Systems (GIS)

GIS combines topologic geometry, capable of manipulating geographic information, with automated cartography, and enables users to compile digital or hard-copy maps. GIS plays a central role in integrating data into coherent databases that connect different attribute data (for example, exposure and health attributes) by geographic location. GIS can store and manipulate input data used to derive key exposure attributes, such as road locations and industrial land uses. GIS increasingly serves as the storage and integrative backbone of remote sensing, geolocation technologies, and sophisticated modeling outputs.

An important role of GIS in exposure assessment is the quantification of topologic relationships. For example, buffer functions that measure the distance between a source and a receptor are used to characterize proximity to roadways, factories, water bodies, and other land uses that have either potentially adverse exposures (for example, pesticide transport from agricultural fields) (Gunier et al., 2011) or potentially favorable exposures (for example, parks and healthy food stores in cities) (Morland and Evenson, 2009). Figure 13.3 demonstrates a road buffer that was used to characterize human exposures to traffic-related air pollution in Hamilton, Canada (Jerrett et al., 2005).

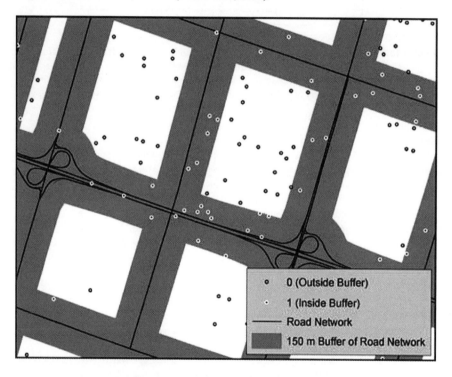

Figure 13.3   Example of a binary buffer overlay showing people likely to experience traffic-related air pollution exposure

Note: The circles represent people. People assigned a "0" are outside a pre-specified distance, while people assigned a "1" are within a given distance. Adapted from Jerrett et al., 2005

Web-based GIS is becoming more common (Evans and Sabel, 2012) and can serve as a tool in policy-making and in educating and empowering communities to understand and better manage their environmental exposures. For example, to promote active commuting, Metro Vancouver collaborated with the University of British Columbia to develop a cycling route planner which allows cyclists to select routes that have the most vegetation, the least traffic pollution, and the least or greatest elevation. That empowers cyclists to choose the routes that best suit their fitness levels, minimize their exposure to traffic pollution, and reduce their carbon dioxide output (Su et al., 2010). The web site runs on the backdrop of Google Maps—an illustration of the potential synergies between new private sector technologies and public-health protection.

Remote Sensing

Remote sensing (RS) has emerged as a key innovation in exposure science. RS encompasses the capture, retrieval, analysis, and display of information collected from a distance by satellite, aircraft, or other technologies that sense energy, light, or optical properties on subsurface, surface, and atmospheric conditions (Short and Robinson, 1998). RS improves human exposure assessment by collecting information on the earth's surface, water, and atmosphere globally, allowing estimation of exposure in regions without dense ground observation systems.

For example, aerosol optical depth (AOD) is an RS product that quantifies the extinction of electromagnetic radiation from aerosols in an atmospheric column at a given wavelength (Emili et al., 2010). Different measures of AOD can be obtained by many primary satellite sensors such as MODIS (MODerate resolution Imaging Spectroradiometer), MISR (Multiangle Imaging SpectroRadiometer), Landsat, IKONOS, Orbview, SPOT (Satellite Pour l'Observation de la Terre), and GOES (Geostationary Operational Environmental Satellite). A review of more than 30 papers that examined the relationship between total column AOD and surface $PM_{2.5}$ (particulate matter with aerodynamic diameter less than 2.5 microns, also known as fine particulate matter) measurements by ground monitoring station through simple linear regressions and correlations found a wide range of uncertainty among the studies (Hoff and Christopher, 2009). In some cases the correlations were strong and AOD served as a predictor of ground level pollution, but in others, AOD was a poor predictor of surface level $PM_{2.5}$ either because the satellite product itself was not sufficiently accurate or because the particles observed in the total column were in layers aloft. The authors suggested conducting a study of extrinsic factors for each region that would aid in understanding the $PM_{2.5}$–AOD relationship. The literature continues to grow with efforts to combine AOD with auxiliary information, such as meteorological data (Pelletier et al., 2007) or boundary layer height (Engel-Cox et al., 2006). Lee et al. (2011) hypothesized that the inherent variability in the particulate matter (PM)–AOD relationship is due to changes in particle size and composition, earth surface properties, vertical distribution of particle concentrations, and other factors. To account for the variability of these factors, they proposed a daily calibration technique that is based on the spatial variability of ground PM measurements and would make it possible to obtain quantitative estimates of PM concentrations by using AOD measurements.

To deal with the limitations of the full-column measurement of AOD to the ground level $PM_{2.5}$ measures that are important for epidemiological investigations, van Donkelaar et al. (2010) created annual global estimates of $PM_{2.5}$ by combining information from

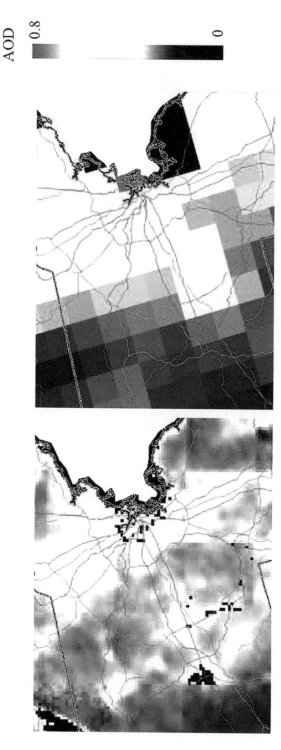

Figure 13.4   MAIAC 1 km (left column) and MOD04 10 km (right column), representing moderate pollution day as measured by PM2.5
Note the loss of AOD variability on conventional images (right column). Consider left column only next. Image A and B show dynamic variation for a given region, regardless of pollution. Adapted from Chudnovsky et al. (2013). Reprinted with permission; copyright © 2013, *Environmental Pollution*.

MODIS and MISR AOD and multiplying by the ratio of estimated full-column AOD and ground level $PM_{2.5}$ from the Geos-Chem global chemical transport model. Epidemiological analyses have found these long-term $PM_{2.5}$ exposure estimates are associated with incidence of diabetes (Chen et al., 2013), diabetes mortality (Brook et al., 2013), and cardiovascular mortality (Crouse et al., 2012) all in studies across Canada. Globally, no association was found with asthma prevalence (Anderson et al., 2012).

Most RS measurements are at too coarse a resolution for use in environmental epidemiological studies that aim at understanding small-area variations in exposure. To deal with this, several studies are now employing 'hybrid' modeling that combines the concepts of land use regression models of exposure with RS measurements (Beckerman et al., 2013a, Beckerman et al., 2013b, Kloog et al., 2012). Through statistical modeling, proxy information about likely locations of pollution at smaller spatial resolution than AOD pixels can essentially distribute the $PM_{2.5}$ estimated from the AOD to its most likely locations within its pixel. These hybrid exposure estimates have been used in a number of epidemiological studies (Jerrett et al., 2013; Kloog et al., 2012; Kloog et al., 2012; Kloog et al., 2013; Madrigano et al., 2013). Another way to increase the spatial resolution needed for epidemiological studies is through the use of the 3 km and 1 km AOD products that are being developed.

An example of efforts to increase spatial resolution is the application of the Multi-Angle Implementation of Atmospheric Correction (MAIAC) algorithm to attain 1 km resolution AOD from MODIS (Lyapustin et al., 2011). The 1 km product was generated for the New England area during 2003. Figure 13.2 compares the 10 km and 1 km retrievals. It clearly shows that considerably more detail is obtained with the 1 km product. Chudnovsky et al. (2013) found that the 1 km AOD product had much higher correlation with ground-level measurements of $PM_{2.5}$.

Several studies and reviews (for example, Maxwell et al. (2010)) have suggested that higher resolution data enhance efforts to identify time–space patterns that are the basis of many risk assessments for diseases (Wilson, 2002). In many studies, remotely-sensed data were used to derive three variables: vegetation cover, landscape structure, and water bodies. The ability to sense vegetation remotely from space is important in that nearly all vector-borne diseases are linked to the vegetative environment during their transmission cycle. Furthermore, crop type information may be important for studying the effects of pesticides (for example, vector resistance and illnesses caused by exposure to toxins) (Beck et al., 2000). Ward et al. (2006), Ward et al. (2000), and Maxwell et al. (2010) used crop location to identify where pesticides were applied in relation to residential locations. Remote sensing of vegetation cover combined with GIS has also been used to develop management strategies to reduce herbicide application (Gómez-Casero et al., 2010) and to assess potential exposure of fish and wildlife to pesticides and metals (Focardi et al., 2006). Green cover is also associated with higher levels of physical activity, and RS has been used with geolocation technologies to show associations between physical activity of children and their exposure to green spaces (Almanza et al., 2012).

Another RS tool relevant to exposure science and health risk assessment is hyperspectral imaging, which collects and processes information from a wide portion of the electromagnetic spectrum. It has been used to assess human health risks associated with infectious diseases and environmental hazards. Researchers have used this method to rapidly assess potential asbestos hazards from dust from the World Trade Center towers collapse in New York (Clark et al., 2001; Swayze et al., 2006), to identify mercury contamination in suburban agricultural soils in China (Wu et al., 2005), and to separate dust aloft from a dust storm in the Sahara from the underlying surface (Chudnovsky et al., 2009).

Global Positioning System and Geolocation Technologies

Launched in the 1980s for defense applications, the global positioning system (GPS) offers exposure scientists a simple means of tracking a person's geographic position. GPS receivers are now embedded into many cellular telephones, vehicle navigation systems, and other instruments (Goodchild et al., 2007). The GPS, owned by the U.S. government, consists of three components: a space segment with at least 24 satellites that transmit one-way signals to the earth; a control segment that maintains ground stations to track the satellites, reset their clocks, and maintain their positions; and a user segment that consists of individual devices that receive signals and calculate three-dimensional positions and times (GPS, 2011). GPS signals can be augmented or complemented by land-based navigation systems that use cellular telephone triangulation to provide positions when satellite signals are unavailable because of, for example, topographic obstruction or weather conditions (Shoval and Isaacson, 2006). Radiofrequency identification can also be used for local tracking of goods, animals, or people (Goodchild et al., 2007). Collectively, these systems are referred to here as geolocation technologies.

Geolocation technologies have been used extensively to improve environmental exposure assessment. For example, they have been used to inform simulation models of potential pesticide exposures (Leyk et al., 2009), and to assess exposure to infectious disease vectors (Vazquez-Prokopec et al., 2009) and air pollution (Paulos et al., 2007). Combining GPS data with accelerometers worn on children, a study in southern California found that children living in a smart growth community were more physically active in their neighborhood than matched controls in neighboring communities (Jerrett et al., 2013).

Geolocation technologies reduce exposure measurement error by providing a more reliable "time–geography of exposure" (compare to Briggs, 2005 and Hägerstraand, 1970). Geolocation technologies can improve our understanding of exposures by defining a person's location in time and space, often revealing important limitations of survey-based assessments of location. For example, Elgethun et al. (2007) compared time–activity diaries with actual measurements from GPS and found severe under-reporting of the amount of time spent outdoors at home in the diaries. Such errors may result in substantial exposure misclassification to pollutants such as ozone that have low penetration ratios from outdoors to indoors. Geolocation technologies also provide a better understanding of likely sources of error when points are used to represent large structures, such as schools and day care facilities (Houston et al., 2006).

GPS data can be combined with pollution monitors carried on study subjects as they walk, ride bicycles, or drive (Mead et al., 2013). When data obtained on environmental intensities (for example, contaminant concentration in air or water) are combined with geolocation information and physical activity measurements (obtained with accelerometers), more detailed estimates of potential chemical, biologic, or physical exposures can be made by using data on inhalation rate, ingestion potential, or dermal contact. Steinle et al. (2013) reviewed many studies that combine GPS devices, personal monitoring, and time-activity diaries to estimate personal exposures and then propose combining this information with health indicators and population information in a GIS to estimate population exposures. Geolocation technologies have already made important contributions to the understanding of exposures at the point of contact between source and receptor, and they appear poised to play an increasingly integral role in widespread population-based individual sensing.

Personal Samplers

Personal exposure measurement devices can be used to obtain information on exposures to volatile organic chemicals (VOCs) and other gases as well as particles, and gases, along with time–activity information.

Miniature microsensor systems have been developed for detecting VOCs with detection limits below parts per billion (Kim et al., 2011). A portable ("wearable") VOC monitor called the Volatile-Organic-Chemical Monitor (Chen et al., 2013; Iglesias et al., 2009) uses a "hybrid chemical sensor" for near real-time measurement and wireless transmission of data via a cellular telephone. Benzene, Toluene, Ethylbenzene, and Xylene (BTEX) components can be individually measured at concentrations as low as 1 ppb. The components and operating characteristics can be varied to measure other types of VOCs.

Many types of personal particle samplers have been developed. A small, portable, dual-chamber, data-logging particle monitor that can sense micrometer and submicrometer particles have been developed (Chowdhury et al., 2007; Edwards et al., 2006; Litton et al., 2004). Because its detection limit is 50 $\mu g/m^3$, it is not sensitive enough for typical ambient concentrations of particles found in many developed countries, but was designed for use in locations with much higher concentrations of particles, such as kitchens with wood stoves which are used in many households in developing countries. Simplicity of operation, low cost, battery operation, low weight, small size, and quiet operation are potentially useful features. A small, lightweight, portable, and battery-operated monitor of real-time black carbon called microAeth is now available (AethLabs, 2012).

Mead et al. (2013) describe the capability of miniature electrochemical sensors to detect carbon monoxide (CO), nitrogen dioxide (NO_2), and nitrogen oxide (NO) at low cost and with a size small enough to be carried on a person or installed at nodes to create a high-density monitoring network to capture small-area within city heterogeneity of pollutants. Previously only able to detect at the parts per million level, the researchers demonstrate the ability to detect parts per billion and with proper post-processing, the sensors are highly sensitive, have low bias, and high linearity. The sensors have been combined with GPS and have the ability to log data *in situ* or transmit data to another location for analysis. A pilot study of 46 of these combined sensors deployed at fixed sites within and around Cambridge, England for 2.5 months demonstrated the increased understanding of spatiotemporal variability in gaseous pollutants (Mead et al., 2013).

The University of California, Berkeley Time-Activity Monitoring System was developed by Allen-Piccolo et al. (2009) to record accurately when and how long each participant spends at a location more reliably than is typically available in diaries maintained by participants. The time–activity information can be used to estimate overall personal exposures on the basis of exposure measurements made with fixed monitors in each microenvironment if personal monitors cannot be used.

Simultaneous active collection of particles on filters and passive sampling of pollutant gases can be made with the Personal Multipollutant Sampler, a personal and microenvironmental monitoring system, developed by Chang et al. (1999) and Demokritou et al. (2001). Different configurations allow measurement of PM_{10} (particulate matter with aerodynamic diameter less than 10 microns) and $PM_{2.5}$ mass, $PM_{2.5}$ trace elements, black carbon, elemental and organic carbon, and passive ozone, nitrogen dioxide, and sulfur dioxide. The sampler can be mounted on a backpack strap on the subject's chest, and the battery-operated pump is carried in the backpack.

Although there have been advances in personal monitoring as described above, there is still a need for the development of portable monitors that can measure multiple pollutants. There is also a need for increasingly smaller sensors that people will carry with them less obtrusively, such as the size of a cell phone. Small monitors that can log, process, and then transmit the data wirelessly would improve the development of ubiquitous monitoring networks that could more rapidly identify populations or individual sources, factors influencing exposures, individuals who are highly exposed and ways to reduce exposures (NRC, 2012).

Cellular Telephone Technology

Although the last 20 years have seen substantial technical advances in personal environmental monitoring, personal sensors are still inadequate in their capacity to obtain highly selective, multistressor measurements in out-of-the-laboratory environments. Many personal monitors are unacceptable to some users because of their size, weight, noise level or appearance. These limitations shorten monitoring times, lower compliance, and may introduce bias. Current techniques also do not consider the physical activity of the subject during exposure, which may have substantial influences on the contact and intake of pollutant (de Nazelle et al., 2011).

Sensing of the environment through modification of cellular telephones, which are carried routinely by billions of people around the world, can ameliorate these shortcoming by allowing large-scale personalized monitoring as people move through time and space. Smart phones come with many embedded sensors, such as accelerometers, compasses, gyroscopes, GPS, dual microphones, dual cameras, ambient light detectors, proximity detectors, WiFi, and Bluetooth radio that can be harnessed for ubiquitous and participatory sensing (Lane et al., 2010). Several software applications have been written to exploit the onboard sensors such as motion, audio (for noise), visual, and location sensors through cellular or wireless networks (for example, Seto et al., 2010). Although there is at least one example of using the camera on a cellular telephone to photograph black carbon on a filter for processing elsewhere (Ramanathan et al., 2011), future endeavors could put devices, such as pollution monitors, into the cellular telephones.

Cellular telephones, supporting software, and expanding cellular and WiFi networks potentially can be used to form "ubiquitous" sensing systems to collect personal exposure information on millions of individuals and large ecosystems using citizen–scientists (Crall et al., 2010). Such networks can also take advantage of "embedded" sensors installed in existing infrastructure, for example, weight-in-motion sensors installed on roadways or sensors installed in public vehicles, such as buses, that provide anonymous, continuous data collection. Cell phones can be used in participatory sensing. Below are examples of embedded, ubiquitous, and participatory sensing that use cellphone technology.

Given the role of traffic in air pollution, noise, and accident risk, better information on traffic and other modes of transportation are needed. Telecom Italia's Localizing and Handling Network Event Systems (LocHNESs) software platform, which uses anonymous information on location from cell phone users in combination with embedded location tracking from public transit vehicles, is being used to manage traffic flows and to track mobility patterns of pedestrians, bicyclists and vehicles. The system is being tested to supply near real-time traffic monitoring and management information (Calabrese et al., 2011). Such information on traffic could be combined with models to estimate noise or air pollution

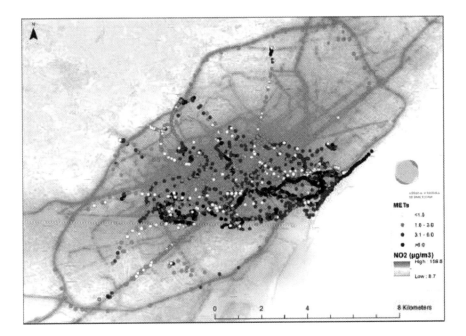

Figure 13.5   Selected 31 days of CalFit data
Note: (Circles represent GPS tracking of participants, color intensity represents energy intensity); background shading is NO$_2$ concentration. Adapted from de Nazelle et al. (2013). Reprinted with permission; copyright © 2013, *Environmental Pollution.*

levels throughout the city on a highly resolved spatiotemporal basis. LocHNESs illustrates how ubiquitous mobile phones can supply anonymous information on location that can be combined with embedded location tracking networks on public infrastructure to deliver real-time data on environmental exposures.

Increasingly, researchers are harnessing the ability to program smartphones with apps that utilize sensors embedded within the phone to develop tools and methods to study and intervene to address sedentary lifestyles, obesity, and ambient risk factors, such as air pollution, noise, or ultraviolet radiation. One example is the CalFit software that runs on mobile telephones that use the Android operating system. The software uses the accelerometer and GPS sensors that are built into smartphones to record activity counts and energy expenditure and the time and location in which an activity occurs. The software program has a single on–off switch for data-logging that continuously collects data as a background service when turned on (Seto et al., 2010). A study that compared the use of cellphone accelerometers with CalFit software to Actigraph accelerometers, the current gold standard device used in epidemiological studies, demonstrated the validity of using smartphones accelerometers with CalFit software to estimate light to moderate physical activity (Donaire-Gonzalez et al., 2013).

A pilot study tested CalFit with 36 human volunteers in Barcelona, Spain. Most of the participants (n=31) complied with the protocol and reported that using the cellphone to collect time-activity information was unobtrusive on their lives. From the data, the researchers could determine that the participants spent, on average, half of their daily time at home, a third at work, six percent in transit, and the rest of their time in other indoor or outdoor locations. By overlaying the GPS and accelerometer data on a map of estimated annual NO_2 exposure derived from a land use regression and adjusted for temporal variations with values obtained from government monitors (Figure 13.5), the researchers determined that transit accounted for 24% of participants inhaled dose of pollution even though only six percent of their time was in transit due to high pollution and high activity levels in transit. Using the CalFit data on smartphones, they were able to estimate time-activity information to improve exposure estimates (de Nazelle et al., 2013). This system can also serve as a base station for other sensors that operate via Bluetooth radio to collect data on air pollution, light, and noise.

CalFit and other cellular-telephone-based systems can also be used to implement context-specific ecologic momentary assessment (CS-EMA). CS-EMA measures real-time exposures and outcomes with sensors that are inside and outside the telephone. The system can communicate information, telling a person to respond to a survey when particular events are observed, such as a period of physical activity, exposure to air pollution, use of steroid inhalants, or consumption of particular food. Responding to these surveys provides opportunities to obtain important information about an exposure or outcome, such as mood, stress, or behaviors at the point of contact between the stressor and receptor (Dunton et al., 2011; Intille, 2007).

In participatory sensing, participants decide on what, where, and when to monitor in their environments (Mun et al., 2009). Participatory sensing can operate on scales from individual to group to regional to global, depending on what is being measured (Burke et al., 2006). Participatory sensing systems often combine embedded and ubiquitous systems with web-based applications that allow participants to share information on their exposures.

The Personal Environmental Impact Report (PEIR) system, in Los Angeles, California, measures four main outcomes: exposure to fine particulate matter ($PM_{2.5}$), exposure to fast food outlets, output of transportation-related greenhouse-gas emissions, and output of transportation-related $PM_{2.5}$ emissions near sensitive receptors. PEIR relies on information from cellular telephones, GPS, and data on land use and traffic to determine location and speed and classify activities as walking or driving with a Markov chain algorithm (Mun et al., 2009). Once activities are classified, $PM_{2.5}$ levels from a near real-time dispersion model are combined with a person's location to assign a likely exposure concentration.

The PEIR system has been piloted by 30 volunteers. Respondents, using a Facebook application, can review their exposures and emissions in comparison with those of others. They can see continuous information on their exposures and contributions such as a weekly impact report and a locational trace of where they have been that demonstrates an innovative fusion of new media and mobile sensing platforms.

An example of using small sensors for ubiquitous monitoring is the Berkeley Atmospheric CO_2 Observation Network ($BEACO_2N$) project (Sanders, 2012). Researchers at the University of California, Berkeley, are installing a combination of sensors on an instrument the size of a shoe box to the rooftops of 30 buildings, most of them schools, which comprise a 27-square-mile grid in Oakland, CA. The goal is to understand how carbon dioxide (CO_2) changes over time in an urban setting to assess California's efforts to decrease climate-acting pollutants. Each instrument contains sensors for CO_2, NO_2, CO, ozone, temperature, pressure, and relative humidity. The data are transmitted to Berkeley through WiFi and can be accessed by anyone online (Sanders, 2012). The researchers are also creating an educational component that can be incorporated into the science curricula of local schools (Teige et al., 2011).

Use of onboard cell phone sensors in combination with existing environmental data or other embedded sensors leads to many challenging issues involving the large data, biased samples, and the integration of input data with different temporal/spatial resolution and errors in measurement. Together these issues emphasize the need for novel modeling approaches to sort through and interpret observations and measurements.

Modeling

As ubiquitous measurements become more widely used, the ability of models to provide a repository for exposure knowledge has been and will continue to be a cornerstone of exposure science. Models have informed epidemiological investigations and other health studies by interpreting data and observations, and providing tools for predicting trends. Key to the future of exposure models is how they incorporate the ever-increasing amounts of observations of natural and human processes and environmental effects. Vast new measurement programs in fields as diverse as genomics and earth-observation systems, ranging from the nanoscale to global dimensions, present important opportunities and challenges for modeling. The interdependence of models and measurements is complex and iterative (EPA, 2009; NRC, 2007). Historically, the cycle of models and observations tended to begin with observations used to build a model, then a second set of observations to calibrate the model, and a third set to "validate" or evaluate the model performance. But in place of validation or even calibration with observations, an alternative approach is to use models and observations (such as biomarker data and environmental samples) as independent tools to evaluate hypotheses about source–receptor relationships (McKone et al., 2007).

An important element of any exposure assessment is quantifying the degree of uncertainty about the exposure estimates that arise from both statistical variability in the finite set of measurements available on which to build the model and their inherent measurement errors, and misspecification of the form of the statistical model used—both in prediction of exposure and in the sampling and measurement error distributions. Well-established techniques are available to estimate the standard errors of parameter estimates and model predictions. In the "classical error model," the measured value is assumed to be distributed randomly around the true value with some deviations caused by instrument or model error. In the "Berkson error model," individual exposures are assumed to be distributed around some applied exposure of a group (for example, ambient pollution around a monitor). The "classical error model" is appropriate when personal measurements that are subject to some error are used. It tends to induce an attenuation bias toward the null. The "Berkson error model" is appropriate when individuals within a group differ because of unmeasured factors (for example, time–activity or household characteristics); it might not produce any bias but will tend to inflate the variance of the estimated regression coefficient (Berkson, 1950). In many instances, estimated coefficients are affected by a combination of classical and Berkson errors (Zeger et al., 2000b).

The second type of uncertainty, model misspecification, is more difficult (Leamer and Leamer, 1978) because the true form of the relationships being modeled can never be known and the most one can hope for is to investigate the robustness of the predictions in a reasonable group of plausible alternative models. This is generally known as sensitivity analysis, but more formal alternatives, such as model averaging (Hjort and Claeskens, 2003, Hoeting et al., 1999), are available that essentially provide an average of a wide variety of models, each weighted in some fashion according to goodness-of-fit. These methods also create an overall estimate of uncertainty that combines the statistical variability within each model and the differences between models to provide a more "honest" assessment of overall uncertainty than simply quoting the standard error of some "best" model. With increased computing powers, however, machine learning or data mining methods in which k-fold cross validation is used to select from a wide library of possible statistical models is now becoming possible. In addition to models as tools to interpret, predict, and evaluate source, concentration, and receptor relationships, there are informatics models—the emerging tools for managing and exploring massive amounts of information from diverse sources and in widely different formats.

Conclusion

This chapter highlights recent advances in the field of exposure science with applicability for spatial epidemiology in the 21st Century. These huge leaps in terms of the ability to sense from afar in remote sensing, or sense with small-scale sensors that we wear on our person or within devices such as smartphones that we rely on for other everyday tasks, or that are embedded into our current infrastructure allow for estimation of exposure on space and time scales that have previously not been possible. Going forward, there are many advances that will continue to improve the space and time scales upon which we measure exposure, thus limiting the measurement errors that have historically constrained the power of epidemiological investigations.

GIS continues to evolve, with many new open-source platforms for analysis of spatial data including spatial packages for use in the R programming language which is commonly used in spatial epidemiology. This could allow the modeling of exposure and the epidemiological analysis within the same computer programming environment. With the growing availability of spatial data, there also needs to be continued support for the GIS "infostructure" to ensure that public health agencies and researchers have access to the wealth of geographic exposure data while protecting the privacy of individuals. Additional efforts to support and enhance web-based exposure mapping is needed to improve access to data on and understanding of potential exposures.

The geolocation technologies, particularly those onboard the cellular phone, will continue to expand possibilities for linking location to specific exposures. Numerous other applications of the kind piloted in Italy will harness anonymous information on the movements of mass populations to better understand patterns of emissions and potential exposures.

Many new satellite products are also coming online in the next decade. A 3 km MODIS AOD product has become publicly available this past year. In 2015, two new hyperspectral sensors will be launched: the National Aeronautics and Space Administration (NASA) HyspIRI (NASA, 2011) and European EnMAP (EnMAP, 2011) missions. With their improved hyperspectral and multispectral capabilities, these sensors will increase the ability to monitor the effects of urbanization on the environment and to assess land cover characteristics that could indicate the presence of risks posed by vector-borne and animal-borne diseases on a global scale.

The increased use of sensors onboard cellular telephones provides an opportunity not just for ubiquitous sensing in the developed world, where there is a relatively large amount of information available on environmental epidemiology, but also in the developing world, where there is increasing use of mobile phones. To use the onboard sensors accurately, there is a need to assess measurement error in cellular telephones in comparison with high-caliber instruments by using either laboratory or field analyses.

With all of these data collection and data management advances in exposure science come difficulties with validation of new instruments and algorithms, storage of large amounts of data, improved ability to analyze large data sets through statistical advances and computing power, and sharing of data sources while also maintaining privacy protections for personal data. Without comparable investments in the development of new modeling analytic techniques to address correlated data on many more variables than subjects and without advances in computation, such as parallel-processing techniques, the analysis of the mountains of data threatens to become the new limiting factor in further progress. As is the case with all models, exposure models must balance the need for transparency with the need for fidelity and credibility. That requires concurrent development of model performance evaluation efforts in any model development program.

In addition to new advances, a need exists for continued support for access to existing exposure data that were collected with public funding for use in scientific research. Efforts by government agencies and universities that are involved in exposure science research should work to foster cooperation with the private sector to encourage data collection, sharing of geographic and exposure information, and the formation of partnerships among exposure scientists with the goal of improving public health protection.

References

AethLabs. 2012. *Microaeth model ae41 [online]*. San Francisco, CA: AethLabs. Available at: www.aethlabs.com [Accessed January 11 2012].

Allen-Piccolo, G., Rogers, J.V., Edwards, R., Clark, M.C., Allen, T.T., Ruiz-Mercado, I., Shields, K.N., Canuz, E., and Smith, K.R. 2009. An ultrasound personal locator for time-activity assessment. *International Journal of Occupational and Environmental Health*, 15(2): 122–132.

Almanza, E., Jerrett, M., Dunton, G., Seto, E., and Pentz, M.A. 2012. A study of community design, greenness, and physical activity in children using satellite, GPS and accelerometer data. *Health Place*, 18(1): 46–54.

Anderson, H.R., Butland, B.K., van Donkelaar, A., Brauer, M., Strachan, D.P., Clayton, T., van Dingenen, R., Amann, M., Brunekreef, B., Cohen, A., Dentener, F., Lai, C., Lamsal, L.N., Martin, R.V., and One, I.P. 2012. Satellite-based estimates of ambient air pollution and global variations in childhood asthma prevalence. *Environmental Health Perspectives* 120(9): 1333–9.

Beck, L.R., Lobitz, B.M., and Wood, B.L. 2000. Remote sensing and human health: New sensors and new opportunities. *Emerging Infectious Diseases*, 6(3): 217–227.

Beckerman, B.S., Jerrett, M., Martin, R.V., van Donkelaar, A., Ross, Z., and Burnett, R.T. 2013a. Application of the deletion/substitution/addition algorithm to selecting land use regression models for interpolating air pollution measurements in california. *Atmospheric Environment* 77: 172–7

Beckerman, B.S., Jerrett, M., Serre, M., Martin, R.V., Lee, S.J., van Donkelaar, A., Ross, A., Su, J., and Burnett, R.T. 2013b. A hybrid approach to estimating national scale spatiotemporal variability of PM2.5 in the contiguous united states. *Environmental Science & Technology*, 47(13): 7233–41.

Berkson, J. 1950. Are there two regressions? *Journal of the American Statistical Association*, 45(250): 164–80.

Briggs, D. 2005. The role of GIS: Coping with space (and time) in air pollution exposure assessment. *Journal of Toxicology and Environmental Health-Part a-Current Issues*, 68(13–14): 1243–61.

Brook, R.D., Cakmak, S., Turner, M.C., Brook, J.R., Crouse, D.L., Peters, P.A., van Donkelaar, A., Villeneuve, P.J., Brion, O., Jerrett, M., Martin, R.V., Rajagopalan, S., Goldberg, M.S., Pope, C.A., and Burnett, R.T. 2013. Long-term fine particulate matter exposure and mortality from diabetes mellitus in canada. *Diabetes Care*, 36(10): 3313–20.

Burke, J.A., Estrin, D., Hansen, M., Parker, A., Ramanathan, N., Reddy, S., and Srivastava, M.B. 2006. Participatory sensing. ACM SenSys, the 4th ACM Conference on Embedded Networked Sensor Systems, Boulder, CO.

Calabrese, F., Colonna, M., Lovisolo, P., Parata, D. and Ratti, C. 2011. Real-time urban monitoring using cell phones: A case study in Rome. *IEEE Transactions on Intelligent Transportation Systems*, 12(1): 141–51.

Chang, I-T., Sarnat, J., Wolfson, J., Rojas-Bracho, L., Suh, H., and Koutrakis, P. 1999. Development of a personal multi-pollutant exposure sampler for particulate matter and criteria gases. *Pollution atmosphérique* 41: 31–9.

Chen, C., Tsow, F., Campbell, K.D, Iglesias, R., Forzani, E., and Tao, N.J. 2013a. A wireless hybrid chemical sensor for detection of environmental volatile organic compounds. *IEEE Sens J,* 13(5): 1748–55.

Chen, H., Burnett, R.T., Kwong, J.C., Villeneuve, P.J., Goldberg, M.S., Brook, R.D., van Donkelaar, A., Jerrett, M., Martin, R.V., Brook, J.R., and Copes, R. 2013b. Risk of incident diabetes in relation to long-term exposure to fine particulate matter in ontario, canada. *Environmental Health Perspectives* 121(7): 804–10.

Chowdhury, Z., Edwards, R.D., Johnson, M., Shields, K.N., Allen, T., Canuz, E., and Smith, K.R. 2007. An inexpensive light-scattering particle monitor: Field validation. *Journal of Environmental Monitoring*, 9(10): 1099–106.

Chudnovsky, A., Ben-Dor, E., Kostinski, A., and Koren, I. 2009. Mineral content analysis of atmospheric dust using hyperspectral information from space. *Geophysical Research Letters*, 36(15).

Chudnovsky, A.A., Kostinski, A., Lyapustin, A., and Koutrakis, P. 2013. Spatial scales of pollution from variable resolution satellite imaging. *Environmental Pollution (Barking, Essex: 1987)* 172: 131–8.

Clark, R.N., Green, R.O., Swayze, G.A., Meeker, G., Sutley, S., Hoefen, T.M., Livo, K.E., Plumlee, G., Pavri, B., and Sarture, C. 2001. *Environmental Studies of the World Trade Center Area after the September 11, 2001 Attack*: U.S. Geological Survey.

Crall, A.W., Newman, G.J., Jarnevich, C.S., Stohlgren, T.J., Waller, D.M., and Graham, J. 2010. Improving and integrating data on invasive species collected by citizen scientists. *Biological Invasions*, 12(10): 3419–28.

Crouse, D.L., Peters, P.A., van Donkelaar, A., Goldberg, M.S., Villeneuve, P.J., Brion, O., Khan, S., Atari, D.O., Jerrett, M., Pope, C.A., Brauer, M., Brook, J.R., Martin, R.V., Stieb, D., and Burnett, R.T. 2012. Risk of nonaccidental and cardiovascular mortality in relation to long-term exposure to low concentrations of fine particulate matter: A Canadian national-level cohort study. *Environmental Health Perspectives* 120(5): 708–14.

de Nazelle, A.; M.J. Nieuwenhuijsen, J.M. Anto, M. Brauer, D. Briggs, C. Braun-Fahrlander, N. Cavill, A.R. Cooper, H. Desqueyroux, S. Fruin, G. Hoek, L.I. Panis, N. Janssen, M. Jerrett, M. Joffe, Z.J. Andersen, E. van Kempen, S. Kingham, N. Kubesch,

K.M. Leyden, J.D. Marshall, J. Matamala, G. Mellios, M. Mendez, H. Nassif, D. Ogilvie, R. Peiro, K. Perez, A. Rabl, M. Ragettli, D. Rodriguez, D. Rojas, P. Ruiz, J.F. Sallis, J. Terwoert, J.F. Toussaint, J. Tuomisto, M. Zuurbier, E. Lebret.. 2011. Improving health through policies that promote active travel: A review of evidence to support integrated health impact assessment. *Environment International*, 37(4): 766–77.

de Nazelle, A., Seto, E., Donaire-Gonzalez, D., Mendez, M., Matamala, J., Nieuwenhuijsen, M.J., and Jerrett, M., 2013. Improving estimates of air pollution exposure through ubiquitous sensing technologies. *Environmental Pollution (Barking, Essex: 1987)* 176: 92–9.

Demokritou, P., Kavouras, I.G., Ferguson, S.T., and Koutrakis, P. 2001. Development and laboratory performance evaluation of a personal multipollutant sampler for simultaneous measurements of particulate and gaseous pollutants. *Aerosol Science & Technology*, 35(3): 741–52.

Donaire-Gonzalez, D., de Nazelle, A., Seto, E., Mendez, M., Nieuwenhuijsen, M.J., and Jerrett, M. 2013. Comparison of physical activity measures using mobile phone-based CalFit and actigraph. *Journal of Medical Internet Research*, 15(6): e111.

Dunton, G.F., Liao, Y., Intille, S., Wolch, J., and Pentz, M.A. 2011. Physical and social contextual influences on children's leisure-time physical activity: An ecological momentary assessment study. *Journal of Physical Activity & Health,* 8(Suppl 1): 103–8.

Edwards, R., Smith, K.R., Kirby, B., Allen, T., Litton, C.D., and Hering, S. 2006. An inexpensive dual-chamber particle monitor: Laboratory characterization. *Journal of the Air and Waste Management Association*, 56(6): 789–99.

Elgethun, K., Yost, M.G., Fitzpatrick, C.T.E., Nyerges, T.L., and Fenske, R.A. 2007. Comparison of global positioning system (GPS) tracking and parent-report diaries to characterize children's time-location patterns. *Journal of Exposure Science and Environmental Epidemiology*, 17(2): 196–206.

Emili, E., Popp, C., Petitta, M. 2010. PM10 remote sensing from geostationary SEVIRI and polar-orbiting MODIS sensors over the complex terrain of the European alpine region. *Remote Sensing of Environment*, 114(11): 2485–99.

Engel-Cox, J.A., Hoff, R.M., Rogers, R., Riffler, M., Wunderle, S., and Zebisch, M. 2006. Integrating lidar and satellite optical depth with ambient monitoring for 3-dimensional particulate characterization. *Atmospheric Environment*, 40(40): 8056–67.

EnMAP. 2011. *Enmap hyperspectral imager* [online]. Available at: www.enmap.org [Accessed October 28, 2011].

EPA 2009. Guidance on the development, evaluation, and application of environmental models. *In* C. f. R.E.M. Office of the Science Advisor, U.S. Environmental Protection Agency ed. Washington, D.C.

Evans, B. and Sabel, C.E. 2012. Open-source web-based geographical information system for health exposure assessment. *International Journal of Health Geographics*, 11(1): 2.

Focardi, S., Corsi, I., Mazzuoli, S., Vignoli, L., Loiselle, S.A., and Focardi, S. 2006. Integrating remote sensing approach with pollution monitoring tools for aquatic ecosystem risk assessment and management: A case study of Lake Victoria (Uganda). *Environ Monit Assess* 122(1–3): 275–87.

Gómez-Casero, M., Castillejo-González, I., García-Ferrer, A., Peña-Barragán, J., Jurado-Expósito, M., García-Torres, L., and López-Granados, F. 2010. Spectral discrimination of wild oat and canary grass in wheat fields for less herbicide application. *Agronomy for Sustainable Development*, 30(3): 689–99.

Goodchild, M.F., Nusser, S.M., Pickle, L.W., Lane, J., Williamson, P., Mulry, M.H., Davison, A., Sardy, S., Feskens, R., and Hox, J. 2007. The Morris Hansen lecture 2006: Statistical perspectives on spatial social science. *Journal of Official Statistics-Stockholm*, 23(3): 269.

GPS. 2011. *The Global Positioning System* [online]. Available at: www.gps.gov/systems/gps/ [Accessed October 28 2011].

Gunier, R.B., Ward, M.H., Airola, M., Bell, E.M., Colt, J., Nishioka, M., Buffler, P.A., Reynolds, P., Rull, R.P., Hertz, A., Metayer, C., and Nuckols, J.R. 2011. Determinants of agricultural pesticide concentrations in carpet dust. *Environmental Health Perspectives* 119(7): 970–76.

Hägerstraand, T. 1970. What about people in regional science? *Papers in Regional Science*, 24(1): 7–24.

Hjort, N.L. and Claeskens, G. 2003. Frequentist model average estimators. *Journal of the American Statistical Association*, 98(464): 879–99.

Hoeting, J.A., Madigan, D., Raftery, A.E., and Volinsky, C.T. 1999. Bayesian model averaging: A tutorial. *Statistical Science*: 382–401.

Hoff, R.M. and Christopher, S.A. 2009. Remote sensing of particulate pollution from space: Have we reached the promised land? *Journal of the Air & Waste Management Association*, 59(6): 645–75, discussion 642–4.

Houston, D., Ong, P., Wu, J., and Winer, A. 2006. Proximity of licensed child care facilities to near-roadway vehicle pollution. *American Journal of Public Health*, 96(9): 1611–17.

Iglesias, R.A., Tsow, F., Wang, R., Forzani, E.S., and Tao, N. 2009. Hybrid separation and detection device for analysis of benzene, toluene, ethylbenzene, and xylenes in complex samples. *Analytical Chemistry*, 81(21): 8930–35.

Intille, S.S. 2007. Technological innovations enabling automatic, context-sensitive ecological momentary assessment. *The Science of Real-Time Data Capture. Self-Reports in Health Research*: 308–37.

Jerrett, M., Arain, A., Kanaroglou, P., Beckerman, B., Potoglou, D., Sahsuvaroglu, T., Morrison, J., and Giovis, C. 2005. A review and evaluation of intraurban air pollution exposure models. *Journal of Exposure Analysis and Environmental Epidemiology*, 15(2): 185–204.

Jerrett, M., Gale, S., and Kontgis, C. 2009. An environmental health geography of risk, in T. Brown, S. McLafferty and G. Moon eds. *A Companion to Health and Medical Geography*. Oxford, Wiley-Blackwell, 418–45.

Jerrett, M., Gale, S., and Kontgis, C. 2010. Spatial modeling in environmental and public health research. *International Journal of Environmental Research and Public Health,* 7(4): 1302–29.

Jerrett, M., Almanza, E., Davies, M., Wolch, J., Dunton, G., Spruitj-Metz, D., and Ann Pentz, M. 2013a. Smart growth community design and physical activity in children. *American Journal of Preventive Medicine*, 45(4): 386–92.

Jerrett, M., Burnett, R.T., Beckerman, B.S., Turner, M.C., Krewski, D., Thurston, G., Martin, R.V., van Donkelaar, A., Hughes, E., Shi, Y., Gapstur, S.M., Thun, M.J., and Pope, C.A. 2013b. Spatial analysis of air pollution and mortality in California. *American Journal of Respiratory and Critical Care Medicine* 188(5): 593–9.

Kim, S.K., Chang, H., and Zellers, E.T. 2011. Microfabricated gas chromatograph for the selective determination of trichloroethylene vapor at sub-parts-per-billion concentrations in complex mixtures. *Analytical Chemistry*, 83(18): 7198–206.

Kloog, I., Coull, B.A., Zanobetti, A., Koutrakis, P., and Schwartz, J.D. 2012a. Acute and chronic effects of particles on hospital admissions in New England. *PloS One*, 7(4):e34664.

Kloog, I., Melly, S.J., Ridgway, W.L., Coull, B.A., and Schwartz, J. 2012b. Using new satellite based exposure methods to study the association between pregnancy PM2.5 exposure, premature birth and birth weight in massachusetts. *Environmental Health* 11: 40.

Kloog, I., Nordio, F., Coull, B.A., and Schwartz, J. 2012c. Incorporating local land use regression and satellite aerosol optical depth in a hybrid model of spatiotemporal PM2.5 exposures in the mid-atlantic states. *Environmental Science & Technology*, 46(21): 11913–21.

Kloog, I., Ridgway, B., Koutrakis, P., Coull, B.A., and Schwartz, J.D. 2013. Long- and short-term exposure to PM2.5 and mortality: Using novel exposure models. *Epidemiology*, 24(4): 555–61.

Lane, N.D., Miluzzo, E., Lu, H., Peebles, D., Choudhury, T., and Campbell, A.T. 2010. A survey of mobile phone sensing. *IEEE Communications Magazine*, 48(9): 140–50.

Leamer, E.E. and Leamer, E.E. 1978. *Specification Searches: Ad hoc Inference with Nonexperimental Data*: Wiley New York.

Lee, H., Liu, Y., Coull, B., Schwartz, J.D., and Koutrakis, P. 2011. A novel calibration approach of MODIS AOD data to predict PM2.5 concentrations. *Atmospheric Chemistry and Physics*, 11(15): 7991–8002.

Leyk, S., Binder, C.R., and Nuckols, J.R. 2009. Spatial modeling of personalized exposure dynamics: The case of pesticide use in small-scale agricultural production landscapes of the developing world. *International Journal of Health Geographics* 8.

Litton, C.D., Smith, K.R., Edwards, R., and Allen, T. 2004. Combined optical and ionization measurement techniques for inexpensive characterization of micrometer and submicrometer aerosols. *Aerosol Science and Technology*, 38(11): 1054–62.

Lyapustin, A., Wang, Y., Laszlo, I., Kahn, R., Korkin, S., Remer, L., Levy, R., and Reid, J. 2011. Multiangle implementation of atmospheric correction (MAIAC): 2. Aerosol algorithm. *Journal of Geophysical Research: Atmospheres (1984–2012)* 116 (D3) :

Madrigano, J., Kloog, I., Goldberg, R., Coull, B.A., Mittleman, M.A., and Schwartz, J. 2013. Long-term exposure to PM2.5 and incidence of acute myocardial infarction. *Environmental Health Perspectives* 121(2): 192–6.

Maxwell, S.K., Meliker, J.R., and Goovaerts, P. 2010. Use of land surface remotely sensed satellite and airborne data for environmental exposure assessment in cancer research. *Journal of Exposure Science and Environmental Epidemiology*, 20(2): 176–85.

McKone, T.E., Castorina, R., Harnly, M.E., Kuwabara, Y., Eskenazi, B., and Bradman, A. 2007. Merging models and biomonitoring data to characterize sources and pathways of human exposure to organophosphorus pesticides in the salinas valley of california. *Environmental Science & Technology*, 41(9): 3233–40.

Mead, M.I., Popoola, O.A.M., Stewart, G.B., Landshoff, P., Calleja, M., Hayes, M., Baldovi, J.J., McLeod, M.W., Hodgson, T.F., Dicks, J., Lewis, A., Cohen, J., Baron, R., Saffell, J.R., and Jones, R.L. 2013. The use of electrochemical sensors for monitoring urban air quality in low-cost, high-density networks. *Atmospheric Environment* 70: 186–203.

Molitor, J., Jerrett, M., Chang, C-C., Molitor, N-T., Gauderman, J., Berhane, K., McConnell, R., Lurmann, F., Wu, J., Winer, A., and Thomas, D. 2007. Assessing uncertainty in spatial exposure models for air pollution health effects assessment. *Environmental Health Perspectives* 115(8): 1147–53.

Morland, K.B. and Evenson, K.R. 2009. Obesity prevalence and the local food environment. *Health Place*, 15(2): 491–5.

Mun, M., Reddy, K., Shilton, N., Yau, N., Burke, J.M., Estrin, D., Hansen, M., Howard, E., West, R., and Boda, P. 2009. PEIR, the personal environmental impact report, as a platform for particpatory sensig system research. 7th Annual International Conference on Mobile Systems, Applications and Services-MobiSys '09, Krakow, Poland.

NASA. 2011. *HyspIRI mission study* [*online*]. Available at: http://hyspiri.jpl.nasa.gov [Accessed October 28 2011].

NRC 2007. *Models in Environmental Regulatory Decision Making*. Washington, D.C.: National Academies Press.

NRC 2012. *Exposure Science in the 21st Century: A Vision and a Strategy*. Washington, D.C.: The National Academies Press.

Patel, C.J., Bhattacharya, J., and Butte, A.J. 2010. An environment-wide association study (EWAS) on type 2 diabetes mellitus. *PloS One*, 5(5):e10746.

Paulos, E., Honicky, R.J., and Goodman, E. 2007. Sensing atmophere. The 5th ACM Conference on Embedded Network Sensor Systems—AMC SenSys, Sydney, Australia.

Pelletier, B., Santer, R., and Vidot, J. 2007. Retrieving of particulate matter from optical measurements: A semiparametric approach. *Journal of Geophysical Research: Atmospheres (1984–2012)* 112 (D6):D06208.

Ramanathan, N., Lukac, M., Ahmed, T., Kar, A., Praveen, P.S., Honles, T., Leong, I., Rehman, I.H., Schauer, J.J., and Ramanathan, V. 2011. A cellphone based system for large-scale monitoring of black carbon. *Atmospheric Environment* 45: 4481–7.

Sanders, R. 2012. UC Berkeley installing first CO2 sensor network in Oakland. *UC Berkeley News Center* [*online*]. Available at: http://newscenter.berkeley. edu/2012/06/27/uc-berkeley-installing-first-co2-sensor-network-in-oakland/ [Accessed: September 30, 2013].

Seto, E., Martin, E., Yang, A., Yan, P., Gravina, R., Lin, I., Wang, C., Roy, M., Shia, V., and Bajcsy, R. 2010. Opportunistic strategies for lightweight signal processing for body sensor networks. 3rd International Conference on Pervasive Technology Related to Assistive Environments-PETRA, Samos, Greece.

Short, N.M. and Robinson, J. 1998. *The remote sensing tutorial*: Goddard Space Flight Center, NASA.

Shoval, N. and Isaacson, M. 2006. Application of tracking technologies to the study of pedestrian spatial behavior. *The Professional Geographer*, 58(2): 172–83.

Steinle, S., Reis, S., and Sabel, C.E. 2013. Quantifying human exposure to air pollution—moving from static monitoring to spatio-temporally resolved personal exposure assessment. *Science of the Total Environment*, 443: 184–93.

Su, J.G., Winters, M., Nunes, M., and Brauer, M. 2010. Designing a route planner to facilitate and promote cycling in metro vancouver, canada. *Transportation Research Part A: Policy and Practice*, 44(7): 495–505.

Swayze, G.A., Clark, R.N., Sutley, S.J., Hoefen, T.M., Plumlee, G.S., Meeker, G.P., Brownfield, I.K., Livo, K.E., and Morath, L.C. 2006. Spectroscopic and x-ray diffraction analyses of asbestos in the World Trade Center dust: Asbestos content of the settled dust, in J.S. Gaffney and N.A. Marley eds. *Urban Aerosols and their Impact: Lessons Learned from the World Trade Center Tragedy*. Oxford: Oxford University Press, 40–65.

Teige, V.E., Havel, E., Patt, C., Heber, E., and Cohen, R.C. 2011. Berkeley atmospheric CO_2 network (BEACON)—bringing measurements of CO_2 emissions to a school near you. AGU Fall Meeting, San Francisco, CA.

van Donkelaar, A., Martin, R.V., Brauer, M., Kahn, R., Levy, R., Verduzco, C., and Villeneuve, P.J. 2010. Global estimates of ambient fine particulate matter concentrations from satellite-based aerosol optical depth: Development and application. *Environmental Health Perspectives* 118(6): 847–55.

Vazquez-Prokopec, G.M., Stoddard, S.T., Paz-Soldan, V., Morrison, A.C., Elder, J.P., Kochel, T.J., Scott, T.W., and Kitron, U. 2009. Usefulness of commercially available GPS data-loggers for tracking human movement and exposure to dengue virus. *International Journal of Health Geographics*, 8(1): 68.

Ward, M.H., Nuckols, J.R., Weigel, S.J., Maxwell, S.K., Cantor, K.P., and Miller, R.S. 2000. Identifying populations potentially exposed to agricultural pesticides using remote sensing and a geographic information system. *Environmental Health Perspectives* 108(1): 5–12.

Ward, M.H., Lubin, J., Giglierano, J., Colt, J.S., Wolter, C., Bekiroglu, N., Camann, D., Hartge, P., and Nuckols, J.R. 2006. Proximity to crops and residential exposure to agricultural herbicides in Iowa. *Environmental Health Perspectives* 114(6): 893–7.

Wilson, M.L. 2002. Emerging and vector-borne diseases: Role of high spatial resolution and hyperspectral images in analyses and forecasts. *Journal of Geographical Systems*, 4(1): 31–42.

Wu, J., Jiang, C., Liu, Z., Houston, D., Jaimes, G., and McConnell, R. 2010. Performances of different global positioning system devices for time-location tracking in air pollution epidemiological studies. *Environmental Health Insights* 4: 93–108.

Wu, Y.Z., Chen, J., Ji, J.F., Tian, Q.J., and Wu, X.M. 2005. Feasibility of reflectance spectroscopy for the assessment of soil mercury contamination. *Environmental Science & Technology*, 39(3): 873–8.

Zeger, S.L., Thomas, D., Dominici, F., Samet, J.M., Schwartz, J., Dockery, D., and Cohen, A. 2000b. Exposure measurement error in time-series studies of air pollution: Concepts and consequences. Environmental Health Perspectives 108(5): 419–26.

SECTION 5
Accessibility and Health

Chapter 14
Locational Planning of Health Care Facilities

Alan T. Murray and Tony H. Grubesic

Health care is an important component of general health and well-being, even if the type and/or delivery mechanism varies across different political, economic, cultural and organizational perspectives. For example, hospitals, clinics and other facilities where health care expertise is made available are in constant demand, regardless of their geographical setting (for example, urban or rural). For this and other reasons, the location of health care services is vital so that people have access to primary and secondary care, emergency medicine, preventive care, diagnostics and testing, treatment, surgery, physiatry, public health, etc. Further, it is well known that proximity and access are highly correlated with service utilization (Abernathy and Hershey, 1972; Mayhew and Leonardi, 1982; Tanser, Gething and Atkinson, 2009; Alegana et al., 2012; Wang, 2012). The better the access and the closer people are, the more likely they are to utilize health care services.

Health care may be highly subsidized by local, state, federal and provider organizations or may be a for profit enterprise. Either way, there is a need for ensuring access to essential services in the face of limited resources—a trait that the health care system shares with a variety of other public and private services. As a result, this has meant making difficult decisions about where facilities are to be located in order to provide the greatest benefit possible. To support such decision making, geographical information systems (GIS), spatial analysis and location modeling have been important tools in planning and decision making associated with siting, extension and change in a health care facility network.

For example, in rural settings, location modeling has been used to identify optimal locations for a new hospital in the Negev region of Israel (Sinuany-Stern et al., 1995; Mehrez et al., 1996). Specifically, a variety of location models and assumptions were tested and compared, with multicriteria decision analysis used to evaluate the identified locational alternatives. In other work, Chu and Chu (2000) utilized spatial optimization for hospital siting as well as allocation techniques to model the supply and demand of public hospital beds in Hong Kong. The developed decision support system can also be loosely coupled with a geographic information system for the visualization of results. Jia et al. (2007) focused on the location of facilities for providing medical services during large-scale emergencies (for example, smallpox, dirty bombs, etc.). Results suggested that the developed model outperforms traditional p-center, p-median and covering models for dealing with large-scale disasters, particularly when multiple-facility coverage requirements are in place.

For more general overviews of the health care context along with methods to support systematic analysis, readers are referred to the work of Thomas (1992), Gatrell and Elliott (2009) and Cromley and McLafferty (2012), among others.

The focus of this chapter is on health care facility siting and analysis, providing an overview of the criteria, measures and generic optimization models that have been employed to support evaluation and locational decision making for health care facilities. Worth highlighting at the outset is that a location model (or optimization model) may be used in both descriptive and prescriptive ways (Church and Murray, 2009). A location model is unique in that it mathematically reflects goals, objectives, constraining conditions along with decision variables that correspond to selection of where facilities should be placed/sited in geographic space. A descriptive context may use a location model to assess aspects of existing systems. For example, the efficiency of a system is examined in Bennett et al. (1982), Rushton (1984), Kumar (2004) and Yao and Murray (2014), where the current system is compared against the theoretical best case in terms of access, service and other stipulated siting criteria. This essentially provides the capacity to evaluate inherent inefficiencies in the system due to its historic evolution relative to the changing needs, usage, and demographic patterns of a community. Alternatively, a prescriptive (or normative) context uses a location model to site, reconfigure, expand or consolidate a system of health care facilities. There are many examples in the literature along these lines (Abernathy and Hershey, 1972; Calvo and Marks, 1973; Love and Trebbi, 1973; Dokmeci, 1977, etc.). Again, this contrast is noted to indicate that a location model can be effective in both descriptive and prescriptive contexts.

A second meaningful distinction made in this chapter relevant to health care planning is the recognition of non-hierarchical and hierarchical service systems. A non-hierarchical system is one where each facility provides the same basic services. In contrast, a hierarchical system has different levels of services, something discussed in Dokmeci (1977), Brotchie, Dickey and Sharpe (1980), Rushton (1984), Hodgson (1988), Rahman and Smith (2000), Galvao et al. (2006), and Yasenovskiy and Hodgson (2007), among others. For example, a hierarchical health care system may have clinics that provide primary care and treatment, but also hospitals that can carry out more specialized services like sophisticated diagnostics (that is, imagining, scanning, etc.), testing, surgery, treatment, rehabilitation, etc. Thus, a higher level facility (for example, hospital) provides all the services of the lower level facility (for example, clinic) but also provides other more advanced services. Such nuanced differences have proven important to explicitly consider in system analysis and design, requiring unique mathematical care and detail.

In the remainder of this chapter we detail the criteria and measures often employed in health care facility planning and analysis. This is then used to formalize general optimization models reflective of location problems applied in practice. Rather than provide an extensive review of the literature in this area, this chapter instead focuses on the mathematical formalization of the general planning problem. Literature reviews of modeling research may be found in Rushton (1984), Hodgson (1988), Rahman and Smith (2000), Daskin and Dean (2004), Cromley and McLafferty (2012).

Criteria and Measures

A range of criteria has been outlined associated with the locational planning of health care systems, including issues of access, efficiency, budgets, equity and overall system utilization. In the context of health care, the concept of access is multifaceted. For example, access to the health care system can reflect the subset of a population that has health insurance (Ku and Matani, 2001), or the mobility of older adults (Rosso et al., 2013). Other facets of access include cost, safety, demographics and socio-economic status. However, access also commonly refers to the travel time/distance between home/work to get to a health care facility. A priority, therefore, is providing people with the best access possible, making the system as efficient as possible. Of course, given limited funds, efficiency also means doing the best you can within budget constraints. Equity generally speaks to resolving spatial disparities in service provision, but also includes strategies for remediating gaps in health care access for different socio-economic groups. There also could be a concern for equity in service variability and care options among health care facilities. This relates to system utilization as well, where there could be interest in ensuring that facilities are well utilized or evenly balanced across a system. For example, a single, highly popular clinic that is operating over capacity versus another clinic that is relatively idle, presents a number of inequity issues. There is a robust literature on measuring the efficiency of health care delivery and systems, including efficiency (Hollingsworth, 2003), but there is no widely accepted global baseline for system utilization, because each system is different. However, if the system is hierarchically arranged, there might be explicit consideration of system integration in this regard.

A challenge then in systematic analysis is operationalizing these criteria through measures and metrics that can ultimately be included in a mathematical model. With appropriate notation, specification is possible. It should be noted that the measures that follow represent those typically utilized in health care facility analysis and planning. It is possible to conceive of alternative measures and metrics in many cases.

Cost: Since locational planning of health care facilities is being considered, it is assumed that existing or potential facility locations are known. This set is referred to as J, with an individual site being $j \in J$. For any site j there are associated operational costs. Let f_j be the fixed cost to open/operate health care facility j. There are also decisions being made, descriptively or prescriptively, regarding the placement of facilities. This may be denoted using a binary variable, where X_j equals 1 if health care facility j is selected for operation or 0 if not. A measure of system cost can then be structured:

$$\sum_j f_j X_j \quad (1)$$

For a given selection of facilities, the associated costs to establish the system may be computed using equation (1). Based on the particular context, this function could then be minimized, if so desired, or it might even be bounded by a budget/investment limit.

Access: Consider first that the areas where people reside and/or work are known. This set is referred to as I, with an individual area being $i \in I$. For any area i the expected service demand for a health care facility is a_i. Given an area i and a potential facility j, it is possible to derive the distance or travel time, d_{ij}, between them. Finally, there is a question of which service demand areas seek (or would seek) treatment at which health care facility. Assuming this is known,[1] a binary variable can denote service assignment (or allocation), where Z_{ij} equals 1 if demand area i is served by health care facility j or 0 if not. A measure of average access can then be structured:

$$\frac{1}{\sum_i a_i} \sum_i \sum_j a_i d_{ij} Z_{ij} \quad (2)$$

The assumption with equation (2) is that it is known which areas are being served by which facilities. With this, average access distance/time may be derived across all demand areas.

---

1 This type of information is often derived in market analysis studies prior to determining the feasible sites for a new health care facility location (Garnick et al., 1987; Duggan, 2002).

Based on the planning context, it may be desirable to optimize this function in order to determine a good allocation of service.

Equity: As with the above measures, there are many ways to view and quantify equity. A common approach to equity in health care facility siting and analysis is to try to ensure that an area is no further than some maximum travel or distance standard. Assume this maximum proximity standard is γ, then the stipulated condition for any area i would be $d_{ij} \leq \gamma$ for the facility j it is served by. Suppose that all potential facilities are examined. This would allow for the specification of a set of potential facilities, Ω_i, that could serve area i within the maximum standard. Specifically, $\Omega_i = \{j \mid d_{ij} \leq \gamma\}$. A condition could therefore be stipulated regarding equitable service assignment as follows:

$$\sum_{j \in \Omega_i} Z_{ij} = 1 \qquad\qquad (3)$$

This is a formal requirement that a health service demand area must be served by a facility that is no further away than γ. An illustration of such a spatial condition is given in Figure 14.1, where two clinics are within the maximum proximity standard shown. This is interpreted in equation (3) using the pre-defined set of facilities that achieve this service level. Of course, if $\Omega_i = \emptyset$ for any i, then maintaining equity in this manner would not be possible.[2]

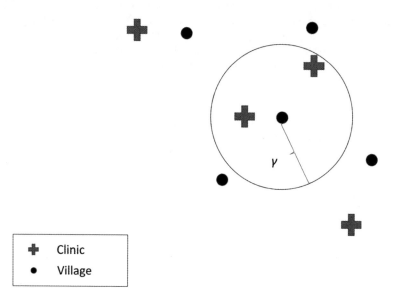

✚	Clinic
●	Village

Figure 14.1   Clinics within maximum distance coverage standard

---

2   Equity measures can be complicated by the distance behavior of hospital patients, contingent upon their particular diagnosis and the availability of medical experts and equipment (Mayer, 1983). However, equity in location decisions must be based on the greater good, not for patients with extreme distance behaviors.

Utilization: A concern in locating health care facilities is that they be sufficiently utilized, but not over utilized. If l_j represents the minimum level of service expected of facility j and u_j indicates the maximum service capacity of facility j, then utilization requirements can be specified as follows:

$$l_j \leq \sum_i a_i Z_{ij} \leq u_j \qquad (4)$$

What is expressed in equation (4) is that the total expected demand assigned (or allocated) for service at facility j must be a least l_j but no more than u_j. The rationale is that system performance remains feasible and justified under these conditions.

Cost, access, equity and utilization are prevailing concerns in locational planning and analysis of health care facilities. The detailed measures and metrics above represent typical approaches that have been relied upon. As noted, it is possible to conceive of alternative measures and metrics in many cases. Nevertheless, the measures and metrics included here serve as representative approaches addressing desired performance criteria in health care planning and analysis.

Non-hierarchical Locational Planning and Analysis

There is a rich history of applying spatial optimization models to support evaluation, analysis and decision making in health care planning. This includes work by Abernathy and Hershey (1972), Calvo and Marks (1973), Love and Trebbi (1973), Dokmeci (1977), Brotchie, Dickey and Sharpe (1980), Bennett, Eaton and Church (1982), Rushton (1984), Hodgson (1988), Kumar (2004), and Yao and Murray (2014), but others as well. In many cases applied approaches represent classic spatial analytical and location models, such as the transportation, the p-median and the maximal covering location problems. Reviews of these classic approaches can be found in Daskin and Dean (2004), Tanser, Gething and Atkinson (2009), and Cromley and McLafferty (2012). Prevalent in the above noted health care siting and analysis work is a mixture of concerns that extend beyond classic location models. Given this, a general model incorporating a range of concerns is presented below. Initial emphasis will be on the case of a non-hierarchical system.

Notation:

i = index of service demand areas
j = index of potential health care center/facility sites
f_j = fixed cost to open/operate health care facility j
a_i = anticipated demand for health care services in area i
d_{ij} = distance or travel time between area i and facility j
l_j = lower service capacity of facility j
u_j = upper service capacity of facility j
Ω_i = set of potential facilities that cover or suitably serve demand area i
$X_j = \begin{cases} 1, & \text{if health care facility } j \text{ is selected for operation} \\ 0, & \text{otherwise} \end{cases}$

$$Z_{ij} = \begin{cases} 1, \text{ if demand area } i \text{ is served by health care facility } j \\ \quad\quad 0, \text{ otherwise} \end{cases}$$

$$Y_i = \begin{cases} 1, \text{ if demand area } i \text{ is suitably covered} \\ \quad\quad 0, \text{ otherwise} \end{cases}$$

This notation is utilized in structuring the generalized spatial optimization model reflecting the goals, objectives and constraining conditions commonly found in health care system siting and analysis efforts.

Minimize	$\sum_j f_j X_j$	(5)
Minimize	$\sum_i \sum_j a_i d_{ij} Z_{ij}$	(6)
Maximize	$\sum_i a_i Y_i$	(7)
Subject to	$\sum_j Z_{ij} = 1 \;\; \forall i$	(8)
	$l_j X_j \leq \sum_i a_i Z_{ij} \;\; \forall j$	(9)
	$\sum_i a_i Z_{ij} \leq u_j X_j \;\; \forall j$	(10)
	$Y_i \leq \sum_{j \in \Omega i} X_j \;\; \forall i$	(11)
	$X_j = \{0,1\} \;\; \forall j$	(12)
	$Z_{ij} = \{0,1\} \;\; \forall i,j$	
	$Y_i = \{0,1\} \;\; \forall i$	

This is a multi-objective optimization problem accounting for cost, access and equity. The first objective, (5), reflects the intent to minimize the fixed costs of building or establishing the system of health care facilities. The second objective, (6), represents the goal of optimizing access to facilities through the minimization of the total demand weighted distance that people are from a sited clinic. This is equivalent to minimizing average distance, specified in equation (2) for a given allocation scheme. The third objective, (7), seeks to maximize the expected service demand that is sufficiently close to a sited health care facility. Constraints (8) specify that each demand area must be assigned, or allocated, to a facility. Constraints (9) indicate that demand allocated to a sited facility must be at least a minimum service level. Constraints (10) indicate that demand allocated to a sited facility cannot exceed a maximum capacity. Constraints (11) track whether a demand area *i* is suitably served by one or more facilities. Constraints (12) impose decision variables associated with site selection, demand allocation and coverage to be binary, zero or one.

This model is an amalgamation of different fundamental criteria, including cost, access, and equity. This is accomplished using a multi-objective optimization framework. As such, it is possible to emphasize certain objectives (Cohon, 1978) as well as omit objectives that are not applicable or important for a particular planning situation. For example, in the context of public hospitals, which are charged with serving all individuals within a

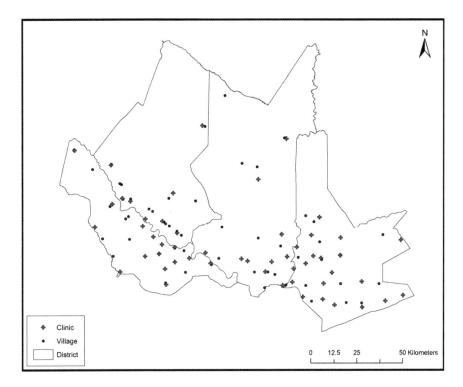

Figure 14.2   Non-hierarchical clinic system

community, a median-based approach (6) might be emphasized for hospital location. For a private hospital, which is trying to maximize market share, a maximum covering approach might be the best strategy. Regardless, changes to the model may mean that constraints would need to be modified or even additional constraints added.

The structured model can be placed in the context of the health care siting and analysis literature. Applications that have detailed cost along the lines of objective (5) include Dokmeci (1977), Achabal et al. (1978), Mehrez et al. (1996), and Griffin, Scherrer and Swann (2008). Work with access distance minimization similar to objective (6) includes Gould and Leinbach (1966), Abernathy and Hershey (1972), Calvo and Marks (1973), Love and Trebbi (1973), Dokmeci (1977), Achabal et al. (1978), Bennett (1981), Hodgson (1988), Mehrez et al. (1996), and Cocking et al. (2006). Studies that have incorporated coverage like in objective (7) include Bennett, Eaton and Church (1982), Mehrez et al. (1996), Cocking et al. (2006), Griffin, Scherrer and Swann (2008), and Leira et al. (2012). Finally, implementation of utilization requirements reflected in constraints (9) and/or (10) can be found in Love and Trebbi (1973), Cocking et al. (2006), and Griffin, Scherrer and Swann (2008).

The non-hierarchical context effectively means that there is no interaction among health care facilities and each facility provides the same basic services. An example of such a system is shown in Figure 14.2, indicating clinics in Chibuto, Chokwe, Guija and Mandlakaze districts of Gaza province in southern Mozambique. Yao and Murray (2014) report each clinic provides the same basic health services.

Hierarchical Locational Planning and Analysis

A hierarchical system explicitly recognizes different facility types, or levels. For example, it is not uncommon to observe three or more levels of an integrated health care system, particularly in rural areas. The lowest service level may be a neighborhood clinic providing general wellness exams. The next level may be a community medical center providing wellness evaluation, basic diagnostic and testing services and other assessment capabilities. The next level might be the regional hospital providing the full range of health care services, including more specialized operations and treatments. Given this basic hierarchy, assume that the set of service levels is K, where $k \in K$ refers to a particular service level. Further, the services are hierarchically ordered, from the most basic to the more advanced. Related to the above example, the service hierarchy would reflect this with $k=1$ corresponding to the clinics, $k=2$ representing the community centers, $k=3$ being the regional hospitals, and so on. A system reflecting a hierarchy of two service levels is depicted in Figure 14.3.

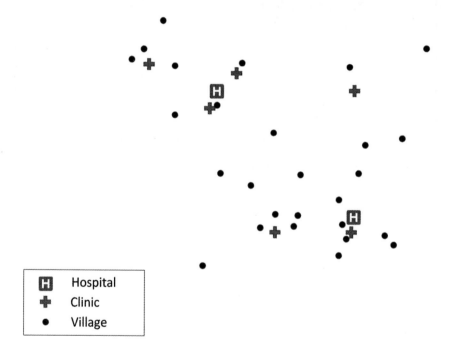

Figure 14.3   Hierarchical health care system

The above general model, (5)–(12) can then be extended to account for hierarchical relationships associated with service levels. Consider the following additional notation:

k = index of health care center/facility levels
$\hat{f}_{jk}$ = fixed cost to open/operate health care facility j at level k
$\hat{a}_{ik}$ = anticipated demand for health care services at level k in area i
$\hat{l}_{jk}$ = lower service capacity of facility j at level k

$\hat{u}_{jk}$ = upper service capacity of facility j at level k

$\hat{\Omega}_{ik}$ = set of potential facilities at level k that cover or suitably serve demand area i

$\hat{X}_{jk} = \begin{cases} 1, \text{ if health care facility } j \text{ at level } k \text{ is selected for operation} \\ \quad 0, \text{ otherwise} \end{cases}$

$\hat{Z}_{ijk} = \begin{cases} 1, \text{ if demand area } i \text{ is served by health care facility } j \text{ at level } k \\ \quad 0, \text{ otherwise} \end{cases}$

This notation has intentionally been used to remain similar to that utilized previously. The basic difference is the inclusion of service levels, both with respect to variability in spatial demand for service as well as decision making regarding where and what level of facility should be sited. The hierarchical model variant for health care planning and analysis is as follows:

Minimize $\quad\quad \sum_k \sum_j \hat{f}_{jk} \hat{X}_{jk}$ $\quad\quad$ (13)

Minimize $\quad\quad \sum_k \sum_i \sum_j \hat{a}_{ik} d_{ij} \hat{Z}_{ijk}$ $\quad\quad$ (14)

Maximize $\quad\quad \sum_i a_i Y_i$ $\quad\quad$ (15)

Subject to $\quad\quad \sum_j \hat{Z}_{ijk} = 1 \; \forall i,k$ $\quad\quad$ (16)

$\quad\quad\quad\quad\quad \hat{l}_{jk} \hat{X}_{jk} \leq \sum_{k'=1}^{k} \sum_i \hat{a}_{ik'} \hat{Z}_{ijk'} \;\; \forall j, k>1$ (17)

$\quad\quad\quad\quad\quad \sum_{k'=1}^{k} \sum_i \hat{a}_{ik'} \hat{Z}_{ijk'} \leq \hat{u}_{jk} \hat{X}_{jk} \;\; \forall j, k>1$ (18)

$\quad\quad\quad\quad\quad Y_i \leq \sum_k \sum_{j \in \hat{\Omega}_{ik}} \hat{X}_{jk} \; \forall i$ $\quad\quad$ (19)

$\quad\quad\quad\quad\quad \hat{X}_{jk} = \{0,1\} \; \forall j,k$ $\quad\quad$ (20)

$\quad\quad\quad\quad\quad \hat{Z}_{ijk} = \{0,1\} \; \forall i,j,k$

$\quad\quad\quad\quad\quad Y_i = \{0,1\} \; \forall i$

As was the case previously, this model remains a multi-objective optimization problem structured to address cost, access and equity issues. The first objective, (13), reflects the intent to minimize the fixed costs in health care facility siting, accounting for different possible levels of service. The second objective, (14), represents the goal of optimizing access to facilities through the minimization of the total demand weighted distance that people are from a sited clinic. The third objective, (15), looks to maximize the expected service demand that is sufficiently close to a sited health care facility. The model constraints are similar, but different in distinct ways. Constraints (16) specify that each demand area must be assigned, or allocated, to a facility. Constraints (17) indicate that demand allocated to a sited facility must be at least a minimum service level. Constraints (18) indicate that demand allocated to a sited facility cannot exceed a maximum capacity. Constraints (19) track whether a demand area is suitably served by one or more facilities. Constraints (20) impose decision variables associated with site and level selection, demand allocation and coverage to be binary, zero or one.

Perhaps the most significant aspect of the hierarchical model to note has to do with the utilization constraints, (17) and (18). Specifically, level k demand for an area can be assigned to any sited level k or higher facility. The model will allocate demand based on optimizing access, but this will be done making use of level k or higher facilities. In addition, to avoid confusion associated with model formulations, we have opted here not to delve into coverage hierarchies. Thus, objective (15) does not make any distinction between levels of service. However, this could readily be accounted for, if so desired, but would necessitate additional constraints to be imposed (see Church and Eaton, 1987; Branas, MacKenzie and ReVelle, 2000).

Worth noting is that this is but one type of system hierarchy. In fact, many are possible, both in terms of allocation as well as coverage: referral, non-referral, parallel, sequential, etc. An overview of different hierarchical relationships can be found in Church and Eaton (1987).

Discussion

The above general models can be modified and extended in various ways to address concerns unique to particular health care circumstances for a given region. This includes adding further constraining conditions, modifying coefficients, changing objectives, other types of decisions, different hierarchical relationships, etc. In the remainder of this section specific examples of modifications and extensions are detailed.

Objectives

One of the more challenging aspects of the above optimization models is contending with multiple objectives. It is possible that not all objectives are of interest for a particular health care planning situation, but rather only one or a specific subset. Beyond this, while the various objectives reflect siting goals and criteria, solving problems with multiple objectives is not straightforward. Objectives can often be conflicting with non-commensurate measurement units, making them difficult to integrate in a comparable manner. Dealing with this leads to the fact that there are typically tradeoff solutions, each representing a valid possible solution with respect to the objective criteria. This means that further analysis and rationale must be employed to choose from among the tradeoff alternatives to make planning decisions or interpret analysis. Of course, this assumes that the tradeoff solutions can be identified, which is often a formidable task. In fact, an entire area of research exists focused on identifying and reaching consensus on multi-objective tradeoff solutions. As a result, different approaches may be employed to solve these problems, including the weighting and constraint methods (see Cohon, 1978). The weighting method combines all objectives into one single objective using weights to reflect the relative importance of each objective. Tradeoffs result by varying the weights, and hence the importance, for each objective. An alternative is the constraint method, where objectives are effectively moved to constraints in some manner and systematically evaluated.

To illustrate the constraint method, consider objective (5) in the non-hierarchical model. This objective could be re-cast as a constraint in the following way:

$$\sum_j f_j X_j \leq \beta \qquad\qquad (21)$$

where β is a total budget limitation. Various a priori budget values could be considered, each time solving the associated problem instance. The range of tradeoff solutions identified could then be further evaluated based on planning and analysis insights. That is, one would solve a series of problems, each with a different value of β. Each unique solution would be considered in some way, then subjective decisions made based on the application results.

Taking the budget issue a bit further, if all fixed costs are the same, then it is actually an issue of how many facilities to be sited. This means that it would be possible to modify constraint (21) are as follows:

$$\sum_j X_j = p \qquad (22)$$

where p is the number of health care facilities to be located. Of course, this is effectively a budget constraint as well because the number of facilities is being fixed. Tradeoff solutions result by examining a range of values for p, or rather different investment levels. This is presented in Figure 14.4 for two objectives, with the x-axis giving the cost, objectives (5) and (13), and the y-axis indicating the percentage of the expected service demand that is sufficiently close to a clinic, objectives (7) and (15).

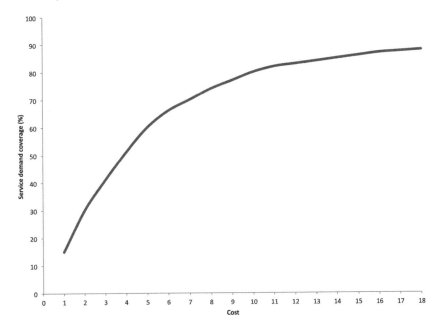

Figure 14.4   Two objective tradeoff curve

The coverage objectives, (7) in the non-hierarchical model and (15) in the hierarchical model, could also be re-cast as constraints as follows:

$$\sum_i a_i Y_i \geq \alpha \qquad (23)$$

where α represents an a priori set amount of demand that must be covered, or suitably served. Again, a range of values for α would be examined, with each associated problem instance solved. This would give tradeoff solutions for subsequent evaluation.

Of course, there are computational considerations in using any multi-objective method as the weights or parameters must be systematically adjusted, then treated and solved as a different problem instance.

Utilization

Addressing issues of utilization in the above models can represent a challenge as well. Minimum and maximum service requirements, while important, can lead to various conflicts in practice. For example, to achieve these requirements may mean that fractional assignments (allocations) are necessary. Specifically, it might be necessary for better overall system efficiency to allocate only a portion of one demand area to be served by one health care facility with the remaining portion allocated to another facility. Often this is the only way to balance utilization among facilities. Fortunately, addressing this in the models is relatively easy as one only need to relax the binary restrictions on the Z_{ij} and $\hat{Z}_{ijk}$ decision variables, allowing them to range between zero and one instead. Specifically, $0 \leq Z_{ij} \leq 1$ and $0\,\hat{Z}_{ijk} \leq 1$, meaning the service demand in area i can be allocated to one or more facilities. An example of this situation is illustrated in Figure 14.5, with the bottom most village (demand) being served by two different clinics.

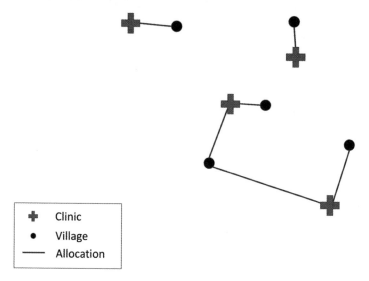

Figure 14.5   Service demand allocation

Beyond fractional allocation, strict utilization limits may actually be too rigid. Consider capacity constraints (10) in the non-hierarchical model. These constraints impose strict maximum demand allocation for each facility. In practice, however, there may well be some flexibility in total allocation. Incorporating this into the model would mean some extension. In particular, the following constraints could replace constraints (10):

$$\sum_i a_i Z_{ij} \leq u_j X_j + O_j \quad \forall j \qquad (24)$$

$$O_j \leq m X_j \quad \forall j \qquad (25)$$

where O_j is the allocation of service demand above/over the maximum capacity of sited facility j and m is a large number. Constraints (24) now allow the allocation capacity to be exceeded. Constraints (25) allow this to happen only if potential facility j is in fact sited. In addition to the constraints, the following additional objective would be necessary:

Minimize $\qquad \sum_j O_j \qquad (26)$

This objective attempts to limit allocations that exceed the maximum limits of facilities. In many ways, this represents a goal programming approach where the bound (minimum or maximum) on allocation is the goal and it is desired that it not be violated. Similar extension is possible for the minimum allocation constraints, (9) and (17).

As noted previously, it is recognized that access and proximity greatly influence whether people will actually use health care services. In particular, it has been observed that utilization declines with distance. Figure 14.6 shows one such pattern of behavior. This behavior can also be included in the above models through the use of a distance decay or other interaction function. Specifically, $a_i d_{ij}$ in objective (6) could be replaced with:

$$\delta_{ij} = a_i e^{-\mu d_{ij}} \qquad (27)$$

where is the distance decay parameter. This would similarly apply to objective (14) as well. Other functional forms reflecting observed/expected behavior could also be used, as could extensions include spatial interaction approaches (see Yasenovskiy and Hodgson, 2007).

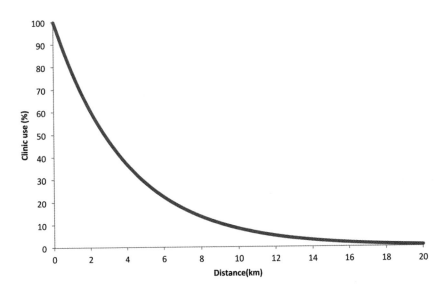

Figure 14.6 **Access distance impacts on health care utilization**

While the above models seek to maximize coverage provided within prescribed standards, it is possible to strictly impose such conditions to address service equity issues. In particular, one could readily restrict assignment/allocation, which would involve the removal of objective (7) and addition of the following condition to constraints (8):

$$\sum_{j \in \Omega_i} Z_{ij} = 1 \qquad\qquad (28)$$

This would similarly apply for objective (15) and constraints (16) in the hierarchical model. Rather than encourage proximal coverage, this approach ensures that each demand area is served by one or more health care facilities that are considered accessible.

The final extension to be mentioned is addressing issues of an expanding system. There is ample evidence to suggest that an existing system is continually in need of re-evaluation in terms of access and accessibility. Both urban and rural environments are continually changing, as are health care needs. The addition of new facilities is often an ideal time to reflect on geographic service performance issues. The above models, therefore, would need to be adapted to account for both existing and new facilities. It may very well make sense to close or move facilities as needs and utilization varies. While this can generally be taken care of through the use of fixed costs, as reflected in objectives (5) and (13), other options may also be appealing. For example, one may want to specify how many existing health care facilities are to remain in an expanded system:

$$\sum_{j \in \Phi} X_j = r \qquad\qquad (29)$$

where Φ is the set of existing health care facilities and r is the number to maintain. An approach along these lines was imposed in Yao and Murray (2014).

Conclusions

Location planning and analysis of health care facilities represents a dynamic and complex process. Balancing core health care objectives (for example, cost, access, equity, and utilization) against a myriad of operational constraints, such as budgets, facility capacity, service hierarchies and local geographic context, requires rigorous and defendable methods of analysis. The purpose of this chapter was to provide a succinct, but detailed overview of these challenges as well as methods that have been relied upon to date.

Two major themes are hopefully evident in the chapter. First, as noted throughout the chapter, the range of mathematical problems and extensions to support location planning and analysis for health care is massive. The core objectives of the planning context, such as equity and utilization, seem relatively simple when taken individually. However, when the facets of equity, utilization, cost and access are co-mingled in a more complex decision making environment, determining appropriate tradeoffs that balance various factors becomes more complicated. Very often, these facets are in direct competition with each other. For example, cost and access are inversely related. While access to health care would certainly be enhanced by placing a clinic in every neighborhood, as an example, the cost of doing so would be prohibitive. As a result, these facets must be evaluated simultaneously and in an unbiased fashion to ensure an equitable and affordable distribution of health care facilities.

A second theme that hopefully is clear is that hierarchical systems and their associated complexities continue to be important in the location planning process for health care facilities. The most operationally efficient health care networks utilize hierarchical systems. For example, the Children's Hospital of Philadelphia, which is based in the University City neighborhood, includes a 430-bed main building devoted to inpatient care, but also includes 24 primary care clinics in Southeastern Pennsylvania, six primary care clinics in Southern New Jersey as well as eight specialty care locations in Pennsylvania and five in New Jersey. The ability to offer the right mix of services across a complex network is exceedingly complex, and no doubt worthy of continued research.

References

Achabal, D.D., Moellering, H., Osleeb, J.P., and Swain, R.W., 1978. Designing and evaluating a health care delivery system through the use of interactive computer graphics. *Social Science & Medicine*, 12: 1–6.

Abernathy, W.J. and Hershey, J.C., 1972. A spatial-allocation model for regional health-services planning. *Operations Research*, 20: 629–42.

Alegana, V.A., Wright, J.A., Pentrina, U., Noor, A.M., Snow, R.W., and Atkinson, P.M., 2012. Spatial modelling of healthcare utilisation for treatment of fever in Namibia. *International Journal of Health Geographics*, 11:6.

Bennett, V.L., Eaton, D.J., and Church, R.L., 1982. Selecting sites for rural health workers. *Social Science & Medicine*, 16: 63–72.

Bennett, W.D., 1981. A location allocation approach to health care facility location: A study of the undoctored population of Lansing, Michigan. *Social Science & Medicine*, 15: 305–12.

Branas, C.C., MacKenzie, E.J., and ReVelle, C.S., 2000. A trauma resource allocation model for ambulances and hospitals. *Health Services Research*, 35: 489–506.

Brotchie, J.F., Dickey, J.W., and Sharpe, R., 1980. *TOPAZ—General Planning Technique and its Applications at the Regional, Urban, and Facility Planning Levels*. Berlin: Springer Verlag.

Calvo, A.B. and Marks, D.H., 1973. Location of health care facilities: An analytical approach. *Socio-Economic Planning Sciences*, 7: 407–22.

Chu, S.C.and Chu, L., 2000. A modeling framework for hospital location and service allocation. *International Transactions in Operational Research*, 7(6): 539–68.

Church, R.L. and Eaton, D.J., 1987. Hierarchical location analysis using covering objectives, in A. Ghosh and G. Rushton, eds. *Spatial Analysis and Location–Allocation Models*. New York: Van Nostrand Reinhold, 163–85.

Church, R.L. and Murray, A.T., 2009. *Business Site Selection, Location Analysis, and GIS*. New York: Wiley.

Cohon, J.L., 1978. *Multiobjective Programming and Planning*. New York: Academic Press.

Cocking, C., Flessa, S., and Reinelt, G., 2006. Locating health facilities in Nouna district, Burkina Faso, in *Operations Research Proceedings*: 431–6. Berlin: Springer.

Cromley, E.K. and McLafferty, S.L., 2012. *GIS and Public Health*. 2nd ed. New York: Guilford Press.

Daskin, M.S. and Dean, L.K., 2004. Location of health care facilities, in M. Brandeau, F. Sainfort and W. Pierskalla, eds. *Operations Research and Health Care: A Handbook of Methods and Applications*. Boston: Kluwer Academic: 43–76.

Dokmeci, V.F., 1977. A quantitative model to plan regional health facility systems. *Management Science*, 24: 411–19.

Duggan, M., 2002. Hospital market structure and the behavior of not-for-profit hospitals. *RAND Journal of Economics*: 433–46.

Galvao, R.D., Espejo, L.G.A., Boffey, B., and Yates, D., 2006. Load balancing and capacity constraints in a hierarchical location model. *European Journal of Operational Research*, 172: 631–46.

Garnick, D.W., Luft, H.S., Robinson, J.C., and Tetreault, J., 1987. Appropriate measures of hospital market areas. *Health Services Research*, 22(1): 69.

Gatrell, A.C. and Elliott, S.J., 2009. *Geographies of Health: An Introduction.* 2nd ed. New York: Wiley.

Gould, P.R. and Leinbach, T.R., 1966. An approach to the geographic assignment of hospital services. *Tijdschrift voor Economische en Sociale Geografie*, 57: 203–6.

Griffin, P.M., Scherrer, C.R., and Swann, J.L., 2008. Optimization of community health center locations and service offerings with statistical need estimation. *IIE Transactions*, 40: 880–92.

Hollingsworth, B., 2003. Non-parametric and parametric applications measuring efficiency in health care. *Health Care Management Science*, 6(4): 203–18.

Jia, H., Ordóñez, F. and Dessouky, M., 2007. A modeling framework for facility location of medical services for large-scale emergencies. *IIE Transactions*, 39(1): 41–55.

Ku, L. and Matani, S., 2001. Left out: Immigrants' access to health care and insurance. *Health Affairs*, 20(1): 247–56.

Kumar, N., 2004. Changing geographic access to and locational efficiency of health services in two Indian districts between 1981 and 1996. *Social Science & Medicine*, 58: 2045–67.

Leira, E.C., Fairchild, G., Segre, A.M., Rushton, G., Froehler, M.T. and Polgreen, P.M., 2012. Primary stroke centers should be located using maximal coverage models for optimal access. *Stroke*, 43: 2417–22.

Love, C.G. and Trebbi, G., 1973. Regional health care planning. *IEEE Transactions on Systems, Man and Cybernetics* SMC-3:10–18.

Mayer, J.D., 1983. The distance behavior of hospital patients: A disaggregated analysis. *Social Science & Medicine*, 17(12): 819–27.

Mayhew, L.D. and Leonardi, G., 1982. Equity, efficiency, and accessibility in urban and regional health-care systems. *Environment and Planning A*, 14: 1479–507.

Mehrez, A., Sinuany-Stern, Z., Tal, A-G., and Shemuel, B., 1996. On the implementation of quantitative facility location models: The case of a hospital in a rural region. *Journal of the Operational Research Society*, 47: 612–25.

Rahman, S. and Smith, D.K., 2000. Use of location-allocation models in health service development planning in developing nations. *European Journal of Operational Research*, 123: 437–52.

Rosso, A.L., Grubesic, T.H., Auchincloss, A.H., Tabb, L.P., and Michael, Y.L., 2013. Neighborhood Amenities and Mobility in Older Adults. *American Journal of Epidemiology*, 178(5).

Rushton, G., 1984. Use of location-allocation models for improving the geographical accessibility of rural services in developing countries. *International Regional Science Review*, 9: 217–40.

Sinuany-Stern, Z., Mehrez, A., Tal, A-G., and Shemuel, B., 1995. The location of a hospital in a rural region: The case of the Negev. *Location Science*, 3: 255–66.

Stummer, C., Doerner, K., Focke, A., and Heidenberger, K., 2004. Determining location and size of medical departments in a hospital network: A multiobjective decision support approach. *Health Care Management Science*, 7(1): 63–71.

Taket, A.R., 1989. Equity and access: Exploring the effects of hospital location on the population served--a case study in strategic planning. *Journal of the Operational Research Society:* 1001–1009.

Tanser, F., Gething, P., and Atkinson, P., 2009. Location-allocation planning, in T. Brown, S. McLafferty and G. Moon, eds. *A Companion to Health and Medical Geography.* New York: Wiley-Blackwell: 540–66.

Thomas, R.W., 1992. *Geomedical Systems: Intervention and Control.* London: Routledge.

Wang, F., 2012. Measurement, optimization, and impact of health care accessibility: A methodological review. *Annals of the Association of American Geographers*, 102: 1104–12.

Yao, J. and A.T. Murray 2014. Locational effectiveness of clinics providing sexual and reproductive health services to women in rural Mozambique. *International Regional Science Review* 37, 172–93.

Yasenovskiy, V. and Hodgson, J., 2007. Hierarchical location-allocation with spatial choice interaction modeling. *Annals of the Association of American Geographers*, 97: 496–511.

Chapter 15

Planning Towards Maximum Equality in Accessibility of NCI Cancer Centers in the U.S.

Fahui Wang, Cong Fu, and Xun Shi

Disparities in Accessibility of Cancer Care and Planning of NCI Cancer Centers

In a larger context, the debate of equity versus efficiency has been around in economics (for example, Savoie and Irving, 1992; Krugman, 2013) and in economic geography (for example, Haveman, 1973; Meyer, 2008), and is also highly relevant in healthcare markets (for example, Light, 1992; Ubel et al., 2000; Bevan et al., 2010; Reidpath et al., 2012). Some argue that the value of equity is beyond this debate, and emphasize that equity is a matter of ethical obligation and needs to be recognized as rights to medical care (Fried, 1975).

Cancer is a leading cause of death in the United States, only second to heart disease (Centers for Disease Control and Prevention, 2010). Spatial access to cancer care can be particularly important to patients' utilization of the services (Onega et al., 2008), and thus the outcomes (Shi et al., 2012; Wang and Onega, 2015). Research suggests that longer travel time to cancer care services increases risk of advanced cancer (for example, Gumpertz et al., 2006), reduces utilization of certain therapy (for example, Celaya et al., 2006), and limits enrollment in clinical trials (for example, Avis et al., 2006). The Cancer Centers designated by the National Cancer Institute (NCI) in the U.S. (hereafter referred to as "NCI Cancer Centers") have demonstrated "scientific excellence and the capability to integrate a diversity of research approaches to focus on the problem of cancer" (National Cancer Institute, 2013), and patients cared by the centers have lower mortality rates in various types of cancer (Onega et al., 2009). A recent study by Shi et al. (2012) reports that there is a great deal of variability in spatial accessibility of the NCI Cancer Centers and other academic medical centers (AMCs), and much demand for quality cancer care is left unfulfilled. Uneven distributions of cancer care facilities and population lead to *geographic disparity* in accessibility, exemplified by presence of ample service in some areas and absence or paucity of service in others. Furthermore, disproportionally higher numbers of racial and ethnic minorities often suffer from poor access to health care including cancer care (National Cancer Institute, 2008), commonly referred to as *racial disparity*. Both geographic and racial disparities contribute to deep gaps in access to care and health outcomes in the U.S.

The NCI Cancer Centers are not the only cancer care providers in the U.S. Other specialized cancer care facilities also include (1) the NCI Cancer Center satellite facilities, (2) the Community Clinical Oncology Programs (CCOPs), and (3) academic medical centers (AMCs) in the Council of Teaching Hospitals and Health Systems (COTH). Nevertheless, the NCI Cancer Centers represent perhaps the cancer care of the highest quality. There are currently a total of 66 NCI Cancer Centers. NCI designation is awarded via a grant using a peer-review process, and an NCI Cancer Center receives substantial financial support from NCI grants and is re-evaluated every 3 to 5 years (National Cancer

Institute, 2013). It is understandable that the current designation criterion focuses on quality of research and care in cancer prevention, diagnosis, and treatment. Given the important role of NCI Cancer Centers as the "backbone" of the cancer care system in the U.S., we argue that when the quality standard is not compromised, improving and promoting equal accessibility as a criterion should be factored into the designation and planning process of NCI Cancer Centers.

On the methodological front, there is a rich collection of models in the study of planning for health care facilities (Wang, 2012), but most follow the line of classic location–allocation problems (Church, 1999). In other words, most models aim to maximize service coverage (Pacheco and Casado, 2005), minimize travel needs of patients (Wang, 2006: 203–11), limit the number of facilities (Shavandi and Mahlooji, 2008), or maximize health outcome (Hemenway, 1982) by adjusting the distribution and supply of healthcare facilities. More recently, Gu et al. (2010) used a bi-objective model to identify optimal locations for healthcare facilities that maximize total coverage of population as well as their total accessibility; and Páez et al. (2013) also considered the accessibility issue in their healthcare allocation problem. However, none of the studies has equity as an objective. Equity in health and health care may be defined as equal access to health care, equal utilization of health care service or equal (equitable) health outcomes among others (Culyer and Wagstaff, 1993). Most agree that equal access is the most appropriate principle of equity from a public health policy perspective (Oliver and Mossialos, 2004).

Most recently, Wang and Tang (2013) formulate the issue of equity in healthcare delivery as equal accessibility of healthcare services, and specifically proposes a new objective of minimizing inequality in accessibility of public services. This chapter further advances this line of work on solving the planning problem of maximal equality in accessibility and applies the method to planning NCI Cancer Centers. Specifically,

1. the objective is to minimize inequality of spatial accessibility of NCI Cancer Centers, and
2. two policy options are considered: allocating additional resources to existing NCI Cancer Centers, and designating new centers.

This research is largely exploratory. Results from the study may inform the public policy decision making process in planning of the NCI Cancer Centers towards equal accessibility.

Data Processing and Measuring Accessibility

Three variables are needed in defining spatial accessibility of cancer care: supply, demand and the geographic link between them. In this case study, the supply is the NCI Cancer Centers and their corresponding capacities (for example, numbers of beds), the demand for cancer care is the estimated cancer patients across the U.S., and the link is travel distance or travel time between them. The study area is the contiguous 48 states in the U.S. (excluding Hawaii and Alaska). In Hawaii or Alaska, the accessibility issue is confined within the state and thus more straightforward to address. The accessibility measures of the 48 contiguous states need to account for complex interaction between supply and demand across state borders.

Besides the NCI Cancer Centers, comprehensive hospitals are the second tier of cancer care facilities in the U.S. Most such hospitals are academic medical centers (AMCs),

Canterbury Christ Church University Library
Canterbury Campus

If the items you have on loan have not been reserved
by another user, they will be automatically renewed
by the library system, on a rolling basis (in
accordance with the loan period of the item).

Item(s) issued:-

Title: Spatial analysis in health geography
Item barcode: 6547176660
Due: 11 December 2018

Total items: 1
13/11/2018 14:47

A monthly statement with details of items on loan will
be sent to you so that you can keep track of the items
you have on loan and decide if you need to return
those that you don't need any longer. And you can
also log into your LibrarySearch library account online
for a list of these items.

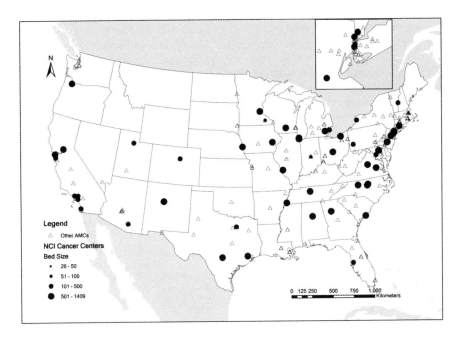

Figure 15.1   NCI Cancer Centers and other AMCs in the contiguous U.S.

which are either independent or integrated with medical schools, and are members of the COTH. The AMCs also provide high quality research and cancer care (Onega et al., 2008), and in fact, there is also considerable overlap between the NCI Cancer Centers and the AMCs. The 243 AMCs that are currently not NCI Cancer Centers thus may be the best candidates for the future designation. This study considers the 58 NCI Cancer Centers that currently provide care to patients in the contiguous 48 states as existing NCI Cancer Centers, and the 243 AMCs (currently not NCI Cancer Centers) as possible candidates for future designations. The combined data set of the 301 hospitals, including their geographic locations and existing beds counts, is shown in Figure 15.1 (the inset is a close-up of New York area, also shown in other Figures). Data of the NCI Cancer Centers are available from the NCI website (NCI, 2013). Data of the number of staffed hospital beds could be found from the American Hospital Directory (http://www.ahd.com/state_statistics.html) and websites of individual hospitals.

For the purpose of illustrating the methodology, this study simply uses population counts from the 2010 Census at the census tract level to represent the demand for cancer care. To more accurately represent the location of each census tract, we calculated population-weighted centroid for each census tract based on the 2010 Census Block data. We are aware of the spatial and temporal variability of cancer rates. More accurate estimate of cancer care demand will require the cancer prevalence data by age, sex, and race and ethnicity at sharper geographic resolutions from the North American Association of Central Cancer Registries (NAACCR) (www.naaccr.org) data and SEER*Stat (http://seer.cancer.gov/seerstat). Future research will provide refined and more in-depth analysis of the issue.

The travel time between each census tract centroid and each hospital is computed using the ArcGIS Network Analyst module. The road network data for this calculation were extracted from the ESRI StreetMap USA data that came with the ArcGIS 10.1 release. We only considered the major roads, including interstate, the U.S. and state highways, due to the computational limitation of both software and hardware. The calculated travel time is the theoretical shortest-path travel time based on the length of each road section and the speed limit information attached to each section. This approach is adequate for capturing the travel impedance between patients and these regional hospitals at the national scale for planning and public policy analysis. For technical detail of travel time estimation in ArcGIS Network Analyst, readers may consult Wang (2015, pp. 38–40). It assumes that patients seeking the specialized cancer care travel by private vehicles. Conceivably, people may also choose other transportation modes such as by air or railway. The former incurs considerably high financial cost and the latter is very limited in the U.S., and neither is considered by this study. The study area has 72,539 census tracts and a combined 301 hospitals, so the computation of the travel time matrix between them is quite time consuming. Specifically, building the road network dataset took over 14 hours and calculating the OD travel time matrix cost about 26 hours for a total of about 40 hours on a HP Pavilion dv7 laptop (2.00GHz CPU and 6.00GB memory).

There are a wide range of models for measuring spatial accessibility, which can be generalized into one model (Wang, 2012):

$$A_i = \sum_{j=1}^{n} [S_j f(t_{ij}) / (\sum_{k=1}^{m} D_k f(t_{kj}))] \qquad (1)$$

where A_i is the accessibility at location i (for example, a census tract); S_j is the capacity of supply (for example, number of hospital beds) at location (hospital) j; D_k is the estimated number of patients (demand) at tract k; t is the travel time between them, and n and m are the total numbers of hospital locations and population locations, respectively. The above measure is similar to the popular *two-step floating catchment area (2SFCA) method* (Luo and Wang, 2003), but uses a generalized function $f(t)$ to capture the distance decay effect, which can be a continuous function as in a gravity model, a discrete variable as in the 2SFCA method, or a hybrid of the two as in a kernel density function.

Our research adopts the gravity model with a power function $f(t) = t^{-\beta}$, where β is the travel friction coefficient. In the following case study, we used $\beta = 2.0$ for illustration. These are certainly arbitrary choices in both the function form and the parameter value. One needs to analyze the actual hospital visitation data to derive the distance decay function and related parameters (Wang, 2012), as attempted by Delamater et al. (2013), which is not feasible in this study. See Páez et al. (2012) for more discussion on selection of distance decay parameter.

In essence, the spatial (geographic) accessibility in equation (1) is the ratio of supply (capacity of cancer care facility) and demand (cancer patients), where their interaction is discounted by the spatial impedance (travel time here) between them. This approach accounts for the reality of patients seeking cancer care across the borders of political units such as county or state but not across countries. In other words, it assumes that the care of NCI Cancer Centers is limited to the patients in the U.S. It also permits patients the options of seeking care in multiple facilities while obeying the rule of distance (travel time) decay.

Optimization Models

An optimization problem is formalized with two components: an objective and a set of constraints that account for limited resources and decision options. As shown in

Figure 15.2, for planning NCI Cancer Centers, we started with an objective with two policy (decision) options. The objective is to minimize geographic disparities, that is, inequality in accessibility across geographic areas. The two decision options are:

1. allocating additional resources to current NCI Cancer Centers (in this case, increasing the total capacity in terms of a fixed number of beds), and
2. designating additional NCI Cancer Centers from existing AMCs.

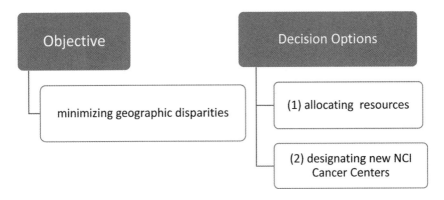

Figure 15.2   Planning goal and decision options

Formulating the Objective

The objective function is formulated as minimizing the variance (that is, least squares) of accessibility index A_i across all demand locations, written as:

$$\min \sum_{i=1}^{m} D_i(A_i - a)^2 \qquad (2)$$

In the above objective function, accessibility gaps $(A_i - a)$ are weighted by corresponding demand D_i. Note the average accessibility (weighted by demand at each tract) defined in eq. (1) across all the census tracts, a, is equal to the ratio of total supply and total demand in the study area. Denoting the total supply as $S = S_1 + S_2 + \ldots + S_n$ and total demand as $D = D_1 + D_2 + \ldots + D_m$, we have the weighted average accessibility a as

$$a = \sum_{i=1}^{m} (D_i/D)A_i = S/D \qquad (3)$$

See Shen (1998) and Wang (2015, pp. 110–11) for the detailed proof.

Decision option (1): Allocating additional resources to current NCI Cancer Centers

Resources can be in various forms and formats. For illustration, this case study examines how to allocate a fixed number of additional beds, denoted by a constant B, to existing NCI Cancer Centers. Another conceivable (though very unlikely) scenario is to relax the constraint further by permitting the reduction of the existing Centers' capacities, as assumed in Wang and Tang (2013), to allocate a fixed amount of resource in existing facilities.

The decision variables are the new capacity (number of beds) of existing NCI Cancer Centers, denoted by X_j. Note that the additional beds increase the total supply to $S+B$, and thus also change the average accessibility to $a = (S + B)/D$. The optimization problem is subject to the following constraints:

$$\sum_{j=1}^{n} X_j = S + B, \text{ and} \qquad (4)$$

$$X_j - S_j \geq 0 \text{ for all } j = 1, 2, ..., n. \qquad (5)$$

In implementation of the optimization problem, the number of beds X_j can be treated as a continuous non-negative real number instead of an integer. The above formulation fits the description of a quadratic programming (QP), where the objective function is a quadratic function of several variables (here X_j) subject to linear constraints on these variables (Wang and Tang, 2013). Matlab (www.mathworks.com/products/matlab) was used to solve the quadratic programming.

Decision option (2): designating additional NCI Cancer Centers

The other planning scenario is to designate new NCI Cancer Centers from the AMCs. Suppose a given number (n_0) of new centers is to be designated. Denote the capacities (number of beds) of all NCI and AMC hospitals as $S_j, j = 1, 2, ..., 301$. In other words, the number of potential supply sites is expanded from only the existing NCI Cancer Centers in the previous decision option to all the 301 hospitals. With a binary decision variable $x_j = 0, 1$, we specified that a hospital designated as an NCI Cancer Center to be 1, and 0 otherwise. For the existing 58 NCI Cancer Centers with patient care, $x_j = 1$ is already the predetermined solution; for the remaining 243 AMCs, x_j is the variable to be solved.

The optimization problem is a 0–1 integer programming subject to the following constraints:

$$\sum_{j=1}^{n} x_j = 58 + n_0, \text{ and} \qquad (6)$$

$x_j = 1$ for the existing 58 NCI Cancer Centers.

$x_j = 0$ or 1 for the other 243 AMCs.

We used Lingo (www.lindo.com/products/lingo) to solve this quadratic integer programming.

Case Studies

This preliminary study includes four experiments to minimize the inequality in spatial accessibility by either allocating new beds or designating new centers. These scenarios are set up in accordance to most feasible policy options: allocating funds currently available only to existing NCI Cancer Centers and the practice of NCI designating new Centers on a periodical basis.

Allocating 500 additional beds to existing NCI Cancer Centers

Currently, the total number of beds at the 58 NCI Cancer Centers is 34,160, and thus each Center has about 500 beds on average. The first case study is to allocate 500 beds (that is, adding the capacity of cancer care equivalent to an average NCI Cancer Center) to the existing 58 NCI Cancer Centers in order to minimize the spatial inequality in accessibility across census tracts. While the objective function in equation (1) remains intact, the optimization constraint in equation (4) is updated as:

$$\sum_{j=1}^{n} X_j = S + B = 34160 + 500 = 34660$$

The solution is presented in Figure 15.3. Most of the additional 500 beds were allocated to southern California, some were also to New England and Indiana. This indicates that the optimal allocation of resources for the purpose of equality is far from a trivial solution that could be obtained by simply map reading or tabulation of cancer care providers. In this case, some of the most needy regions (that is, those receiving additional services) appear to already have multiple facilities on site or nearly. However, they are also major urban areas with a lot more demand. Table 15.1 shows that the additional beds also improve the overall accessibility (a larger mean accessibility score 0.1330 than the mean value 0.1313 in the existing condition) by increasing the total supply capacity, as discussed previously. To normalize the measure of dispersion of accessibility distribution across census tracts, the coefficient of variation (that is, standard deviation divided by mean) is computed to

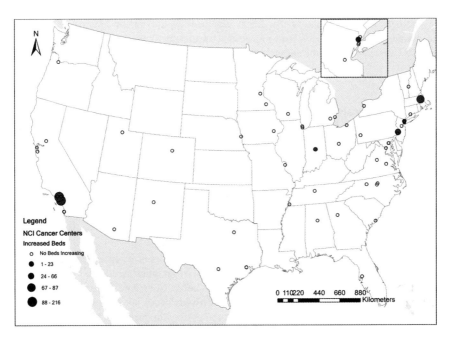

Figure 15.3   Allocation of 500 additional beds towards maximum equality in accessibility

evaluate the impact of optimization. As shown in Table 15.1, the coefficient of variation is reduced from the current 16.3329 to 16.1291, an improvement of 1.25%.

Table 15.1   Basic statistics for accessibility before and after optimization

Scenarios	Min	Max	Mean	St. Dev.	Cof. Var.
Existing condition	0.000150	339.3099	0.1313	2.1440	16.3329
1. Adding 500 beds	0.000151	339.3100	0.1330	2.1454	16.1291
2. Adding 2500 beds	0.000163	339.3103	0.1393	2.1330	15.3084
3. Adding 2500 beds with 25% cap	0.000156	339.3104	0.1384	2.1345	15.4197
4. Designating 5 new centers	0.000168	340.2747	0.1417	2.1365	15.0765

Allocating 2,500 additional beds to existing NCI Cancer Centers

As an experiment, we increased the number of additional beds to 2,500 (that is, adding the capacity of cancer care equivalent to five NCI Cancer Centers):

$$\sum_{j=1}^{n} X_j = S + B = 34160 + 2500 = 36660$$

The result is presented in Figure 15.4. Compared with Figure 15.3, the 2,500 beds were allocated to similar regions but with an expanded list of receiving hospitals. Some hospitals received more than 500 beds, and doubled their sizes, which raises the concern of feasibility in practice. Table 15.1 shows that 2,500 new beds would further improve the overall accessibility mean to 0.1393 and reduce the coefficient of variation down to 15.3084, a 6.27% improvement to the current situation.

Allocating 2,500 additional beds with a 25% expansion cap for each center

In the second study we see unrealistic scenarios that some hospitals had their sizes dramatically increased (in some cases even doubled). To address this problem, we tried another implementation that caps the increase at each hospital. For illustration purpose, we arbitrarily specify this cap to be 25%, and this updates both constraints in equations (4) and (5) to:

$$\sum_{j=1}^{n} X_j = S + B = 34160 + 2500 = 36660$$

$$0.25 S_j \geq X_j - S_j \geq 0 \text{ for all } j = 1, 2, \ldots, 58.$$

Figure 15.5 presents the result. With the additional constraint of an expansion cap, the additional beds are now allocated to a wider range of facilities across the U.S. (note the absence of hospitals with more than 200 increased beds). Compared with the two previous experiments, Table 15.1 shows that this new constraint does not compromise much the optimization, but certainly seems much more feasible in practice. The coefficient of variation is reduced by 5.59% to 15.4197 in comparison to a reduction of 6.27% to 15.3084 in the previous case without the cap.

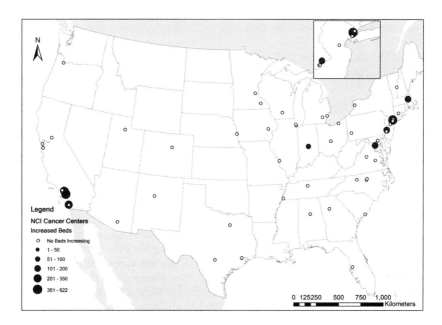

Figure 15.4   Allocation of 2500 additional beds towards maximum equality in accessibility

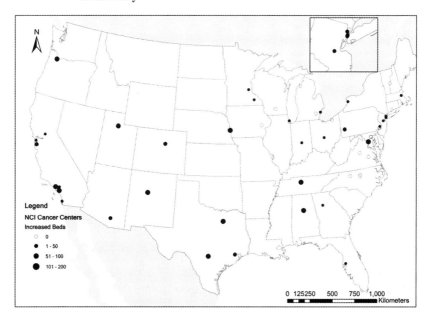

Figure 15.5   Allocation of 2500 additional beds with 25% expansion cap towards maximum equality in accessibility

Designating Five New NCI Cancer Centers

The final experiment was to designate five new NCI Cancer Centers from the 243 AMCs that are currently not NCI Cancer Centers. This updates the constraint in equation (6) to

$$\sum_{j=1}^{n} x_j = 58 + 5 = 63$$

The solution is presented in the last row in Table 15.1 with detail listed in Table 15.2. Among the five hospitals, one in Texas filling the void in the middle of heavily-urbanized Dallas-San Anotio-Houston triangle, and one in Mississippi and one in south Alabama to serve the populous Gulf coastal area; one in Denver to enhance the service capacity in the heartland, and one in the Rhode Island to once again highlight the shortage in the northeast corner. These five hospitals have a combined capacity of 545 beds, similar to the 1st case study of adding 500 beds. Yet this scenario is able to exert the highest impact in equalizing accessibility by reducing the coefficient of variation by 7.69% to the lowest value 15.0765. In other words, the most cost-efficient policy in reducing the geographic disparities in accessibility of NCI Cancer Centers is to go beyond allocation resource on the existing list and designate new centers.

Table 15.2   Designation of five new NCI Cancer Centers in a planning scenario

Hospital	Address	No. beds
University of South Alabama Medical Center	2451 Fillingim St, Mobile, AL 36617	131
Scott & White Hospital	2401 S 31st St, Temple, TX 76508	116
Methodist Rehabilitation Center	1350 E Woodrow Wilson Blvd, Jackson, MS 39216	124
National Jewish Health	1400 Jackson St, Denver, CO 80826	24
Memorial Hospital of Rhode Island	111 Brewster St, Pawtucket, RI 02860	150
Total		545

It is a significant finding. A recent paper by Delmelle et al. (2014) focuses on the decision choices between increasing capacity of existing facilities and adding new facilities in a school location problem. Our study suggests that adding new facilities is a more favorable policy option in terms of its effect of more significant reduction in inequality of accessibility. One likely reason is that the new facilities reduce the travel time for patients in resource-deprived areas in addition to added capacity of service supply.

Conclusions

The Cancer Centers designated by the National Cancer Institute (NCI) form the "backbone" of the cancer care system in the United States. Awarded via a grant using a peer-review process, the current designation criterion focuses on quality of research and care in cancer prevention, diagnosis, and treatment. We argue that when the quality standard is not compromised, an additional criterion for improving and promoting equal accessibility should be factored into the designation and planning process of NCI Cancer Centers. This

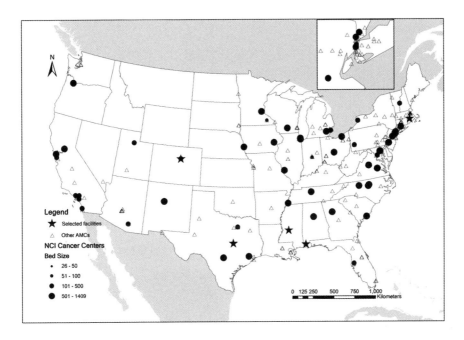

Figure 15.6   Designating five new NCI Cancer Centers towards maximum equality in accessibility

chapter proposes several planning scenarios to minimize the inequalities in accessibility. One is to allocate additional resources to the existing NCI Cancer Centers, and the other is to designate new centers. The issue is formulated as various optimization problems, which can be solved by the quadratic programming and integer programming techniques.

Four case studies were conducted to illustrate the solution and examine the impacts of various policy decisions. Several valuable lessons are learned from the results. Foremost, the optimal solution for maximal equality in accessibility helps identify most needy areas that could benefit from allocation of additional resource or designation of a new center, which would otherwise be hard to detect by conventional visual examination of maps or tabulation. Increasing the resource available for allocation does improve the equality. Imposing an expansion cap on a cancer center makes the policy more feasible without sacrificing much on the overall goal of inequality reduction. The most cost-effective policy is to go beyond the existing centers by designating new NCI Cancer Centers. This finding suggests that the best policy for reducing geographic disparities is "geographic." When new NCI Cancer Center designations are not feasible for financial constraint, the best approach is perhaps for existing ones to expand their geographic coverage by establishing satellite facilities.

References

Avis, N.E., K.W. Smith, C.L. Link, G.N. Hortobagyi, and E. Rivera. 2006. Factors associated with participation in breast cancer treatment clinical trials. *Journal of Clinic Oncology*, 24: 1860–67.

Bevan, G., J-K. Helderman and D. Wilsford. 2010. Changing choices in health care: Implications for equity, efficiency and cost. *Health Econ Policy Law*, 5: 251–67.

Celaya, M.O., J.R. Reese, J.J. Gibson, B.L. Riddle, and E.R. Greenberg. 2006. Travel distance and season of diagnosis affect treatment choices for women with early-stage breast cancer in a predominantly rural population (United States). *Cancer Causes and Control*, 17: 851–6.

Centers for Disease Control and Prevention (CDC). 2010. Deaths and mortality. http://www.cdc.gov/nchs/fastats/deaths.htm [Accessed 21st December 2013].

Church, R.L. 1999. Location modelling and GIS, in *Geographical Information Systems*, ed. P.A. Longley et al., 293–303. New York: John Wiley.

Culyer, A.J. and A. Wagstaff. 1993. Equity and equality in health and health care. *Journal of Health Economics*, 12: 431–57.

Delamater P.L., Messina J.P., Grady S.C., WinklerPrins, V., Shortridge A.M. 2013. Do more hospital beds lead to higher hospitalization rates? A spatial examination of Roemer's Law. *PLoS ONE*, 8(2):e54900.

Delmelle, E.M, J-C. Thill, D. Peeters and I. Thomas. 2014. A multi-period capacitated school location problem with modular equipment and closest assignment considerations, forthcoming in *Journal of Geographical Systems*.

Fried, C. 1975. Rights and health care—beyond equity and efficiency. *The New England Journal of Medicine*, 293: 241–5.

Gu, W., X. Wang and S.E. Mcgregor. 2010. Optimization of preventive health care facility locations. *International Journal of Health Geographics*, 9: 1–16.

Gumpertz, M.L., L.W. Pickle, B.A. Miller, and B.S. Bell. 2006. Geographic patterns of advanced breast cancer in Los Angeles: Associations with biological and socio-demographic factors (United States). *Cancer Causes and Control*, 17: 325–39.

Haveman, R.C. 1973. Efficiency and equity in natural resource and environmental policy. *American Journal of Agricultural Economics*, 55: 868–78.

Hemenway, D. 1982.The Optimal Location of Doctors. *The New England Journal of Medicine*, 306: 397–401.

Krugman, P. 2013. Why inequality matters. *The New York Times* (the Opinion Pages on Dec 15, 2013). http://www.nytimes.com/2013/12/16/opinion/krugman-why-inequality-matters.html?_r=0 (last accessed 21 December 2013).

Light, D.W. 1992. Equity and efficiency in health care. *Social Science & Medicine*, 35: 465–9.

Luo, W. and F. Wang. 2003. Measures of spatial accessibility to health care in a GIS environment: Synthesis and a case study in the Chicago region. *Environment and Planning B: Planning & Design*, 30: 865–84.

Meyer, D. 2008. Equity and efficiency in regional policy. *Periodica Mathematica Hungaria*, 56: 105–19.

National Cancer Institute. 2008. Cancer Health Disparities. http://www.cancer.gov/cancer topics/factsheet/disparities/cancer-health-disparities [Accessed 21 December 2013].

National Cancer Institute. 2013. National Cancer Institute-Designated Cancer Centers. http://www.cancer.gov/researchandfunding/extramural/cancercenters [Accessed 21 December 2013].

Oliver, A. and E. Mossialos. 2004. Equity of Access to Health care: Outlining the Foundations for Action. *Journal of Epidemiology and Community Health*, 58: 655–8.

Onega, T., E.J. Duell, X. Shi, D. Wang, E. Demidenko, and D. Goodman. 2008. Geographic access to cancer care in the U.S. *Cancer*, 112(4): 909–18.

Onega, T., E.J. Duell, X. Shi, E. Demidenko, D. Gottlieb, and D. Goodman. 2009. Influence of NCI-Cancer Center attendance on mortality in lung, breast, colorectal, and prostate cancer patients. *Medical Care Research and Review*, 66(5): 542–60.

Pacheco J.A. and S. Casado. 2005. Solving two location models with few facilities by using a hybrid heuristic: A real health resources case. *Computers & Operations Research*, 32: 3075–91.

Páez, A., J. Esita, K.B. Newbold, N.M. Heddle and J.T. Blake 2013. Exploring resource allocation and alternate clinic accessibility landscapes for improved blood donor turnout. *Applied Geography*, 45: 89–97.

Páez, A., D.M. Scott and C. Morency. 2012. Measuring accessibility: Positive and normative implementations of various accessibility indicators. *Journal of Transport Geography*, 25: 141–53.

Reidpath, D.D., A.E. Olafsdottir, S. Pokhrel and P. Allotey. 2012. The fallacy of the equity-efficiency trade off: Rethinking the efficient health system. *BMC Public Health*, 12 (Suppl 1): S3. doi:10.1186/1471–2458–12-S1-S3

Savoie, D.J. and I. Brecher (editors). 1992. *Equity and Efficiency in Economic Development Essays in Honour of Benjamin Higgins*. Montreal: McGill-Queen's University Press.

Shavandi, H. and H. Mahlooji. 2008. Fuzzy hierarchical queuing models for the location set covering problem in congested systems. *Scientia Iranica*, 15: 378–88.

Shen, Q. 1998. Location characteristics of inner-city neighborhoods and employment accessibility of low-income workers. *Environment and Planning B*, 25: 345–65.

Shi, X., J. Alford-Teaster, T. Onega and D. Wang. 2012. Spatial access and local demand for major cancer care facilities in the United States. *Annals of the Association of American Geographers*, 102: 1125–34.

Ubel, P.A., J. Baron, B. Nash and D.A. Asch. 2000. Are Preferences for Equity over Efficiency in Health Care Allocation "All or Nothing"? *Medical Care*, 38: 366–73.

Wang, F. 2012. Measurement, optimization, and impact of health care accessibility: A methodological review. *Annals of the Association of American Geographers*, 102:1104–12.

Wang, F. 2015. *Quantitative Methods and Socio-Economic Applications in GIS* (2nd ed.). Boca Raton, FL: CRC Press

Wang, F. and T. Onega. 2015. Accessibility of cancer care:disparities, outcomes and mitigation. *Annals of GIS*. DOI: 10.1080/19475683.2015.1007893.

Wang, F. and Q. Tang. 2013. Planning toward equal accessibility to services: A quadratic programming approach. *Environment and Planning B*, 40: 195–212.

Chapter 16
Spatial Dimensions of Access to Health Care Services

Daniel J. Lewis

Health and medical geographers are not distinct in valuing access as a foundational component of a spatial system. Historically, notions on the accessibility of places have: underpinned classical forms of spatial arrangement, pattern, and hierarchy; highlighted social and environmental inequalities or injustices; and, helped shape the past, present, and future dynamics of location-based resources and services (Meade and Emch, 2010). However, the study of healthcare systems is a context in which consideration of access is ubiquitous, and has provided a great many developments in how access is conceptualized, qualified, quantified and modeled.

Providing healthcare is undoubtedly a geographic problem; all care takes place somewhere: whether in a doctor's office, at home, at the site of an accident, at a hospital or affiliated treatment center, or elsewhere. The differing dimensions of access have fueled debate the world over as to what constitutes an equitable system of healthcare, and at what point inequalities in access become unjust. Barnett and Copeland (2010) suggest that health systems are under mounting pressure to effectively meet the changing health needs of their populations; the implication of which being that failure to understand a population's access to healthcare services could have profound impacts upon fairness, social justice and healthcare equity. Accessibility has been core to the specification of universal service defined by the UK's National Health Service (NHS) at its inception, and is identified as a right in the current NHS contract. Similarly, in the US, the Patient Protection and Affordable Care Act (known as: "Obamacare") seeks to, among other things, improve access to health insurance, and hence access to care.

In this chapter, access will be constrained to a consideration of its explicitly spatial dimensions, discussed as a subset of a more holistic approach to evaluating access. In taking a strictly spatial approach to access, we will further focus in on quantitative methodologies, particular those that stem from spatial analysis and Geographic Information Science. An overview of existing spatial methods for capturing access is an important but commonly encountered section, and as such the prospects for research on access will take priority after an illustrated recap of the status quo. In doing so, the chapter will emphasize the need to move away from seeing spatial analysis as a discrete and isolated function in the analysis of a healthcare system, to a process which both informs, and is informed by the greater research effort.

Naturally, this chapter will have to make some compromises, therefore it is important to note that the context of the chapter is largely urban, and largely developed world. The conceptualization and application of access to healthcare services will necessarily differ in some developing contexts. For instance, Perry and Baker (2000) discuss the constraints of mountainous Andean Bolivia in terms of accessing care, while Royal (2013) asserts the importance of nomadic herding routes in Niger in understanding disease spread. Kruk and

Freedman (2008) provide a useful overview of health system performance in developing countries for a range of policy relevant areas.

My own interest in access to healthcare comes from studying the provision of primary care in London, UK, and examining how a densely populated urban environment challenges assumptions of how individuals might, through choice or constraint, access healthcare. I believe that studies are required that deal with the dimensionality of access at fine spatial scales; scales relevant to local communities, and small urban neighborhoods, in addition to studies that have a more regional purview. As such, examples from Southwark, a borough in South London, UK, will be used to highlight the importance of local context in furthering our knowledge of healthcare accessibility.

A Practical Definition of Spatial Access

There are several models of access to healthcare services in health research, all of which value geography, primarily by positing the existence of a spatial interaction between a patient and a medical professional (Ricketts, 2010). The dominant model is the "behavioral model of health services use" articulated by Andersen and Newman (1973), who set out a framework for health service utilization which focuses on three factors: the individual; the healthcare system; and, changes in how illness is perceived by society, and treated by doctors. Andersen and Newman's intent was to create a method for evaluating the use of health services so as to provide the necessary evidence to achieve their equitable distribution.

In the model, individuals are differentiated in their likely utilization of healthcare services according to 3 sets of criteria: predisposing characteristics; enabling characteristics; and, need for care. Predisposing characteristics, including demographic and socio-economic factors, define the types of individuals who are more or less likely to use healthcare services. Enabling characteristics, including characteristics of the family and local neighborhood, define whether an individual is able to utilize care, allowing for social support, economic security, or simply through convenient location of healthcare services. Finally, need for care captures whether or not an individual has a perceived, or clinically diagnosed, requirement to use healthcare services. Opposing these individual criteria are the nature of healthcare services themselves, from basic assessments of whether a given service has adequate capacity and resource to treat patients, to more sophisticated views of the organizational role of healthcare providers. This could include the commissioning of care, and provision of appropriate investment in training, equipment, infrastructure, and so on. Finally, changing satisfaction with services, cultural attitudes, and advancing technologies are dynamically linked to the assessment of individual and systemic factors in the model. This kind of framework tallies with wider discourses around social justice (Harvey, 1973) and welfare (Smith, 1977) that emerged at the time.

Aday and Andersen (1974) further developed the behavioral model of health service use, picking up many of the points that Penchansky and Thomas (1981) would later distill into five relatively discrete dimensions: accessibility, availability, accommodation, affordability, and acceptability. Crucial to a spatial definition of access to healthcare services are availability and accessibility, which capture whether there exists enough capacity in an area to meet local need for healthcare, and whether that capacity is suitably located with respect to the distribution of people. Additionally, accommodation asks whether the capacity provided by a healthcare service is suitable to meet the needs of the area served, while affordability and acceptability deal with a patient's ability to pay and with their satisfaction as to the service provided. These five dimensions, discussed further in Cromley and McLafferty (2011), represent a popular interpretation of the behavioral model of health service use; each dimension is readily quantifiable, and allows for simple models of health service used to be iteratively refined to capture additional complexity if required.

In 1995, Andersen was able to reflect on the circa 25 year development of the behavioral model of health services use and could point to its continued widespread adoption, application, criticism and adaptation. An interesting expansion of the existing behavioral model dealt with the notion of a spatial interaction between a doctor and a patient as defining access to healthcare; initially seen as solely important to the health of individuals, this spatial interaction is now viewed as equally important to the politics of healthcare and as instrumental in influencing policy-making (Aday et al., 2004). We only have to look to the politics that surround the closure or relocation of healthcare services to verify this observation.

Aday and Andersen (1974) discuss the way in which socio-organizational and geographic components of access can be reasoned as separable, or at least distinguishable. In a sense, accessibility is more than simply the presence of a healthcare service; it is also a function of the geographic separation between someone in need of care and someone able to provide care. Attempts to isolate the geographic components of access are key in furthering research on spatial inequalities in health. The idea of distributional fairness, also commonly encountered as spatial equity (Talen and Anselin, 1998; Truelove, 1992), is a relatively longstanding attempt to do this, and derives its specification from the wider definition of healthcare equity. The World Health Organization (WHO; Whitehead, 1992) is instructive in its definition of healthcare equity as: equal access to available care for equal need; equal utilization for equal need; and, equal quality of care for all (p. 432).

While a geographic approach on its own could be viewed as reductive, a considered approach to space has produced startling work on the past unevenness of care in urban contexts (Knox, 1978; Knox and Pacione, 1980) as well as helping to define one of the most enduring representations of inequality in health: the inverse care law. Hart (1971) defines this law as the availability of health services varying inversely with patient need, a feature of the health landscape only emphasized in past years by media scrutiny on the existence of "postcode lotteries." That the quality of care you receive might be a function of where you live is a distinctly spatial characteristic of access to, and use of, healthcare. The Marmot Review (2010) has most recently recast this in terms of a "social gradient" in health in the UK.

The field of theories and models of access and use of healthcare services is large, and covers a number of sub-disciplines. My approach is health-geographic, and aims to focus in on the inequalities brought about by the spatial arrangement of health services and their patients. This spatial focus is an important subset of the behavioral model of health utilization, and picks up particularly on the accessibility and availability components of Penchansky and Thomas's (1981) dimensions. It is aligned with established geographic ideas about injustice (Dorling, 2010), and crucially, it is an examination of the question of: "Who gets what, where, and how?" (Smith, 1977).

Measuring Spatial Access

In this section we will discuss a range of existing measures of access implemented using a Geographic Information System (GIS), a tool for capturing, storing, manipulating, analyzing and visualizing geographic information (Longley et al., 2011). The successful use of GIS in an analytical capacity in public health relies on the application of a spatial-scientific framework to allow for methodological transparency and reproducibility (Longley et al., 2011; Goodchild, 1992; Goodchild, 1990). Cromley and McLafferty (2011) show that only since the 1990s has GIS been used with any regularity in health research, suggesting that it is a field yet to mature in terms of the commonplace usage and creation of geographic information and application of spatial analysis. Despite this, Meade and Emch (2010) highlight that quantitative health research involving location has a longer history in the medical geography literature, and it is also more pervasive in the broader public service arena (Longley, 2005).

In a study of car and bus travel to primary care services in the East of England, UK, Lovett et al. (2002) remark on the "appreciable effort" in creating useful geographic information from patient registers, timetables, and census tables, and that "[g]etting the best from a GIS analysis is, however, sometimes far from straightforward" (p. 110). The rationale for using GIS, despite its difficulty, is the systematic way in which large volumes of spatial data can be processed at a high level of detail. What though is the rationale for creating such measures of access to healthcare services?

Andersen (1995) confronts this question in his reflections on access, subtitled: "Does it matter?" In summary, measures of access can tell us a variety of important things, from descriptive information about how and why individuals, families and communities use healthcare services, to the adaptability, capacity and flexibility of healthcare services to meet those needs. Further, such descriptive measures can be used to monitor performance in delivery of healthcare, and assess changes in that performance. Most importantly, measures of access form a basis for the analysis of uneven distributions of healthcare, and contribute to the discussion of whether such distributions are unjust. In this last case, measures of access can contribute politically to the development of healthcare policy, interventions and improvements to the existing distribution of resources and services, and provide evidence for the efficacy and equity of reforms.

The next four subsections will cover established GIS approaches to measuring access. Cromley and McLafferty (2011) provide a more expansive resource covering these methods, while Higgs (2009) provides a recent overview of the literature of the use of GIS in healthcare utilization studies.

Potential and Revealed Measures of Access

Generally, a measure or model of access to healthcare will fall into one of two broad categories depending on whether the output is intended to capture: the *potential* (opportunity) to access a healthcare service; or, the *revealed* (actual) access based upon healthcare service registration or utilization information.

Rich (1980) defines potential as the intensity of the possibility of spatial interaction between different social or economic groups, resources or services at different locations. What is crucial to this definition is that there is an implied spatial interaction, a flow, between those locations. Measures of potential are commonly used to assess inequality in access to healthcare services because they are generalizable to populations; at their simplest all that is required is knowledge of where healthcare services are located with respect to populations, and if confounding factors are introduced it is generally the case that these are readily obtainable from administrative data resources such as a population census.

Revealed access, by comparison, is a measure of access based upon the healthcare services that an individual patient, or group of patients, actually use, regardless of the opportunity offered by their relative location. As Higgs (2009) notes, these are less common than potential measures, simply owing to the difficultly of obtaining the specialized data that links individual or neighborhood healthcare utilization with appropriate spatial information. Further, while studies of revealed access can tell us about the conditions experienced by a particular sample of the population, often for a distinct disease, disease group, or procedure, they may not necessarily be generalizable to wider contexts, particularly across national systems of healthcare. However, revealed access may be much more valuable in informing the importance of access in the causal pathway of a disease from diagnosis, to treatment and outcome than measures of potential access. This is because potential access is not necessarily a good predictor of revealed access, particularly when applied to individuals.

Network and Density measures

Joseph and Phillips (1984) discuss geographical proximity to services, contrasting potential and revealed measures of access in order to say something about the relative, or realized accessibility of an individual or group. In a study of children with spin bifida in Florida, US, Delmelle et al. (2013) suggest that geographical barriers and parental choice play an important role in explaining clusters of high revealed travel times of some patients, despite the apparent proximity of available hospitals. In describing how patients accessed primary care general practitioner (GP) services in Southwark, London, UK, road network distance and public transport travel times between individuals' home addresses and their nearest GP, and the GP with whom they were registered, were calculated.

Southwark (Figure 16.1) is a densely populated (circa 10,000 people km^{-2}) Inner London borough, with a highly diverse population both ethnically (circa 50% of the population is White British, and over 20% are either Black African or Black Caribbean), and economically (the index of multiple deprivation (IMD, 2010; DCLG, 2011) positions Southwark in

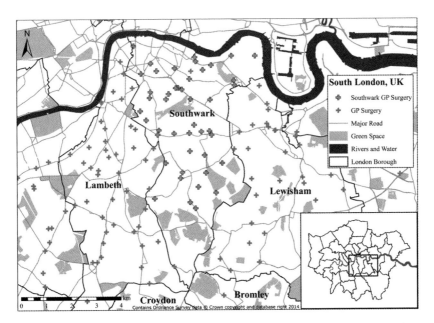

Figure 16.1   Map of Southwark, Lambeth and Lewisham. Inset: Greater London
Source: Edina Digimap/JISC. Crown Copyright 2014 Ordnance Survey.

most deprived decile of English districts). There is little to distinguish Southwark from its immediate neighbors Lambeth and Lewisham, this area of South London is continuously urban. Attention has to be paid to borough edges as there are significant in- and out-flows of patients across boundaries.

It was important to test both distance and time travel on the public transport network in Southwark and its neighbors, as bus services provide frequent local transportation to a population with low rates of car ownership. There is a strong linear relationship between bus travel time and road network distance to GP surgeries; a fitted line gives an R^2 goodness-of-fit of 0.847. This is likely due to the widespread coverage of the public transport network, and the relatively short distances traveled. Figure 16.2 shows the distribution of public transport travel times for the c.300,000 people in Southwark who use GP surgeries. It is clear that there are differences between the potential and revealed accessibility of patients, however it also demonstrates the density of services in Southwark. If all patients used their nearest GP surgery (assuming unconstrained capacities in this case) everyone could be served within 20 minutes of travel time (1,500 m). The fact that the revealed distribution of travel time to GP services differs from the potential tells us that there are intervening factors in how patients access healthcare.

These intervening factors could either be a product of choice or constraint: patients either actively choose to use a particular GP surgery because it offers something that better fits their preferences than the nearest GP surgery; or, patients are constrained by the availability of services and thus have to use GP surgeries other than their nearest. While constraints cannot be completely ruled out, we do know that all GP surgeries studied were willing to accept new patients. Repeating the analysis assuming GP capacities are constrained to current levels using an allocation approach (the Transportation Problem—see Cromley and McLafferty, 2011) reveals similar patterns of choice or constraint in patterns of patient access. An average additional distance of 790m (median: 479m) distinguishes between patients who use their nearest GP surgery, and those who do not. As only 40% of patients in Southwark use their nearest GP surgery, this may indicate that the majority of patients are willing to tradeoff relatively small additional distances in order to satisfy particular preferences in terms of the healthcare they receive.

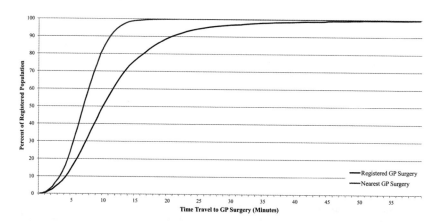

Figure 16.2   Cumulative frequency of public transport travel time to nearest and registered GP surgeries

Computing distances to healthcare services requires individualized data on patients, detailed, routable road and transport networks, and even timetabled public transport data. In addition, sophisticated routing software, such as ESRI ArcGIS Network Analyst, is required to create outputs for proposed journeys between known origins and destinations. Delamater et al. (2012) demonstrate that distance measure (straight-line or network) and data model (vector or raster) both have implications in explaining inequalities in spatial access to healthcare. However, Boscoe et al. (2012) suggest that in most cases, the use of straight-line or network distances, or travel times is largely inconsequential in terms of accessing US hospitals. Spatial scale is a likely influence on these kinds of results, at regional scales space can be treated as isotropic making straight-line methods no more or less useful than network methods. However, at more local scales, the effect of natural barriers and the built environment are likely to make an assumption of isotropy less reliable, and hence network methods more appropriate.

Undoubtedly, it is a luxury in researching access to healthcare services to be able to use individualized patient data with a fine spatial disaggregation. Many studies, such as those that rely on secondary data on hospital admissions, have to make do with areal administrative geographies. In these cases, caution should be exercised when considering distance or travel time calculations owing to aggregation effects such as the modifiable areal unit problem (MAUP: Openshaw, 1984). Instead, densities are often calculated per area, or per capita, in order to account for the heterogeneity of service availability, Ricketts et al. (1994) and Guagliardo (2004) detail a range of such metrics.

Increasingly, researchers have used continuous space measures of density, such as those derived using kernel density estimation (KDE), to estimate access to health services. McLafferty and Grady (2005) demonstrate this in the case of prenatal clinics in Brooklyn, NY, USA, finding access inequalities according to the country of birth of expectant mothers. Lewis and Longley (2012) use KDE to delineate service areas for Southwark GP surgeries based on the known distribution of patients, this allows for the number of different healthcare service options that an individual has to be counted as an intervening factor in health seeking behavior. In Southwark, the number of accessible GP surgeries significantly affects patient enrollment with their nearest GP surgery, controlling for distance and demographic factors; patients living in areas of higher GP surgery density are less likely to use their nearest GP surgery.

Lewis and Longley (2012) suggest that because the KDE-derived service areas are based upon observed distributions of patients, they capture some of the inherent complexities of access better than distance measures alone. This is evident in Figure 16.3, which shows the density surface of patient registration with a GP in Southwark, London calculated using an adaptive KDE method (Davies et al., 2011). While there are core areas of service defined by the residential locations of patients, the pattern is complex, and reflects the underlying spatial structure. This density surface can be delineated by drawing percent volume contours (Gibin et al., 2007) which capture the smallest area that effectively encloses a given percentage of the volume of the surface, creating defined, often multi-nucleated, areas of service. Historically, GP surgeries operating in the UK NHS, had to provide a boundary within which they agreed to provide care to a specified number of patients; a measure intended to regulate the workload of doctors within a GP surgery. However, when service areas based upon Southwark patient registration (as derived from Figure 16.3), are compared to the catchment areas that Southwark GP surgeries are required to draw, there is little congruence. This suggests that accessing healthcare services in dense

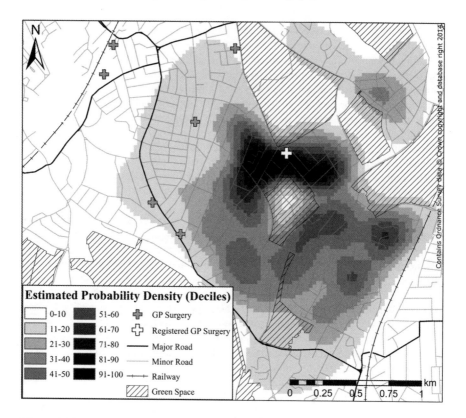

Figure 16.3   An Adaptive Kernel Density Estimate of Patient Registration for a GP in Southwark, London.

Source: Edina Digimap/JISC. Crown Copyright 2014 Ordnance Survey.

urban environments such as Southwark is a discrete, local behavior which is not directly observable by standard administrative processes, such as the drawing of catchment areas.

Gravity Models

Numerous modeling approaches have been developed that aim to both account for the geographical proximity of patients to care, and the characteristics of those patients and the capacity of each health service to deliver care. Point-to-point distances, and spatial aggregations representing densities of population or service, are common in health research, however, they tend to overlook the idea of capacity, or availability of healthcare services in meeting the needs of a population. The distance from a patient to a healthcare service will remain constant, assuming constant geographical factors, just as the density of health services does for a particular cross-section of administrative data, assuming that a service is not closed or relocated. In reality, the potential to access a service is more than just the friction of distance, or the density of services in an area; such measures may be irrelevant if a patient cannot be treated by a health service, or the patient does not have the means to make a particular journey.

Gravitational potential models are commonly used to measure spatial equity, and are so-called because they are analogous to Newtonian models of gravitational interaction. These models are intended to introduce some confounding factors in the opportunity to access healthcare services, and are formulated by dividing a "mass," recast as a measure of capacity or availability of care, by distance. Often such models are disaggregated to attribute different availabilities of healthcare, or different travel costs to different population groups. Talen and Anselin (1998 p. 600) specify a potential model for a population as:

$$Z_i = \sum_j \frac{s_j}{d_{ij}^\alpha} \qquad (1)$$

In which the potential accessibility Z at location i is equal to the capacity s of a service j over the distance between location i and service j subject to a distance decay α, summed for all services.

A simple potential model for Southwark can be fitted using the above model. Capacities are set as the number of full time doctors at each GP surgery, and the value of α is estimated to be 2.2 ($R^2 = 0.992$) based on fitting a curve to the observed distribution of distances of patients to their registered GP surgery. Figure 16.4 demonstrates the modeled output interpreted on a 100m grid, with values divided into quintiles representing different bands of accessibility. These potential values can then be interpreted in terms of the characteristics of the population distribution to reveal inequalities in opportunity to access healthcare. In Southwark, a range descriptive statistics for demographic, family and neighborhood-level variables were created to disaggregate the spatial equity measure. Households with dependent children, people of African ethnicity and people living in social housing all had a lesser opportunity to access GP surgeries. However, these were not necessarily inequitable; for instance, in terms of realized access the African group did not have significantly different spatial access to patients of other ethnicities.

Generally speaking, models are employed in an attempt to accurately simulate the observed or revealed patterns of utilization of healthcare. Such information is important to policy makers, both in assessing how different populations access services, but also as a way of testing policy scenarios, such as opening new healthcare centers, closing others, changing operating capacities or investigating the effect of changing population density and composition through area regeneration. In these cases, gravitational potential models, such as Figure 16.4 are fairly mediocre predictors, and researchers will employ more sophisticated models such as the floating catchment model, or the spatial interaction model.

The floating catchment model, most often implemented as a two or three-step floating catchment area (2SFCA or 3SFCA) procedure, is a development of the gravity model discussed above, and summarizes a range of pertinent information about a population and their health services into a single index. In the two-step model, a measure of availability is first computed at each health service location based upon an a priori researcher specification of their catchment area (often involving a distance decay function as in Luo and Qi's (2009) enhanced method), and then summing the computed availability according to a given zoning of population. Langford et al. (2008) provide an interesting account of how alternative zonings of population distribution (census tracts vs. dasymetric mapping) can influence 2SFCA estimations of service accessibility.

Wan et al. (2012) introduce a three-step model to deal with perceived inadequacies in the two-step methodology related to the overestimation of healthcare demand, a factor pertinent to contexts with high densities of healthcare services such an inner-urban areas. To resolve this, a third step can be added that accounts for differences in the local availability

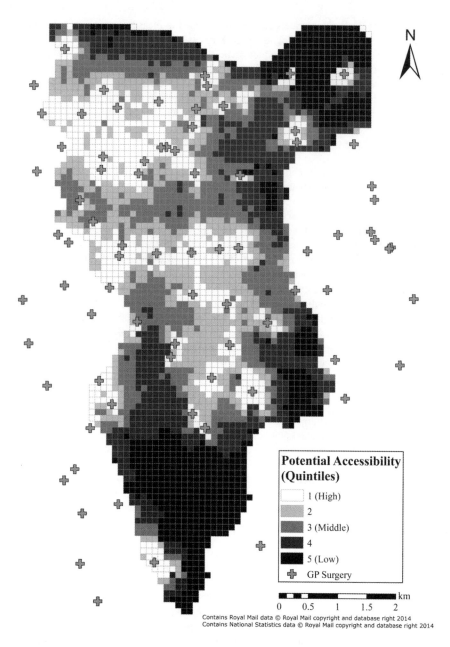

Figure 16.4   Modeled Potential Access to GP surgeries in Southwark, London, UK
Source: Edina Digimap/JISC. Crown Copyright 2014 Ordnance Survey.

of medical sites. Delamater (2013) offers a broader commentary on the role of competing services in this context, adapting a floating catchment model to deal with the fact that healthcare resources are rarely optimally distributed across a population.

Again, the quality of the data used in estimating such models is important, with Bell et al. (2012) suggesting that positional error could play a role in assessing access to primary healthcare, although their implementation of differently geocoded data using a floating catchment method did not produce statistically significant differences. Bissonnette et al. (2012) give a practical demonstration of the use of a floating catchment model, demonstrating that access to primary healthcare in Mississauga, Ontario, Canada is reduced for linguistic minorities and recent immigrant populations, and that neighborhood definition is important in this assessment.

Finally, the spatial interaction model (SIM) allows for one of the most detailed and complex views of access to, and particularly utilization of, healthcare. SIMs can also effectively integrate the relative contributions of the spatial and aspatial factors that influence the patterns of patient utilization. Wilson (2000) heralds "tremendous opportunities" in the use of SIMs for understanding and developing healthcare services, although there is a relative paucity of work in this area comparative to floating catchment models or network and density models. Clarke and Wilson (1985) and Martin and Williams (1992) show early attempts to formulate models of health services in the UK with implications for spatial planning and patterns of health service use, however there has been little since (although see inter alia: Morrissey et al., 2008). The fact that such models attempt to reconstruct realized patterns of care, coupled with their mathematical complexity (evidenced in Fisher and Wang, 2011) and need for specialized data on actual patient utilization, have made them less desirable than generalizable models of potential access, although arguably of much greater value to the geographical areas of their implementation.

Frameworks: Epidemiological and Spatial

What is clear from the discussion of methods employed to capture spatial access to, and utilization of, healthcare services, is that our efforts are governed by a range of frameworks that define the kind of analysis we can do. These frameworks can influence the kind of metrics we are able to produce, as well as the kind of analysis techniques we choose to employ.

Undoubtedly, access to data is a significant issue in studies of healthcare, in the first instance it might help define your research questions based upon whether or not you are capable of framing your research in terms of potential access or realized access. Secondly, the amount of spatial disaggregation in any data you might be able to obtain will define the likely methods you are able to use; data at small area or address level might allow for distance and travel time estimations using transport networks, whereas neighborhood or regional aggregations may require a density-based approach, or the implementation of a gravity or catchment-based model. In this case, the spatial discretization of your data frames your analysis.

Finally, planned analyses may be framed by disciplinary background; an epidemiological approach might be to compute a measure of access and then use a global statistical model, such as regression analysis, to test a series of hypotheses related to access, socio-economic factors and health outcomes. Conversely, spatial analysts are increasingly using spatially-explicit methods that account for spatial dependence, either by using a series of local regressions as in Geographical Weighted Regression (GWR: Brunsdon et al., 1998), employing a spatial weights matrix, as in spatial econometrics (Anselin et al., 2004), or by accounting for spatial autocorrelation by using spatial filters (Getis and Griffith, 2010).

Prospective Developments in Spatial Access

Clearly there are a wide range of existing and emergent methods with which the spatial dimensions of access to healthcare can be evaluated. These techniques are only as good as the assumptions we place upon them though, and so rather than expand upon the prospect for methods themselves, this section will focus on improvements to how we conceptualize and apply existing models.

Uncertain Geographic Assumptions

Key to the consideration of any research is the role of uncertainty, and this is certainly no more or less true in spatial analysis. The spatial dimension adds additional complexity, and potential for error beyond that of tabulated data, however of greater concern than the accuracy of location in data, are the assumptions we make about space and place. Many studies of spatial access to healthcare services apply estimates of how far patients are willing to travel, or arbitrary definitions of individual neighborhoods or living spaces. Not only can the use of inappropriate areal-delineations lead to MAUP, but as Kwan (2012) elaborates, inherent uncertainties over geographic context which can cause inferential errors and confounding effects.

The uncertain geographic context problem (UGCoP) arises because there are no natural units to account for a neighborhood or a place (Longley et al., 2011). Instead we must make assumptions about the validity of existing neighborhood delineations, such a census tracts and threshold travel distances, in our models. In spite of our assumptions, as Diez Roux and Mair (2010) note, the "true causally relevant" spatial context is often unknown and difficult to observe. Kwan (2012) expands this further, noting that social contexts, such as family and friendship networks are not spatial in their nature, and hence difficult to delineate. Similarly, everyday activities can alter the relevance of a fixed neighborhood to an individual dependent upon time of day. Many models of spatial access privilege so-called "night time" distributions of populations; census and residential data can only tell us where people live, they actually tell us relatively little about the everyday contexts experienced by individuals.

While models will often explain differential access to healthcare services in terms of socio-economic or demographic niches, less has been done to account for the differential mobility of populations. Gatrell (2011) calls for more work to capture the dynamic and mobile elements of geographic contexts, and the need to move away from a "static" view of the world. Mobility can be expressed in terms of the individual, as well as the aggregate, accounting for individual journeys, as well as population churn, and the rate of change over time of geographically located phenomena.

Spielman and Yoo (2009) develop an interesting set of simulations on the principle that "different people have different neighborhoods," demonstrating that the heterogeneity of individual neighborhoods leads to differing estimates of the effect on an examined health outcome. Chaix et al. (2009) give an excellent overview of this problem, suggesting that what is important to an individual are personal, or "ego-centric" definitions of neighborhood and these should be distinguished from territorial neighborhoods, which may better reflect "social collective entities." Similar conclusions are made by Spielman and Yoo (2009) who council for theoretical, rather than technical fixes to the problem of access and the neighborhood.

Neighborhood Contexts in Access to Healthcare Services

Researchers are only too aware of the importance of including time varying information into their models. Accessibility and availability of care will necessarily vary depending upon the time of day or the day of the week, while realized access may depend on the ability of an individual to access transport. Cummins (2007) suggests that in investigating individuals and their neighborhood we need to avoid falling into the trap of believing that only the "local" matters; certainly in terms of realizing care, the definition of what is "local" will vary by individual mobility and temporal context. Páez et al. (2010) demonstrate that accessing healthcare in Montreal may be more difficult for older residents owing to reduced mobility, and the urban context.

Activity spaces present an interesting framework for capturing temporal and mobility-based contexts in terms of access to care. Timmermans et al. (2002) summarize much of the research into space-time behavior, with perhaps one of the key lessons being that we cannot continue to think of accessing healthcare as a discrete action, unconnected to an individual's wider lived experience. Accessing care might be based upon co-location of other important functions, such as work, school, or retail, and rather than being accessed as an isolated trip, might form part of a trip-chain (Timmermans et al., 2002). In the broader context of public services, Neutens et al. (2010) demonstrate that access can be subject to hourly and daily variations, and that different groups of people may be affected differently by this. Weber and Kwan (2010) are able to further demonstrate that unrealistic measures of travel time diminishes our ability to realistically evaluate individual accessibility to services, and that accounting for congestion at key business hours in accessibility models produces additional inequalities in access. Kwan (1998) sets out a range of space-time accessibility measures for point-based data, arguing that the additional individual heterogeneity in potential access that they bring about make spatio-temporal models a better candidate for capturing differences between population groups.

Our health behaviors are increasing thought to be influenced by a range of individual and environmental factors; Cummins et al. (2007) suggest that we need a relational conceptualization of place in order to better understand the dynamic, multi-scale, mobile, and heterogeneous views of place that individuals experience. To this end, we need to develop representations of neighborhoods that are as broad as they are deep, we cannot simply concern ourselves with health services. We need to consider other hypothesized elements of the social and physical environment that causally affect access to healthcare, particularly characteristics of the built environment: the diversity, clustering and concentration of complementary resources and services, and the transport infrastructure.

Dealing with Complexity

The notion that healthcare is organized as a system, and hence should be understood in terms of its interdependencies and dynamics is not a new idea, Thomas (1992) gives an excellent overview of how systemic thinking applies to the spatial planning and structure of healthcare. Many of the methods used to measure access to healthcare services, such as computation of distance or travel time on transport networks, or gravitational potential models of spatial equity are built on the systems approach. Thomas (1992) discusses the goal of finding "static equilibriums" which can effectively balance supply of, and demand for healthcare services in such a way that access can be (theoretically) satisfied alongside a range of other goals.

However, Batty (2008) elucidates how long held notions about systems in the aggregate equilibrium form are evolving to encompass new arguments about emergence, multiple interactions, and non-linearity. These dynamic systems are largely what we have been discussing in the previous two subsections; accounting for individual space-time effects, the emergent hierarchies of service, and the wider interconnectedness of peoples everyday activities. Sturmberg and Martin (2009) provide an overview of complexity in healthcare, while Diex-Roux (2011) presents a more directed approach to the practical application of complex systems thinking. Such targeted ways of thinking may be important in advancing research into how populations access and utilize healthcare services.

Casas, Delmelle and Varela (2010), for example, assess space-time clustering events in order to better understand the significant complexities of first-time hospital visitation at a hospital in Columbia. Such studies give hospital administrators a much richer idea of their community of service, and help them make better strategic decisions than might be possible with static cross-sectional data.

Discussion

Undoubtedly the study of access to healthcare is an active research area, it has a history of robust theory and modeling behind it, and it has made significant contributions to the evaluation, monitoring and reform of healthcare services. However, the message for the future is that we need to start seeing it as part of a greater whole, there has been a tendency to consider it as an isolated function that can be reasoned about, and analyzed discretely. The kind of research that is being published, and has been published over the last 10 years, suggests that it is important to consider factors such as: time, mobility, uncertainty and the wider effect of neighborhood resources, and contexts. All of these factors greatly complicate the ways in which we can practically analyze access to healthcare and bring to bear a consideration of complex systems.

On a pragmatic level, the best way that detailed studies of spatial access to healthcare can be brought about is through greater integration of space across the relevant disciplinary contexts. This means a wider interaction between quantitative and qualitative researchers on the theme of access; as well as collaborative research that encompasses different spatial scales, as well as comparative studies of different health systems where possible. Naturally, the biggest issue that undermines such suggestions is the availability of appropriate data resources, certainly rigorous administrative data is difficult to acquire, and utilization data even harder. While there are signs that governments are taking open data, and access to administrative and institutional data more seriously, there is a long way to go, particularly in the healthcare arena.

My own work exploring Southwark, London, UK has demonstrated the spatial complexity inherent in looking at how healthcare services are accessed in inner cities. One of the key notions that the work challenged was that patient choice had hitherto not been exercised in primary care in the NHS. While proximity still explains a large part of access to GP surgeries in Southwark, there is good evidence for behaviors that contradict this norm, particularly among ethnic minority patients who might prefer to be treated by doctors with similar ethnic backgrounds. While choice has notionally been a part of primary care provision in the NHS since its birth (Corrigan, 2005; Moon and North, 2000), a lack of information on the differences between GP services and an insistence on their uniformity largely privileged spatial difference in access to care, however, this looks set to change. Recent reforms to primary care in the NHS have been made on the basis of changing the relationship between doctor and patient into a patient-centered experience, putting more choice in the hands of the patient. The belief was that such a move would help reduce health inequalities.

The gold standard for research on healthcare accessibility should no longer be isolated studies that characterize the magnitude of inequalities in the spatial arrangement of services for particular population groups. Instead we should focus on attempting to more fully reason and demonstrate how access to healthcare impacts causally upon health behaviors and health outcomes, and think through how interventions can be formulated to tackle some of the current inequalities in access in order to improve health for all.

Consolidation

This chapter has dealt with spatial access to healthcare from a quantitative standpoint. The dominant model of healthcare utilization, the behavioral model, has been discussed, and from it, the notion of spatial access has been extracted and dealt with more fully. The use of geographic information systems (GIS) in evidencing variation in spatial access to healthcare services has been demonstrated with respect to several existing methodologies. In doing so, we have been careful to consider the differences between potential and realized access, as well as the value of epidemiological and spatial frameworks for analysis.

Having provided some evidence for the status quo, the chapter considers the prospect for research into spatial access to healthcare, highlighting a range of emergent research that suits a more dynamic, mobile, and complex way of looking at healthcare systems. In systems such as the NHS, which is undergoing significant reforms, and attempting to expand the choices that patients have in accessing care, it is clear that more relevant measures of access are required. We require a better conceptual understanding of how patients access care, and how that might have changed, as well as a fuller integration of lived experience so that we might provide the relevant evidence for the efficacy of reforms and interventions. In moving to such a research context, we have suggested that notions of spatial access need to broaden, and open up to contributions from other disciplines, both health-related and from further afield.

References

Aday, L. and Andersen, R. 1974. A Framework for the Study of Access to Medical Care. *Health Services Research*, 9: 208–20.

Aday, L., Begley, C., Lairson, D., and Balkrishnan, R. 2004. *Evaluating the Healthcare System: Effectiveness, Efficiency, and Equity*. 3rd Edition. Health Administration Press, Chicago, IL, USA.

Andersen R. 1995. Revisiting the Behavioral Model and Access to Medical Care: Does It Matter? *Journal of Health and Social Behavior*, 36(1): 1–10.

Andersen, R. and Newman, J. 1973. Societal and Individual Determinants of Medical Care Utilization in the United States. *The Milbank Memorial Fund Quarterly: Health and Society*, 51(1): 95–124.

Anselin, L., Florax, R., and Rey, S. *Advances in Spatial Econometrics: Methodology, Tools and Applications*. Springer, Berlin, Germany.

Barnett, R. and Copeland, A. 2010. Providing Health Care, in Brown T., McLafferty S., and Moon G. (Ed.) *A Companion to Health and Medical Geography*. Oxford: Wiley-Blackwell.

Batty, M. 2008. Cities as Complex Systems: Scaling, Interaction, Networks, Dynamics and Urban Morphologies, in Meyers R. (Ed.) *Encyclopedia of Complexity and Systems Science*. Springer, Berlin, Germany.

Bell, S., Wilson, K., Shah, T., Gersher, S., and Elliot, T. 2012. Investigating Impacts of Positional Error on Potential Health Care Accessibility. *Spatial and Spatio-Temporal Epidemiology*, 3(1): 17–29.

Bissonnette, L., Wilson, K., Bell, S., Shah, T. 2012. Neighbourhoods and Potential Access to Health Care: The Role of Spatial and Aspatial Factors. *Health and Place*, 18(4): 841–53.

Boscoe, F., Henry, K., and Zdeb, M. 2012. A Nationwide Comparison of Driving Distance Versus Straight-Line Distance to Hospitals. *Professional Geographer*, 64(2): 188–96.

Brunsdon, C., Fortheringham, S., and Charlton, M. 1998. Geographically Weighted Regression. *Journal of the Royal Statistical Society* D, 47(3): 431–43.

Casas, I., Delmelle, E., and Varela, A. 2010. A Space-Time Approach to Diffusion of Health Service Provision Information. *International Regional Science Review*, 33: 134–56.

Chaix, B., Merlo, J., Evans, D., Leal, C., and Havard, S. 2009. Neighborhoods in Eco-Epidemiologic Research: Delimiting Personal Exposure Areas. A Response to Riva, Gauvin, Apparicio and Brodeur. *Social Science and Medicine*, 69: 1306–10.

Clarke, M. and Wilson, A. 1985. A Model-based Approach to Planning in the National Health Service. *Environment and Planning* B, 12: 287–302.

Corrigan, P. 2005. Registering choice: How primary care should change to meet patient needs. Social Market Foundation. Available at: http://www.smf.co.uk/assets/files/publications/Choice%20&%20Health.pdf Last Accessed: 15/4/2013.

Cromley, E. and McLafferty, S. 2011. *GIS and Public Health*. Second Edition. Guilford Press, New York, NY, USA.

Cummins, S. 2007. Commentary: Investigating Neighbourhood Effects on Health—Avoiding the 'Local Trap.' *International Journal of Epidemiology*, 36(2): 355–7.

Cummins, S., Curtis, S., Diez Roux, A., and Macintyre, S. 2007. Understanding and representating 'place' in health research: A relational approach. *Social Science and Medicine*, 65(9): 1825–38.

Davies, T., Hazelton, M., and Marshall, J. 2011. sparr: Analyzing Spatial Relative Risk Using Fixed and Adaptive Kernel Density Estimation in R. *Journal of Statistical Software*, 39(1): 1–14.

Delamater, P. 2013. Spatial accessibility in suboptimally configured health care systems: A modified two-step floating catchment area (M2SFCA) metric. *Health & Place*, 24: 30–43.

Delamater, P., Messina, J., Shortridge, A., and Grady, S. 2012. Measuring geographic access to health care: Raster and network-based methods. *International Journal of Health Geographics*, 11(15): 1–18.

Delmelle, E., Cassell, C., Dony, C., Radcliff, E., Tanner, J., Siffel, C., and Kirby R. 2013. Modeling Travel Impedance to Medical Care for Children with Birth Defects Using Geographical Information Systems. *Birth Defects Research* Part A, 97: 673–84.

Department of Communities and Local Government. 2011. The English Indices of Deprivation 2010. Available at: https://www.gov.uk/government/uploads/system/uploads/attachment_data/file/6871/1871208.pdf [accessed: 9/4/2013].

Diez Roux A. 2011. Complex Systems Thinking and Current Impasses in Health Disparities Research. *American Journal of Public Health*, 101(9): 1627–34.

Diez Roux, A. and Mair, C. 2010. Neighborhoods and Health. *Annals of the New York Academy of Sciences*, 1186: 125–45.

Dorling D. 2010. *Injustice: Why Social Inequality Persists*. Bristol: Policy Press.

Fisher, M. and Wang, J. 2011. *Spatial Data Analysis: Models, Methods and Techniques*. London: Springer.

Gatrell A. 2011. *Mobilities and Health*. Farnham: Ashgate.

Getis, A. and Griffith, D. 2010. Comparative Spatial Filtering in Regression Analysis. *Geographical Analysis*, 34(2): 130–40.

Gibin, M., Longley, P., and Atkinson, P. 2007. Kernel Density Estimation and Percent Volume Contours in General Practice Catchment Area Analysis in Urban Areas, in Winstanley A. (ed.) *Proceedings of GIS Research UK*, National University of Ireland, Maynooth, 11th–13th April 2007: 270–77.

Goodchild, M. 1992. Geographical Information Science. *International Journal of Geographical Information Science*, 6: 31–45.

Goodchild, M. 1990. Spatial Information Science. Proceedings of the Fourth International Symposium on Spatial Data Handling, Zurich: 3–12.

Guagliardo, M. 2004. Spatial Accessibility of Primary Care: Concepts, Methods and Challenges. *International Journal of Health Geographics*, 3(3): 1–13.

Hart, J. 1971. The Inverse Care Law. *The Lancet*, 297(7696): 405–12.

Harvey, D. 1973. *Social Justice and the City*. London: Edward Arnold.

Higgs, G. 2009. The Role of GIS for Health Utilization Studies: Literature Review. *Health Services and Outcomes Research Methodology*, 9: 84–99.

Joseph, A. and Phillips, D. 1984. *Accessibility and Utilization: Geographical Perspectives on Health Care Delivery*. London: SAGE.

Knox, P. 1978. The intraurban ecology of primary medical care: Patterns of accessibility and their policy implications. *Environment and Planning* A, 10(4): 415–35.

Knox, P. and Pacione, M. 1980. Locational behaviour, place preferences and the inverse care law in the distribution of primary medical care. *Geoforum*, 11(1): 43–55.

Kruk, M. and Freedman, L. 2008. Assessing health system performance in developing countries: A review of the literature. *Health Policy*, 85(3): 263–76.

Kwan M-P. 2012. The Uncertain Geographic Context Problem. *Annals of the Association of American Geographers*, 102(5): 958–68.

Kwan M-P. 1998. Space-Time and Integral Measures of Individual Accessibility: A Comparative Analysis Using A Point-based Framework. *Geographical Analysis*, 30(3): 191–216.

Langford, M., Higgs, G., Radcliffe, J., and White, S. 2008. Urban Population Distribution Models and Service Accessibility Estimation. *Computers, Environment and Urban Systems*, 32(1): 66–80.

Lewis, D. and Longley, P. 2012. Patterns of Patient Registration with Primary Health Care in the UK National Health Service. *Annals of the Association of American Geographers*, 102(5): 1135–45.

Longley, P. 2005. Geographical Information Systems: A renaissance of Geodemographics for Public Service Delivery. *Progress in Human Geography*, 29(1): 57–63.

Longley, P., Goodchild, M., Maguire, D., and Rhind, D. 2011. *Geographic Information Systems and Science*. 3rd Edition. Wiley, Hoboken, NJ, USA.

Luo, W. and Qi, Y. 2009. An Enhanced Two-Step Floating Catchment Area (E2SFCA) method for measuring spatial accessibility to primary care physicians. *Health and Place*, 15(4): 1100–107.

Marmot Review. 2010. Fair Society, Healthy Lives. Strategic Review of Health Inequalities in England Post-2010. Available from UCL Institute of Health Equity, London, UK.

Martin, D. and Williams, H. 1992. Market-area Analysis and Accessibility to Primary Health-Care Centres. *Environment and Planning* A, 24(7): 1009–19.

McLafferty, S. and Grady, S. 2005. Immigration and Geographic Access to Prenatal Clinics in Brooklyn, NY: A Geographic Information Systems Analysis. *American Journal of Public Health*, 95: 638–40.

Meade, M. and Emch, M. 2010. *Medical Geography*. Third Edition. New York: Guilford Press.

Moon, G. and North, N. 2000. *Policy and Place: General Medical Practice in the UK*. London: Macmillan Press.

Morrissey, K., Clarke, G., Ballas, D., Hynes, S., and O'Donoghue, C. 2008. Examining Access to GP Services in Rural Ireland Using Microsimulation analysis. *Area*, 40(3): 354–64.

Neutens, T., Schwanen, T., Witlox, F., and de Maeyer, P. 2010. Evaluating the Temporal Organization of Public Service Provision Using Space-Time Accessibility Analysis. *Urban Geography*, 31(8): 1039–64.

Openshaw, S. 1984. The modifiable areal unit problem. *Concepts and Techniques in Modern Geography*, 38. Geobooks, Norwich.

Páez, A., Mercado, R., Farber, S., Morency, C., and Roorda, M. 2010. Accessibility to health care facilities in Montreal Island: An application of relative accessibility indicators from the perspective of senior and non-senior residents. *International Journal of Health Geographics*, 9: 52.

Penchansky, R. and Thomas, J. 1981. The Concept of Access: Definition and Relationship to Consumer Satisfaction. *Med Care*, 19(2): 127–40.

Perry, B. and Gesler, W. 2000. Physical access to primary health care in Andean Bolivia. *Social Science & Medicine*, 50(9): 1177–88.

Rich, D. 1980. Potential Models in Geography. *Concepts and Techniques in Modern Geography*, 26. Geo Abstracts, UEA, Norwich, UK.

Ricketts, T. 2010. Accessing Health Care, in Brown, T., McLafferty, S., and Moon, G. (Eds) *A Companion to Health and Medical Geography*. Oxford: Wiley-Blackwell.

Ricketts, T., Savitz, L., Gesler, W., and Osborne, D. 1994. Geographic Methods for Health Servcies Research: A Focus on the Rural—Urban Continuum. University Press of America, Lanham, MD, USA.

Royal, N. 2013. Dracunculiasis, Proximity, and Risk: Analyzing the Location of Guinea Worm Disease in a GIS. *Transactions in GIS*, 17(2): 298–312.

Smith, D. 1977. *Human Geography: A Welfare Approach*. London: Edward Arnold.

Spielman, S. and Yoo E-H. 2009. The Spatial Dimensions of Neighborhood Effects. *Social Science and Medicine*, 68(6): 1098–105.

Sturmberg, J. and Martin, C. 2009. Complexity and Health—Yesterday's Traditions, Tomorrow's Future. *Journal of Evaluation in Clinical Practice*, 15(3): 543–8.

Talen, E. and Anselin, L. 1998. Assessing spatial equity: An evaluation of measures of accessibility to public playgrounds. *Environment and Planning A*, 30: 595–613.

Thomas R. 1992. *Geomedical Systems: Intervention and Control*. London: Routledge.

Timmermans, H., Arentze, T., and Chang-Hyeon, J. 2002. Analysing Space-Time Behaviour: New Approaches to Old Problems. *Progress in Human Geography*, 26(2): 175–90.

Truelove M. 1993. Measurement of Spatial Equity. *Environment and Planning C: Government and Policy*, 11: 19–34.

Wan, N., Zou, B., and Sternberg, T. 2012. A Three-Step Floating Catchment Area Method for Analyzing Spatial Access to Health Services. *International Journal of Geographical Information Science*, 26(6): 1073–89.

Weber, J. and Kwan, M-P. 2010. Bringing Time Back In: A Study on the Influence of Travel Time Variations and Facility Opening Hours on Individual Accessibility. *The Professional Geographer*, 54(2): 226–40.

Whitehead M. 1992. The concepts and principles of equity and health. *International Journal of Health Services*, 22: 429–45.

Wilson, A. 2000. *Complex Spatial Systems: The Modelling Foundations of Urban and Regional Analysis*. Harlow: Pearson.

Chapter 17

Nature and Death: An Individual Level Analysis of the Relationship Between Biophilic Environments and Premature Mortality in Florida

Christopher J. Coutts and Mark W. Horner

There is an increasing amount of empirical work demonstrating the biophilic (Wilson, 1984) linkages between green space and the conditions and behaviors that affect our physical and mental health and well-being. The biophilia hypothesis (Kellert and Wilson, 1993) is the scientific formalization of the innate biological connection between humans and nature and the inclination for humans " ... to affiliate with natural systems and processes instrumental in their health and productivity" (Kellert et al., 2008, p.vii). The wide array of potential health benefits ranges from the most fundamental of human needs to higher levels of cognitive functioning. These benefits accrue through green space being present in one's environment and delivering important ecosystem services, through access and use of green space, and also through simple exposure and viewing greenery. Green space such as parks, forests, and greenways, or collectively green infrastructure,[1] allows these biophilic needs to be met. The body of evidence demonstrating the importance of green spaces to the health and overall quality of life of communities has been growing, but rare among this body of work are epidemiological studies that link green space to mortality.

This exploratory study drew a sample of death certificates (n=143,725) from 2000–2012 in the state of Florida and employed GIS to determine if green space proximity to one's residential location at time of death was predictive of all-cause premature mortality. The two research questions tested in this study are:

1. Does the *distance* from home to the nearest green space help account for all-cause premature mortality?
2. Does the *amount* of green space within four defined distances of home help account for all-cause premature mortality?

In addition to testing these two research questions, we discuss some of the challenges inherent in the acquisition and management of large public health data sets. In the methods section we provide more detail on data acquisition and analysis preparation than is normally offered in a scientific paper. We do this in order to explicate the potential pitfalls and challenges in the spatial analysis of health phenomena. Our discussion also offers extra detail on how this initial analysis can be extended to parse out more nuanced relationships between health

---

1   Green infrastructure is defined as an " ... interconnected network of natural areas and other open spaces that conserves natural ecosystem values and functions, sustains clean air and water, and provides a wide array of benefits to wildlife and people" (Benedict and McMahon, 2006, p.1).

outcomes and green space. First, however, we present a sampling of the literature that shapes the theory behind the relationship between green space and human health.

Green Infrastructure and Public Health

As McCally states, "at least since the time of Hippocrates's essay 'Air, Water, and Places,' humans have been aware of the many connections between health and the environment" (McCally 2002, p.1). In order to protect ourselves and our health, the natural environment, and its salutogenic (Antonovsky, 1987) or health-supporting properties, must be protected from rapidly expanding built environments. The various components of green infrastructure, from small neighborhood parks to expansive national forests, support health and health behaviors in various ways. For example, wetland protection improves water quality and a local park may act as a setting for social interaction and physical activity.

Positive correlations found between self-reported health status and the amount of greenspace in one's immediate living environment (Maas et al., 2006; De Vries et al., 2003; Verheij et al., 2008) support an otherwise tacit knowledge of the benefits of exposure and access to nature. Specific to the outcome measure in this study, mortality, and speaking to a universal human need, green space has proven to be protective against mortality regardless of one's economic status (Mitchell and Popham, 2008). In other words, a greener environment can provide health benefits to a population regardless of their personal financial resources.

Biophilia and Health

The biophilia hypothesis was originally presented to bring to the forefront humans' evolutionary connection to natural elements and processes, and it has subsequently been examined for the many ways that these connections are evident (Kellert and Wilson, 1993). Among the many benefits stemming from the "experience with nature" that Kellert has cataloged, relatively very little research has explored how one's experience with nature reveals itself in health outcomes (Frumkin, 2008).

In *Biophilic Cities*, Timothy Beatley references a handful of studies that confirm the positive relationship between satisfaction of our biophilic needs and improved health outcomes. He concluded that "few elixirs have the power and punch to heal and restore and rejuvenate the way that nature can" (Beatley 2011, p.6). The literature Beatley references and more is divided into the following sections which represent the pathways by which green space supports health.

Presence, Access, and Exposure

Conserved green infrastructure supports health in three ways: 1) by its presence and quality, 2) from accessing and using it, and 3) from simply viewing or being exposed to it. The following subsections covering presence, access, and exposure represent a number of ways in which land conservation can support public health.[2]

---

2   Unless otherwise noted, evidence for the following sections was derived from summaries provided by the Health Council of the Netherlands (Health Council of the Netherlands and Dutch Advisory Council for Research on Spatial Planning, 2004), the National Association of County and City Health Officials (NACCHO, 2013), the Trust for Public Land (Gies, 2006), and from reviews of

Presence

The dependency of public health on ecosystem services—some as fundamental as water and air—is likely the most complex, understudied, and encompassing relationship presented here. Public health can be affected by the mere presence of the green space needed to deliver ecosystem services. Accordingly, due to its complexity, this is an area of intersection for many disciplines including public health, geography, psychology, and urban planning to name but a few. The theoretical space for the consideration of the connection between ecosystem services on health is the "ecological" public health paradigm (Duhl and Sanchez, 1999). This paradigm recognizes the importance of the environment on health but has yet to truly place biological ecology and ecosystems into this ecological framework (Coutts et al., 2014). The World Health Organization report, *Ecosystems and Human Well-Being* (Millennium Ecosystem Assessment, 2005), illustrates the complex way that our health and existence are linked to ecosystem services. With health as an outcome, most of the literature on the importance of these services to health has come from disciplines other than public health. This is despite the recognized importance of our place within ecosystems, our dependence upon these systems and the essential services they deliver, and their role in mitigating climatic variation (Sanesi et al., 2011).

A *public health ecology* approach (Coutts, 2010) attempts to reconcile this conceptual disconnect by making the natural environment the foundation on which other environmental threats to health are based. Supporting a public health ecology approach involves securing the landscape needed to support ecosystem services and the health of the humans that depend on it. As the human signature on the earth expands and the prospect of an increasingly tainted environment leads to greater risks to human health, it will be increasingly important to protect the landscape on which health depends.

The quality and quantity of the most basic of life-supporting elements, water, is dependent upon the ability of green space to filter pollutants and facilitate the recharging of groundwater stores. Ecologically insensitive development practices that replace green space with impervious surfaces, such as asphalt and concrete, increase the levels of non-point source pollution in surface and groundwater (Arnold and Gibbons, 1996). Runoff carried unchecked over impervious surfaces and into surface water bodies carries with it the pollutants accumulated on these surfaces. On the other hand, runoff that passes over permeable surfaces and green space filters non-point source pollution and allows absorption into groundwater reservoirs.

The conservation of green space can reduce airborne pollutants as well. Selected forms of vegetation, most notably trees, have the capacity to filter both gaseous and particulate airborne pollutants (Bealey et al., 2007; Nowak et al., 2006). These pollutants exacerbate asthma, a growing malady among children in the U.S., and are associated with lung cancer and cardiopulmonary mortality (Pope et al., 2002).

Conserving land and vegetation also plays a role in sequestering carbon (Nowak and Crane, 2002). This sequestration not only has an immediate improvement on air quality, but it also helps mitigate climate change. The potential health effects of climate change are now being explored (Coutts and Berke, 2013; McMichael et al. 2006). The interrelated and sweeping effects of climate change on health will not be done justice here as this is not hypothesized to reveal itself in this cross-sectional study, but this effect should certainly be isolated in future research that examines, prospectively, changes in health status. A micro-

---

environmental audit tools and guides which capture the public health impacts of development (Alaimo et al. 2008, San Francisco Department of Public Health, 2008).

climate phenomenon that may be captured in this study is the urban heat island. It has been shown that local temperatures in urban environments can be partially reduced through the cooling effect of urban nature (Gill et al., 2007). Although the evidence of urban heat islands is irrefutable, it is less clear what affect local green space conservation efforts may have on reducing temperatures within the biosphere as a whole.

Green space may also have some benefit to not only the remediation of pollutants but also the prevention of their release. Conserved space in the form of greenways and trails supports walking and biking (Coutts, 2008; Lindsey et al., 2001) which reduces emissions from motorized vehicles. The co-benefit to health, of course, is the physical activity gained through non-motorized forms of transportation and recreation.

The potential risks to health caused by the desertification and diminished biodiversity from global climate change include changing patterns of infectious disease (McMichael et al., 2006; Aron and Patz, 2001; Dobson and Carper, 1993). Although biodiversity has been put forward as " ... essential, not optional, to our lives and health and to our continuing to flourish as a species" (Beatley 2011, p.18), there is still a small pool of evidence from which to support this precautionary claim. Again, similar to global climate change, diminished biodiversity is not hypothesized to have an impact on the localized analysis presented here, but it should be a consideration in future studies which aim to model the comprehensive set of health threats exacerbated by loss of green space.

While the presence of green space is important to support the most fundamental of human needs, accessing green space could help to alleviate the chronic diseases that are the leading causes of death in the developed world.

Access

The predominant pathway by which health and green space has been linked to public health outcomes rests in the ability of people to access green space for physical activity and social interaction.

Environmental supports for the performance of physical activity have been the focus of a considerable amount of research over the past two decades (Frank et al., 2003; TRB, 2005). This thrust in research was spurred by a renewed appreciation for the environmental supports and barriers to physical activity. An example of an environmental support is an attractive public space, such as a park, where physical activity can be performed (Cohen et al., 2007; Bedimo-Rung et al., 2005; Coombes et al., 2010; Giles-Corti et al., 2005; Dyck et al., 2013; Jong et al., 2012). Even modest amounts of physical activity, such as the levels recommended by the Centers for Disease Control and Prevention (CDC), can prevent increases in visceral fat, a major risk factor for many diseases (Slentz et al., 2005). In addition, the regular physical activity that could be achieved by taking a stroll in a park could reduce the risk of cardiovascular disease, type 2 diabetes, selected forms of cancer, and osteoporosis, as well as improving mental health and mood and increasing general longevity (CDC, 2011).

The other benefit derived from accessing green space is the opportunity to improve one's social capital. Social capital is the value of the relationships that exist between people in a community. Creating appropriate spaces for human interaction is essential for creating a sense of community, as " ... community cannot form in the absence of communal space ... " (Duany et al., 2000). Parks and open space can provide gathering places for social interaction, recreation, and civic function. Those bonds fostered by interaction in public spaces have been proven to be important indicators of many health outcomes (Kuo et al., 1998; Coley and Kuo, 1997).

Exposure

Beyond the implications that land conservation may have on physical health, exposure to nature also has mental health benefits. If we are to fully achieve health, defined by the World Health Organization as the "state of complete physical, *mental* [ital added] and social well-being and not merely the absence of disease or infirmity" (WHO, 1946), then mental health must be a component of environmental interventions aimed at improving overall health.

Modern tendencies towards what has been termed "nature deficiency disorder" (Louv, 2011) or a lack of "vitamin G" (vitamin Green) (Maas et al., 2006) run counter to the biophilia hypothesis and our innate connection to nature and natural elements. This "disorder" or "discord" has proven detrimental to our mental health. Research in the environmental psychology field demonstrates that exposure to nature provides psychological regeneration and a wide variety of other mental health benefits (Grinde and Patil, 2009; Kaplan and Peterson, 1993; Maller et al., 2006). In addition to the rather well-developed research into the mental health benefits of exposure to nature, there appears to be synergistic benefits between the mental and physical health benefits of green space. Physical activity performed in natural settings produces greater mental health benefits as compared to activity performed in common urban outdoor spaces (Bodin and Hartig, 2003; Hartig et al., 2003; Barton and Pretty, 2010). It is studies such as this which confirm the complex and interrelated co-benefits of green space conservation.

Clearly, green space is important for public health outcomes because it supports the elements on which human life depends, including air quality and water needs. Green spaces in the form of parks and greenways are also important because they support the behaviors, such as physical activity, associated with numerous positive health outcomes. Although there is increasing evidence of these associations, to date there has been little attention given to the role of the distribution of green spaces on mortality. This exploratory study aims to begin to remedy this by setting the stage for more comprehensive analyses of this phenomenon. Given that these are spatial issues, GIS is a useful tool for helping to model the role of green space distribution on health.

Methods

The analysis employed in this study aims to test if green space in one's environment is predictive of premature mortality from all-causes. It is predicted that having green space in one's immediate environment will decrease premature mortality from all-causes. This is tested by modeling both the distance to the nearest green space and the amount of green space within defined proximities from one's residential location at time of death controlling for one's race, ethnicity, educational attainment, and marital status.

Data

Death certificate records were obtained from the Florida State Department of Health, Bureau of Epidemiology for the years 2000–2012 (n=2,216,641). The records for 2012 extend only to the month of July and were the most recent available at the time of the data request. Data requests of this type are essential for researchers using health and mortality data at levels smaller than the county. Data at the county-level is freely available, but this level of analysis obviously limits the types of conclusions that can be drawn. States may vary in the ease to which county data is made available, but a U.S. national dataset (http://www.

countyhealthrankings.org) that compiles county data has alleviated some of the challenges inherent in navigating state departments of health data acquisition protocols.

Individual level health and mortality data is, rightfully so, much more difficult to obtain. The death certificate mortality data used in this study was obtained by filing a request with the Florida Department of Health (DOH), Bureau of Epidemiology. In the application to acquire these data, the researchers were required to demonstrate that these data would be stored in a secure way and destroyed once the analysis was complete. After approximately four months of review by the DOH, the data were delivered via a secure File Transfer Protocol (FTP) site for a fee of $78. It is very important to note that these data do not come with a unique identification number for each observation. This must be generated by the researcher to ensure ease of potential merges if data are broken apart and analyzed in various software environments. A review of the acquisition and use of mortality data by a researcher's own Institutional Review Board (IRB) may not be required. As was the case with this study, IRB review was not necessary because the data was derived solely from deceased individuals.

The raw data was prepared for analysis by removing observations where: the residential address given on the death certificate was outside the US, outside the state of Florida, or "unknown"; the cause of death was recorded as something other than "natural"; race was "unknown" or noted as "other" and nothing else; or the deceased were under age 18 or over age 125.

A random sample of 10% of the total 1,576,580 remaining cases was drawn for the analysis. The addresses of those in the sample were geocoded using ArcGIS v.10. The geocoding resulted in a 92.7% match leaving 146,254 individual observations. 84 individuals were removed as they were geolocated outside the state of Florida leaving 146,170. A number of records were discovered that had marital status and education denoted as "unknown." These were dropped after geocoding (n=2,445) leaving 143,725 records for the final analysis.

Premature mortality was defined as Years of Potential Life Lost (YPLL) (CDC, 2012). YPLL measures the impact of premature mortality on a population and closely resembles crude mortality rates (Wise et al., 1988). YPLL was calculated here using the sex-specific average life expectancy (or age at death) of the sample. This average life expectancy of the sample was 78 years for females and 74 years for males. This roughly conforms to the average life expectancies in the U.S. as a whole (male=75.5, female=80.5). YPLL was calculated in the following way:

$$\text{YPLL}_{i \, sex} = \text{average life expectancy}_{sex} - \text{age of death}_{i \, sex}$$

The negative values resulting from an individual living beyond average life expectancy were converted to zeroes as these persons had no years of potential life lost.

Data on green space was contained in a 2009 public land file obtained from the Florida Geographic Data Library. These data included all state and national forests and parks, wildlife conservation areas and preserves, and city and county parks. This database is the best approximation of a statewide inventory of green space in Florida. A close review of these data led us to exclude categories of land that were either inaccessible to the public or had no strong theoretical basis for influencing health and mortality. Among the categories of excluded lands was land owned by the Disney Corporation and the 9.2 million acres of agricultural land in Florida (USDA, 2007). Although it could be argued that some types of agricultural lands provide benefits derived from exposure to pastoral views, they were excluded because their type (for example, crop, pasture, animal processing) could not be confirmed. While agricultural land also obviously provides the benefits of sustenance, these would likely not be revealed locally. It was felt that the exclusion of this land category would prove an underestimate of the benefits of health because of the reduction in overall green space in the analysis. The data used in this analysis accounted for many, but not all, public beaches identified as state parks and preserves along the state's extensive coastline.

These data were imported into TransCAD GIS v. 5.0 to compute measures of green space proximity to individual residences. We computed several measures with these data in an attempt to uncover the presence, exposure to, and accessibility of green space in one's environment. Although accessibility was not the sole focus here as it has been in previous work (Coutts et al., 2013; Coutts, 2008; Coutts et al., 2010), the measures we used could be considered among the many ways to quantify accessibility, all of which have their strengths and limitations (Talen and Anselin, 1998; Lei and Church, 2010). First, we calculated the straight-line distances of each residential address to its nearest green space (based on the green space polygon centroid). Second, we calculated the amount of green space within a set of defined distances from each residence. For each residence we computed the amount of green space within a ¼, ½, and 1 mile distance. This was accomplished by using standard buffering techniques. The GIS was used to construct Euclidian polygon buffers around the residences at each of the aforementioned lengths. For each distance interval, an overlay routine was employed to estimate the amount of green space (in square miles) within each residence's buffer.

Analysis

Before attempting to model relationships, the data were examined to determine if any patterns were inherent in the data that may skew the results. The most obvious culprit would be the abnormal patterns in average age of death for males and females, but this was not found to be an issue. The standard deviations were nearly identical for average age of death for males and females. Although the standard deviations were nearly identical, we did find a need to separate sex into two different models in order to accommodate the varied age structure of male and female mortality. The Yule-Simpson Effect fallacy that can arise from not doing so is discussed briefly in the results section. The standard deviations were also nearly identical for average age of death between the years 2000–2012 in the sample.

We employed negative binomial regression to test the effect of the distance to and amount of green space on YPLL in four separate models for males and four separate models for females controlling for education, race, Hispanic ethnicity, and marital status. The models tested in the analysis included the following variables:

$Y_i = \beta_{oi} + \beta_1 Green_i - \beta_2 Education_i + \beta_3 Race_i + \beta_4 Ethnicity_{ij} + \beta_5 Marital_i + e_i$

Y = YPLL

Green = one of four green space measures (distance to nearest green space, amount of green space within ¼ mile, ½ mile, and 1 mile from residential location)

Education = <high school, high school diploma, bachelor degree, graduate degree

Race = white

Ethnicity = Hispanic (white and black)

Marital = divorced, married, never married, widowed

e = error term

A negative binomial regression model was used to accommodate the non-continuous count nature of the dependent variable, YPLL. A Poisson model was considered but rejected due to over-dispersion. This was evident in the variance of YPLL far exceeding its mean. Further tests of goodness-of-fit of the Poisson model and of over-dispersion (alpha being significantly different than 1 in the negative binomial results) confirmed that the negative binomial model was the best choice.

Finally, a Gettis-Ord Gi* statistic was calculated independently for both male and female YPLL. The Gi* statistic identifies statistically significant "hot spots" and "cold spots" of spatial clustering. To be statistically significant, a feature with a high or low value will be surrounded by features with similar high or low values. The results are z-score values that are interpreted as standard deviations. Values over 1.96 or under -1.96 are statistically significant at the $p<0.05$ level. Values over 2.58 or under -2.58 are statistically significant at the $p<0.01$ level. This statistic is useful as it illuminates those regions of the state that had significant clustering of YPLL; these areas therefore can be identified as hot spots (reflecting high levels of YPLL) or cold spots (reflecting low levels of YPLL).

Results

The descriptive summary of the sample data used in this study revealed that the values of the four green space main effects variables were reasonable and behave in an expected fashion (Table 17.1). The average distance from home to the closest green space for a person who died between 2000–2012 in Florida was 2.55 miles, and the amount of green space increased as the radius of the buffered area surrounding one's home increased. The average amount of green space 0.25 miles from home for a person who died between 2000–2012 in Florida was.002 square miles (1.3 acres). As one would expect, the amount of green space from home increased as successively larger areas surrounding one's home were analyzed, the largest area being a.08 square mile (51.2 acre) average within 1 mile from home.

Table 17.1   Descriptive summary of data (n=143,725)

Variable	n (%)	Mean	σ
YPLL			
Male			
0 (died ≥ 74)	41,580 (29.0)		
1–10 (died 73–64)	13,874 (9.7)		
11–20 (died 63–54)	9,164 (6.4)		
21–30 (died 53–44)	4,806 (3.3)		
31–40 (died 43–34)	1,582 (1.1)		
41–50 (died 33–24)	488 (0.3)		
51–56 (died 23–18)	203 (0.1)		
Total	71,697 (49.9)		
Female			
0 (died ≥ 78)	50,844 (35.5)		
1–10 (died 73–64)	10,145 (7.0)		
11–20 (died 63–54)	5,933 (4.1)		
21–30 (died 53–44)	3,314 (2.3)		

Variable	n (%)	Mean	σ
31–40 (died 43–34)	1,188 (0.8)		
41–50 (died 33–24)	448 90.3)		
51–56 (died 23–18)	156 (0.1)		
Total	72,028 (50.1)		
Main Effects			
Distance to nearest green space (miles)		2.55	1.90
Amount green space (sq miles) within defined distances from residential location			
0.25 mile		.002	.01
0.50 mile		.012	.05
1.0 mile		.080	.22
Controls			
Education			
<HS	32,899 (22.9)		
HS	87,932 (61.2)		
BA	14,937 (10.4)		
Graduate	7,957 (5.5)		
Race/ethnicity			
White	128,281 (89.3)		
Black	14,711 (10.2)		
Other (Asian, Am Indian, Pacific Islander)	831 (0.6)		
Hispanic (white and black	9,442 (6.6)		
Marital Status			
Divorced	60,572 (42.1)		
Married	19,267 (13.4)		
Never married	9,857 (6.9)		
Widowed	54,029 (37.6)		

The regression results revealed that the distance to the nearest green space from home was a significant predictor of YPLL for both males (Table 17.2) and females (Table 17.3). The interpretations of these coefficients are: for every one mile of increased distance from the nearest green space, the average change in the mean of the YPLL is .029 years (10.6 days) for males and .047 years (17.2 days) for females. The only other significant green space variables were the amount of green space within one-half mile and one mile of home for females (Table 17.3). The interpretations of these coefficients are: for every 1 square mile of green space within one-half mile of home, the average change in the mean of YPLL is .452 years (165.2 days) for females; and, for every 1 square mile of green space within 1 mile of home, the average change in the mean of YPLL is .156 years (56.9 days) for females. Note that this was in the positive direction meaning that more green space equated to more YPLL. Also note that the one-half mile measure for females was significant at p=.04, giving us only cautious confidence in its significance.

Table 17.2   Regression results from analysis of green space and YPLL for males (n=71,697)

Variable	Model 1 β (SE)	Model 2 β (SE)	Model 3 β (SE)	Model 4 β (SE)
Distance to nearest green space (miles)	.029*** (.004)			
Amount green space (sq miles) within three defined distances from residential location				
0.25 mile		1.14 (.853)		
0.50 mile			.268 (.170)	
1.0 mile				.060 (.036)
Education <High School (reference)	X	X	X	X
High School	.245*** (.022)	.236*** (.021)	.236*** (.021)	.237*** (.021)
Bachelor	−.008 (.030)	−.025 (.030)	−.025 (.031)	−.024 (.031)
Graduate	−.017*** (.037)	−.185*** (.037)	−.185*** (.037)	−.185*** (.037)
Race/ethnicity White				
	−.719*** (.028)	−.715*** (.028)	−.716*** (.028)	−.717*** (.028)
Hispanic	.282*** (.034)	.287*** (.034)	.288*** (.034)	.289*** (.034)
Marital Status Divorced (reference)	X	X	X	X
Married	.753*** (.024)	.752*** (.024)	.751*** (.024)	.752*** (.024)
Never-married	1.29*** (.031)	1.28*** (.031)	1.28*** (.031)	1.28*** (.031)
Widowed	−1.57*** (.023)	−1.58*** (.023)	−1.58*** (.023)	−1.58*** (.023)

Note: *p<0.05, **p<0.01, ***p<.001

The control variables that behaved in the expected fashion for both males and females were: being white significantly reduced YPLL, and being Hispanic significantly increased YPLL. The control variable for education behaved in the expected fashion for males (graduate education reducing YPLL as compared to those without a high school diploma), but this was not the case for females. Higher levels of education for females resulted in higher YPLL. Also unexpected was, for both males and females, being married resulted in increased YPLL as compared to those who were divorced. Having once been married (widowed) behaved more predictably in reducing YPLL as compared to those who were

Table 17.3   Regression results from analysis of green space and YPLL for females (n=72,028)

Variable	Model 1 β (SE)	Model 2 β (SE)	Model 3 β (SE)	Model 4 β (SE)
Distance to nearest green space (miles)	.047*** (.006)			
Amount green space (sq miles) within three defined distances from residential location				
0.25mile		1.58 (1.12)		
0.50 mile			.452* (.220)	
1.0 mile				.156** (.047)
Education <High School (reference)	X	X	X	X
High School	.235*** (.026)	.222*** (.026)	.222*** (.026)	.222*** (.026)
Bachelor	.208*** (.042)	.186*** (.042)	.186*** (.042)	.188*** (.042)
Graduate	.179** (.059)	.156** (.059)	.155** (.059)	.156** (.059)
Race/ethnicity				
White	−.977*** (.033)	−.966*** (.033)	−.967*** (.033)	−.971*** (.033)
Hispanic	.120** (.042)	.120** (.042)	.121** (.042)	.124** (.042)
Marital Status Divorced (reference)	X	X	X	X
Married	.187*** (.034)	.182*** (.034)	.183*** (.034)	.184*** (.034)
Never-married	.607*** (.046)	.599*** (.046)	.600*** (.046)	.601*** (.046)
Widowed	−2.12*** (.024)	−2.13*** (.024)	−2.13*** (.024)	−2.13*** (.024)

Note: *p<0.05, **p<0.01, ***p<.001

divorced. What the married versus widowed results revealed was that, compared to those who were divorced at time of death, the first one to die in a married relationship had an increased YPLL and the second to die had a decreased YPLL.

We also ran a model that combined the YPLL of males and females into a single dependent variable and included a control variable for sex. The results of this model were significant for all the green space measures, but this speaks to a potential fallacy that can occur when aggregating data that vary significantly on a third explanatory variable (in this case, age) that has a significant influence on the dependent variable. This fallacy is referred to as Simpson's Paradox, or the Yule-Simpson Effect, and explains the statistical phenomena of trends appearing in the aggregated data that disappear when the same data

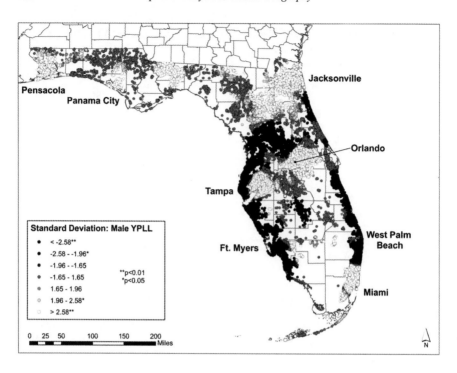

Figure 17.1   Gettis-Ord Gi* measure of clustering of Years of Potential Life Lost, Males

are then analyzed separately (Yule, 1902). This could be caused by an important variable not being considered or the numerical values of the data itself. In this case, aggregating the data ignores the varied age structure of the time of death between males and females (Table 17.1). When sex is aggregated, there appears to be a relationship between YPLL and all of the green space measures, but when sex is disaggregated, most of these relationships disappear. This speaks to the importance of carefully examining the descriptive data before modeling commences.

Figures 17.1 and 17.2 represent the results of the Gettis-Ord Gi* analysis of YPLL clustering. In general, the patterns for males and females were quite similar. There was statistically significantly clustering of higher levels of YPLL in Florida's largest metro areas (Orlando, Jacksonville, Miami), but there were exceptions to this potential urban clustering of increased YPLL. There was statistically significant clustering of lower levels of YPLL along Florida's coasts where most of Florida's population centers and development are located. While this may lead one to consider a potential urban effect that occurs only in Florida's largest metro areas, there was a stark exception to this hypothesis. Tampa, another very large metro area, is split evenly east to west into significantly increased and decreased YPLL even though both areas are highly developed and densely populated. This discrepancy between east and west Tampa also cannot be explained neatly by discrepancies in income distribution in these two parts of the metro area.

It would seem that the clustering of increased female YPLL in the rural north-central part of the state may help explain the regression results that revealed females were significantly

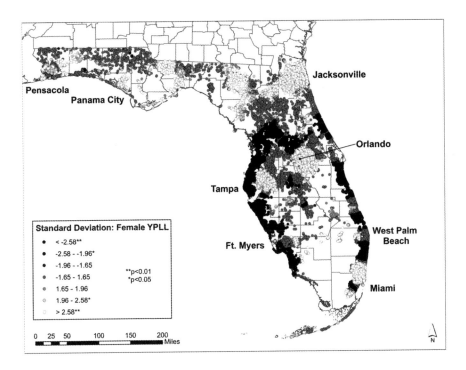

Figure 17.2   Gettis-Ord Gi* measure of clustering of Years of Potential Life Lost, Females

more likely to have increased YPLL as the amount of green space within one-half mile and one mile of home increased. The problem with this explanation was a rural clustering of increased male YPLL just east of the female cluster that would seem to counteract any uniquely female rural effect.

Conclusions

In the state of Florida, for both males and females, as distance from place of residence at the time of death to the nearest green space increases so too does one's years of potential life lost. For every mile further away from the nearest green space, males die a little less than two weeks earlier than their sex-specific average life expectancy and females die a little over two weeks earlier than their sex-specific average life expectancy.

Somewhat surprising are the positive relationships between an increased amount of green space within one-half mile and one mile of home and years of potential life lost in females. In other words, more green space makes females die sooner. For every one square mile (640 acres or about 320 soccer fields) increase in the amount of green space within one-half mile of one's home at time of death, females die approximately six months earlier than their average life expectancy. Recall that this result was considered at the borderline of significance and should therefore be interpreted very cautiously. The squarely significant results reveal that for every one square mile increase in the amount of green space within one mile of one's home at time of death, females die approximately two months earlier than their average life expectancy. The results of the Gettis-Ord Gi* analysis do not seem to support any rural/urban explanation for this peculiar result.

These overall results are somewhat contrary to a previous study of green space and mortality performed at the county-level (Coutts et al., 2010). Coutts et al. found that only the amount of green space within defined distances from home, and not distance to the nearest green space, was predictive of county-level mortality rates in 67 Florida counties. YPLL closely resembles crude mortality rates (used in Coutts et al., 2010) calculated for populations (CDC, 2012). The largest factor which reduces the ability to compare these studies directly is the fact that the previous study did not examine males and females separately and was therefore susceptible to the aforementioned Simpson-Yule effect. In fact, a more recent study examining the city as the unit of analysis and that did separate males and females found a positive relationship between high levels of city greenness and all-cause mortality (Richardson et al., 2012). Without performing a more in-depth case study that would complement these results with more qualitative information, an explanation for the differences found in studies employing various scales of analysis will likely to remain difficult to develop.

Discussion

The results of this exploratory study set the stage to address its most significant limitation; it is unknown how long a person has been at the address given on the death certificate and, therefore, the potential health benefits accrued by the presence, access, and exposure to green space. A more in-depth lifecourse approach can remedy this limitation by incorporating data on inter- and intra-state migration and the influence of environments that vary in their level of greenness. Of course, to examine the potential cumulative, lifecourse effects of green space on health or mortality, data beyond what is present on a death certificate are needed. The collection of these data could occur in a retrospective fashion by interviewing persons familiar with the deceased on past places of residence. This could also occur in a prospective fashion by following those not yet deceased and collecting data on their environments over the lifecourse.

In addition to employing a lifecourse approach, there are a number of other ways that this line of biophilic epidemiology research can be expanded to remedy the pressing need " … to establish a tradition of health research within the community of scientists interested in nature contact" (Frumkin, 2003). Three examples of possible directions for this research are:

1. Creating a typology of green space. This is fundamental to differentiating the types of benefits derived from various forms of green space. Heavily landscaped parks may be important for the benefits derived from access but may not be significant in the benefits derived from presence. Conversely, the presence of large national forests isolated from urban centers could be essential to protect air and water quality, but could be poor at supporting the health benefits of access. Parsing out public health outcomes by type of green space would add much needed specificity.
2. Analyzing the connectivity of green space. Landscape ecology has long recognized the importance of connected green space systems to functioning ecosystems. It could be possible that creating and maintaining a connected system of green

infrastructure, both at the local and regional level, could provide not only animal habitat but also health benefits to people. Are people more likely to perform physical activity on an extensive greenway system as compared to a neighborhood park? A connected system of green space may increase accessibility both by reducing proximity and increasing the number of potential destinations once the system has been reached. Techniques used in transportation analysis and visualization (Roorda et al., 2010; Morency et al., 2011) could be adapted to examine the potential spatial variation in mortality and health outcomes caused by the accessibility of green space. Other questions might include: Does a green space network increase the likelihood of exposure to green space, and does improved ecosystem functioning translate into human health benefits?

3. Lastly, a case-control design that compares cities or regions with high and low occurrences of a health outcome of interest may help us in understanding the qualitative differences in the relative importance of presence, access, and exposure. A careful comparison of two places that are similar demographically, but that vary in the quality and abundance of their green infrastructure, would allow us to isolate the influence of green infrastructure on health. A survey conducted in the case and control communities would also allow comparisons in behaviors that lead to specific health outcomes of interest. This type of design could more rigorously address the possible rural effect of increased YPLL in females in regionally greener environments.

This area is ripe for research, and there is arguably no more fundamental of a connection than that between humans and the environment they depend upon to satisfy both basic and higher order needs.

References

Antonovsky, A. 1987. *Unraveling the Mystery of Health: How People Manage Stress and Stay Well*, Hoboken, NJ: Jossey-Bass.

Arnold, C. and J. Gibbons. 1996. Impervious surface coverage: The emergence of a key environmental indicator. *Journal of the American Planning Association*, 62(2): 243–58.

Aron, J.L. and J. Patz. 2001. *Ecosystem Change and Public Health: A Global Perspective*, Baltimore: Johns Hopkins University Press.

Barton, J. and J. Pretty. 2010. What is the best dose of nature and green exercise for improving mental health? A multi-study analysis. *Environmental Science & Technology*, 44(10): 3947–55.

Bealey, W.J. et al. 2007. Estimating the reduction of urban PM10 concentrations by trees within an environmental information system for planners. *Journal of Environmental Management*, 85(1): 44–58.

Beatley, T. 2011. *Biophilic Cities: Integrating Nature into Urban Design and Planning*, Washington, D.C.: Island Press.

Bedimo-Rung, A.L., Mowen, A.J., and D.A. Cohen. 2005. The significance of parks to physical activity and public health: A conceptual model. *American Journal of Preventive Medicine*, 28(2S2): 159–68.

Benedict, M. and E.T. McMahon. 2006. *Green Infrastructure: Linking Landscapes and Communities*, Washington, D.C.: Island Press.

Bodin, M. and T. Hartig. 2003. Does the outdoor environment matter for psychological restoration gained through running? *Psychology of Sport & Exercise*, 4(2): 141.

CDC. 2011. Physical activity for everyone: The importance of physical activity. Available at: http://www.cdc.gov/physicalactivity/everyone/health/ [Accessed November 19, 2013].

CDC. 2012. Principles of epidemiology in public health practice. Available at: http://www.cdc.gov/osels/scientific_edu/ss1978/lesson3/Section3.html [Accessed January 4, 2014].

Cohen, D. et al. 2007. Contribution of public parks to physical activity. *American Journal of Public Health*, 97(3): 509–14.

Coley, R.L. and F.E. Kuo. 1997. Where does community grow? The social context created by nature in urban public housing. *Environment & Behavior*, 29(4): 468.

Coombes, E., Jones, A.P., and M. Hillsdon. 2010. The relationship of physical activity and overweight to objectively measured green space accessibility and use. *Social Science & Medicine*, 70(6): 816–22.

Coutts, C. et al. 2013. County-level effects of green space access on physical activity. *Journal of Physical Activity and Health*, 10: 232–40.

Coutts, C. 2008. Greenway accessibility and physical activity behavior. *Environment and Planning B: Planning and Design*, 35(3): 552–63.

Coutts, C. 2010. Public health ecology. *Journal of Environmental Health*, 72(6): 53–5.

Coutts, C. and T. Berke. 2013. The extent and context of human health considerations in London's spatial development and climate action strategy. *Journal of Urban Planning and Development*, 139(4): 322–30.

Coutts, C., Forkink, A., and J. Weiner. 2014. The portrayal of nature in the evolution of the ecological public health paradigm. *International Journal of Environmental Research and Public Health*, p.in press.

Coutts, C., Horner, M., and T. Chapin. 2010. Using geographical information system to model the effects of green space accessibility on mortality in Florida. *Geocarto International*, 25(6): 471–84.

De Vries, S., Verheij, R.A., Groenewegen, P.P., and P. Spreeuwenberg. 2003. Natural environments healthy environments ? An exploratory analysis of the relationship between greenspace and health. *Environment & Planning A*, 35(10): 1717–32.

Dobson, A. and R. Carper. 1993. Biodiversity. *The Lancet*, 342(8879): 1096–9.

Duany, A., Plater-Zyberk, E., and J. Speck. 2000. *Suburban Nation*. New York: North Point Press.

Duhl, L.J. and A.K. Sanchez. 1999. *Healthy Cities and the City Planning Process: A Background Document on Links between Health and Urban Planning*, Copenhagen: WHO Publication No. EUR/ICP/CHDV 03 04 03.

Dyck, D. Van et al. 2013. Associations of neighborhood characteristics with active park use : An observational study in two cities in the USA and Belgium. *International Journal of Health Geographics*, 12(26): 1–9.

Frank, L.D., Engelke, P.O., and T.L. Schmid. 2003. *Health and Community Design: The Impact of the Built Environment on Physical Activity*, Washington D.C.: Island Press.

Frumkin, H. 2003. Healthy places: Exploring the evidence. *American Journal of Public Health*, 93(9): 1451–6.

Frumkin, H., 2008. Nature contact and human health: Building the evidence base, in J. Heerwagen, S.R. Kellert, and M. Mador, eds. *Biophilic Design: The Theory, Science, and Practice of Bringing Buildings to Life*. Hoboken, N.J: Wiley: 107–18.

Gies, E. 2006. *The Health Benefits of Parks: How Parks Help Keep Americans and their Communities Fit and Healthy*. San Francisco, CA: The Trust for Public Land.

Giles-Corti, B. et al. 2005. Increasing walking: How important is distance to, attractiveness, and size of public open space? *American Journal of Preventive Medicine*, 28(2 Suppl 2): 169–76.

Gill, S. et al. 2007. Adapting cities for climate change: The role of the green infrastructure. *Built Environment*, 33(1): 115–33.

Grinde, B. and G.G. Patil. 2009. Biophila: Does visual contact with nature impact on health and well-being? *International Journal of Environmental Research and Public Health*, 6(9): 2332–43.

Hartig, T. et al. 2003. Tracking restoration in natural and urban field settings. *Journal of Environmental Psychology*, 23(2): 109–23.

Health Council of the Netherlands and Dutch Advisory Council for Research on Spatial Planning. 2004. *Nature and Health: The Influence of Nature on Social, Psychological and Physical Well-being*, The Hague: Health Council of the Netherlands, RMNO.

Jong, K. De, Albin, M., and E. Sk. 2012. Perceived green qualities were associated with neighborhood satisfaction, physical activity, and general health: Results from a cross-sectional study in suburban and rural Scania, southern Sweden. *Health & Place*, 18: 1374–80.

Kaplan, S. and C. Peterson. 1993. Health and environment: A psychological analysis. *Landscape and Urban Planning*, 26(1): 17–23.

Kellert, S.R., Heerwagen, J., and M. Mador. 2008. *Biophilic Design: The Theory, Science and Practice of Bringing Buildings to Life*. New York: John Wiley & Sons.

Kellert, S.R. and E.O. Wilson. 1993. *The Biophilia Hypothesis*, Washington, D.C: Island Press.

Kuo, F.E. et al. 1998. Fertile ground for community: Inner-city neighborhood common spaces. *American Journal of Community Psychology*, 26(6): 823–51.

Lei, T.L. and R.L. Church. 2010. Mapping transit based access: Integrating GIS, routes and schedules. *International Journal of Geographical Information Science*, 24(2): 283–304.

Lindsey, G. et al. 2001. Access, equity, and urban greenways: An exploratory investigation. *The Professional Geographer*, 53(3): 332.

Louv, R. 2011. *The Nature Principle: Human Restoration and the End of Nature-deficit Disorder*, Chapel Hill, NC: Algonquin Books of Chapel Hill.

Maas, J. et al. 2006. Green space, urbanity, and health: How strong is the relation? *Journal of Epidemiology and Community Health*, 60(7): 587–92.

Maller, C. et al. 2006. Healthy nature healthy people: "contact with nature" as an upstream health promotion intervention for populations. *Health Promotion International*, 21(1): 45–54.

McCally, M. 2002. Environment, health, and risk, in M. McCally, ed. *Life Support: The Environment and Human Health*. Cambridge, MA: MIT Press: 1–14.

McMichael, A.J., Woodruff, R., and S. Hales. 2006. Climate change and human health: Present and future risks. *The Lancet*, 367(9513): 859–69.

Millennium Ecosystem Assessment. 2005. *Ecosystems and Human Well-being: Synthesis*, Washington D.C.

Mitchell, R. and F. Popham. 2008. Effect of exposure to natural environment on health inequalities: An observational population study. *The Lancet*, 372(9650): 1655–60.

Morency, C. et al. 2011. Distance traveled in three Canadian cities: Spatial analysis from the perspective of vulnerable population segments. *Journal of Transport Geography*, 19(1): 39–50.

NACCHO. 2013. Public health in land use planning and community design checklist. Available at: http://www.naccho.org/toolbox/tool.cfm?id=604 [Accessed January 4, 2014].

Nowak, D. and D.E. Crane. 2002. Carbon storage and sequestration by urban trees in the USA. *Environmental Pollution*, 116(3): 381–9.

Nowak, D., Crane, D.E., and J.C. Stevens. 2006. Air pollution removal by urban trees and shrubs in the United States. *Urban Forestry & Urban Greening*, 4(3–4): 115–23.

Pope, C.A., Burnett, R.T., Thun, M.J., Calle, E.E., Krewski, D., and K. Ito. 2002. Lung cancer, cardiopulmonary mortality and long-term exposure to fine particulate air pollution. *Journal of the American Medical Association*, 287(9): 1132–41.

Richardson, E., Mitchell, R., Hartig, T., de Vries, S., Astell-Burt, T., and H. Frumkin.. 2012. Green cities and health: A question of scale? *Journal of Epidemiology and Community Health*, 66(2): 160–5.

Roorda, A., Páez, A., Morency, C., Mercado, R., and S. Farber. 2010. Trip generation of vulnerable populations in three Canadian cities: A spatial ordered probit approach. *Transportation*, 37: 525–48.

Sanesi, G., Gallis, C., and H.D. Kasperidus. 2011. Urban forests and their ecosystems services in relation to human health, in K. Nilsson, ed. *Forests, Trees and Human Health*. New York: Springer Verlag: 23–40.

Slentz, C.A., Aiken, L., Houmard, J.A., Bales, C.W., Johnson, J.L., Tanner, C.J., Duscha, B.D., and W.E. Kraus. 2005. Inactivity, exercise, and visceral fat. STRRIDE: A randomized, controlled study of exercise intensity and amount. *Journal of Applied Physiology*, 99(4): 1613.

Talen, E. and L. Anselin. 1998. Assessing spatial equity: An evaluation of measures of accessibility to public playgrounds. *Environment and Planning A*, 30(4): 595–613.

TRB. 2005. *Does the Built Environment Influence Physical Activity? Examining the Evidence*, Washington D.C.

USDA. 2007. Census of agriculture. Available at: http://www.agcensus.usda.gov/Publications/2007/index.php [Accessed January 4, 2014].

Verheij, R. a., Maas, J., and P.P. Groenewegen. 2008. Urban--Rural Health Differences and the Availability of Green Space. *European Urban and Regional Studies*, 15(4): 307–16.

WHO. 1946. Preamble to the Constitution of the World Health Organization. In Geneva: 100.

Wilson, E.O. 1984. *Biophilia*, Cambridge, MA: Harvard University Press.

Wise, R., Livengood, J., Berkelman, R., and R. Goodman. 1988. Methodologic alternatives for measuring premature mortality. *American Journal of Preventive Medicine*, 4: 268–73.

Yule, G.U. 1902. Notes on the theory of association of attributes in statistics. *BIOMETRIKA*, 2(1902/3): 121–34.

Appendix

Chapter 5

Transportation Network Sources available online:

Peter   Robins: http://   pilgrim.peterrobins.co.uk/itineraries/list.html   [Accessed November 2010]:

http://en.wikipedia.org/wiki/File:The_Ancient_Roads_of_Italy_and_Sicily_nopng.svg [Accessed December 2010]

http://www.spainthenandnow.com/spanish-history/roman-roads-in-hispania/default_88.aspx [Accessed December 2010]

http://www.paradoxplace.com/Perspectives/Maps/Roads%20to%20Santiago.htm [Accessed December 2010].

http://www.paradoxplace.com/Perspectives/Maps/Via%20Francigena.htm   [Accessed December 2010].

http://istrianet.org/istria/index.html [Accessed December 2010].

Index

adjustment 16, 184–7, 189, 190, 191, 192,
 see also standardization
Aedes aegypti 89, *see also* dengue fever
aggregation 1, 2, 3, 4, 19–20, 32, 35, 48, 85,
 163, 176, 211–12, 282, 305–6, *see also*
 ecological analysis
 data errors 20–21
 dis- 2, 5, 19, 281, 283, 285
antibiotic paradox 102
autocorrelation 4, 18, 168
 spatial 4, 8, 18, 129, 139, 144, 145, 147,
 151, 155, 204, 211, 212–13, 215, 216,
 286
 eigenvector 210, 211, 212, 213
 global univariate 144
 local univariate 145
 Moran's I statistic in 4, 144, 145, 147,
 151

barriers 4, 52, 72, 279, 281
barrier analysis 107–9, 112, *see also* disease
 Monmonier's algorithm 108
 Womble's method 108–9
Bayesian estimation 24–5, 48, 105
 Bayesian Information Criterion (BIC) 105
 Bayesian spatially varying coefficient
 (SVC) 109
 curvilinear approach 109
 wombling 109
bias 3, 32, 34, 111, 122, 124, 126, 163, 183,
 184, 186, 220, 228, 231
 ecological 15, 24, 26, 121
binomial kriging 165–6, 168, 170, 174,
 see also noise
biophilia hypothesis 295, 296, *see also* green
 infrastructure; mortality; public health
 public health and 296–9
Black Death, *see* Medieval Black Death
blood lead levels (BLLs) 197, 205–7, 211–12
Box-Cox power transformation 201–2, 204,
 206, 208, 215, *see also* transformation
bubonic plague 71, 72, 79, 80, *see also* Medieval
 Black Death
 peak mortality 71–2
 transmission velocities 71, 77

buffers 75, 77, 140, 222, 301, 302

cancer 47, 161, 261
 carcinogens 47, 56, 60, 62
 breast 161
 covariates, and 161, 163–5
 late-stage diagnosis 161–2, 163–5,
 172–4
 environmental histories 121–2, *see also*
 environmental factors; risk
 leukemia 48, 58, 60–62, *see also* leukemia
 non-Hodgkin lymphoma (NHL) 131–3
 pediatric 47, 56–9
 environmental histories (PEH) 56–7,
 60–62, *see also* PEH
 risk 121–2, *see also* risk
 treatment centres 261–2, *see also* health
 care
 accessibility 262–4
canonical correlation analysis (CCA) 202,
 206–7, 208–10, 212
cartography, *see* mapping
census data 2, 19, 21, 22, 26, 34, 165, 171, 198,
 286
 census blocks 205–6, 208, 210–11, 215,
 263
 census tracts 32, 35, 36, 39, 164, 165, 211,
 212, 263
civilizational collapse-level epidemics,
 periodicity of 79–80
cluster analysis 2, 18–19, 31, 47, 48, 49, 105,
 168, 175–6
 geocoding, and 42–3
 Mantel tests 106–7, 112
 model-based 105, *see also* modeling
 Monte Carlo simulations 3, 18, 124, 126,
 see also Monte Carlo permutation test
 multiple diseases, of 48
 significance 54
 space-time 3–4, 73, 85, 90, 95–6, 163, 176
clustering 48, 49, 62–3, 85, 96, 177
 cancer, of 47
 –co-occurrence distinction 49–50, 63
 K-means 105